Applications and Investigations in Earth Science

Applications and Investigations in Earth Science

EDWARD J. TARBUCK
FREDERICK K. LUTGENS
Illinois Central College

KENNETH G. PINZKE
Belleville Area College

Macmillan College Publishing Company
New York

Maxwell Macmillan Canada
Toronto

Maxwell Macmillan International
New York Oxford Singapore Sydney

Cover illustrations by Dennis Tasa, Tasa Graphic Arts, Inc.
Editor: Robert A. McConnin
Production Editor: Rex Davidson
Art Coordinator: Peter A. Robison
Text Designer: Jill E. Bonar
Cover Designer: Russ Maselli
Production Buyer: Pamela D. Bennett
Illustrations by Dennis Tasa, Tasa Graphic Arts, Inc.

This book was set in Souvenir by York Graphic Services, Inc. and was printed and bound by
R. R. Donnelley & Sons Company. The cover was printed by Phoenix Color Corp.

Macmillan College Publishing Company
866 Third Avenue
New York, New York 10022

Macmillan College Publishing Company is part of the
Maxwell Communication Group of Companies.

Maxwell Macmillan Canada, Inc.
1200 Eglinton Avenue East, Suite 200
Don Mills, Ontario M3C 3N1

ISBN: 0-02-419011-X

Printing: 1 2 3 4 5 6 7 8 9   Year: 4 5 6 7 8

I hear and I forget.
I see and I remember.
I do and I understand.

—Ancient Proverb

Preface

The earth is a very small part of a vast universe, but it is our home. It provides the resources to carry on a modern society and the ingredients necessary to support life. Therefore, a knowledge and understanding of our planet is critical to our economic and social welfare and, indeed, vital to our survival.

In recent years, media reports have made us increasingly aware of our place in the universe and the forces at work in our physical environment, as well as how human interaction with natural systems can often upset a delicate equilibrium. News stories inform us of new discoveries in the solar system and beyond. Daily reports remind us of the destruction created by hurricanes, flooding, mudflows, and earthquakes. We have been made aware of the depletion of ozone, an increase in carbon dioxide in the atmosphere caused by human activities, and growing environmental concerns over the fate of the oceans. To comprehend, prepare for, and solve these and other concerns requires a full awareness of how science is done and the scientific principles that govern the operation of the earth, atmosphere, and oceans.

To achieve scientific understanding requires not only a knowledge and appreciation of broad scientific theories but also the ability to gather scientific data and to solve problems creatively with critical reasoning skills. As part of the educational process, laboratory experience is essential to developing these competencies. In the laboratory, nature can be explored, principles examined, and hypotheses developed, tested, and compared with experimental data. It is often in the laboratory that understanding begins.

Applications and Investigations in Earth Science is a college-level laboratory manual consisting of twenty-two exercises designed to examine many of the basic principles of geology, meteorology, oceanography, and astronomy. The manual is written to accompany and supplement the text *Earth Science* by Edward J. Tarbuck and Frederick K. Lutgens (Macmillan College Publishing Company). All textbook chapters, figures, and table references in the manual are for the seventh edition of the text.

The format of each of the exercises is the same. Each investigation includes

Statement of Purpose that outlines the major area(s) to be investigated in the exercise.

Objectives that clearly state what should be learned by completing the exercise.

Textbook Reference that indicates the chapter(s) and/or appendix in the text *Earth Science* where the material is discussed in detail. The reference material should be read and thoroughly reviewed prior to beginning each exercise.

Materials other than the text, paper, and pencil that are needed and supplied by the student. Also indicated are those materials that are supplied by the instructor and should be located in the laboratory.

Terms are the key words that are shown in bold type in order of appearance in the exercise. In some cases the terms are defined within the exercise. Those that are not defined can be located in the text and/or class notes.

Introduction that briefly discusses the material to be investigated. In most instances, a more detailed discussion can be found in the appropriate chapter of the text. This material should be read and reviewed prior to beginning the exercise.

Questions and Problems that investigate and apply the principles examined in the exercise. The questions require various levels of understanding and build from basic comprehension to more demanding applications. When necessary, adequate space has been provided after each question so that the answer can be recorded for future reference.

Review that presents the main points of the exercise that have been investigated and developed.

Summary/Report Page located at the end of each exercise to be filled out after each exercise is completed. Most of the information requested will have been examined and answered in the exercise. Your laboratory instructor may require that a copy of the Summary/Report page be submitted for evaluation.

A Note to the Student. The purpose of *Applications and Investigations in Earth Science* is to assist you in understanding the basic concepts of earth science that are presented in most college-level courses by providing the experience of participation. Often, simply being told that something is true is not enough. However, the opportunity to collect facts, examine information, and draw conclusions in a scientific manner frequently results in comprehension. Learning is much more than the accumulation of knowledge; it is also the understanding that comes from "doing." In the laboratory we hope you will refine the critical and creative thinking skills that are vital to your effective participation in a modern society.

Each of the manual's twenty-two exercises is self-contained and will require approximately two hours to complete. However, your instructor may not follow the exercises in the order presented in the manual. Also, exercises may be assigned for completion *outside* of the regularly scheduled laboratory session.

Prior to beginning each exercise you should

1. Read the required text assignment, paying particular attention to the key terms that are listed in the exercise.
2. Examine the exercise and read the purpose and list of objectives.
3. Gather any additional materials listed at the beginning of the exercise. Bring these materials along with your laboratory manual, text, course notes, paper, and pencils to the laboratory session.

After you have completed each exercise,

1. Examine the review at the end of the exercise.
2. Complete the Summary/Report page at the end of the exercise. Your instructor may require that you submit this report for evaluation by a specified date.

A Note to the Instructor. *Applications and Investigations in Earth Science* is designed to supplement an introductory general education earth science course. We have attempted to include a general survey of most of the major topics in earth science, building upon questions that require various levels of understanding to answer. The first type of question is concerned with the basic concepts and fundamental principles of earth science. These can be answered by reflective thought and careful reading of class notes and textbook materials. Other sections involve conducting experiments, gathering data, and drawing conclusions. The third type of investigation entails the application of principles to achieve solutions.

Recognizing that different aspects of earth science are emphasized by each of us, we have included twenty-two exercises that cover a wide variety of subjects. Since each exercise is basically self-contained and covers a specific topic, individual exercises or whole units may be omitted or introduced in a different sequence without difficulty. Furthermore, with only minor modification, several exercises, or portions of exercises, may be assigned for completion outside the regularly scheduled laboratory sessions.

The last two exercises in the manual investigate the basic skills of determining location and distance on the earth, working with the metric system, and conducting a scientific inquiry. Should you decide that completing all, or any part, of these exercises would be of benefit to your students, we suggest that you introduce them near the beginning of the course. In addition, Exercise Seventeen, "Astronomical Observations," will take the student several weeks to complete.

Many of the exercises require that equipment and materials be supplied and present in the laboratory. When needed, a general list has been included at the beginning of each exercise. Realizing that some laboratory settings have limited access to materials such as minerals, rocks, and fossils, we have left the specific choices of these specimens to the instructor.

In summary, we have attempted to put together a versatile and adaptable collection of laboratory experiences that investigate many of the topics in introductory earth science. We sincerely feel that each exercise has merit and, through active student participation, the learning process is carried one step closer to complete understanding.

ACKNOWLEDGEMENTS

The authors wish to express their sincere thanks to each of the many individuals, institutions, and agencies that provided photographs and illustrations for the manual. Further, we would like to acknowledge the aid of our many students, past and present. Their comments have helped us maintain our focus on readability and understanding.

A special debt of gratitude goes to the following colleagues who prepared in-depth reviews of this first edition of *Applications and Investigations in Earth Science.* Their critical comments and thoughtful input helped strengthen the manual.

Craig A. Chesner, Eastern Illinois University
Beverly Hardin, Jackson State Community College, TN
Jerry M. Hoffer, University of Texas, El Paso

Keenan Lee, Colorado School of Mines
John Licari, Cerritos College, CA
Barbara M. Manner, Duquesne University
C. Nicholas Raphael, Eastern Michigan University
Gerald E. Schultz, West Texas State University
Henry I. Snider, Eastern Connecticut State University
Peter W. Whaley, Murray State University

A very special acknowledgement goes to our typist, Ms. Marilae Perkins, whose even temper and steady hand always managed to keep the project moving forward. Our thanks also go to the members of the Macmillan Publishing production team who skillfully transformed our manuscript into a finished product.

All lives are precious, but those that end suddenly and too soon
are remembered forever in a special place in our minds and
hearts. This manuscript is dedicated in loving memory to one of
those lives. His name is Tristan. He will be missed.

Contents

ASTRONOMY

EARTH SCIENCE SKILLS

Applications and Investigations in Earth Science

EXERCISE ONE

The Study of Minerals

The ability to identify minerals using the simplest of techniques is a necessity for the earth scientist, especially those scientists working in the field. In this exercise you will become familiar with the common physical properties of minerals and learn how to use these properties to identify minerals (Figure 1.1). In order to understand the origin, classification, and alteration of rocks, which are for the most part aggregates (mixtures) of minerals, you must first be able to identify the minerals that comprise them.

OBJECTIVES

After you have completed this exercise, you should be able to

1. Recognize and describe the physical properties of minerals.
2. Use a mineral identification key to name minerals.
3. Identify several minerals by sight.
4. List the uses of several minerals that are mined.

TEXTBOOK REFERENCE

Chapter 1 and Appendix B

MATERIALS

Materials Supplied by Your Instructor

mineral samples	dilute hydrochloric acid
streak plate	set of quartz crystals (various sizes)
magnet	contact goniometer
glass plate	

|← 5 cm →|

FIGURE 1.1 Quartz crystals. Slender six-sided, transparent crystals that will scratch glass. The shape of the crystals and their hardness are two physical properties used to identify this mineral. (Photo by E. J. Tarbuck)

TERMS

mineral	translucent	cleavage plane
rock-forming	transparent	direction of
mineral	hardness	cleavage
luster	color	fracture
metallic luster	streak	specific gravity
nonmetallic	crystal form	magnetism
luster	contact goniometer	striations
opaque	cleavage	tenacity

1

INTRODUCTION

A **mineral** is a naturally occurring, inorganic solid with an orderly internal arrangement of atoms (called *crystalline structure*) and a definite, but not fixed, chemical composition. Some minerals, such as gold and diamond, are single chemical elements. However, most minerals are compounds consisting of two or more elements. For example, the mineral halite is composed of the elements sodium and chlorine. The distinctive crystalline structure and chemical composition of a mineral give it a unique set of physical properties such as its luster, its hardness, and how it breaks. The fact that each mineral has its own characteristic physical and chemical properties can be used to distinguish one mineral from another.

Minerals, especially gems and precious metals, are among the oldest objects used and treasured by society. Today our industrial economy relies extensively upon the metals extracted from many minerals. For example, the metal iron is refined from the mineral hematite (iron oxide), using a heating process to separate the iron from the oxygen. Often the refining processes of metal-bearing minerals produce potentially harmful by-products that must be monitored and controlled to ensure that they are not released into the environment. In addition to the minerals that contain metals, many others are also useful to society. An example would be the mineral gypsum, used for making drywall and wallboard. Table 1.1 lists a few of the minerals that are mined as well as their uses.

Of the nearly four thousand known minerals, only a few hundred have any current economic value. Of the remaining minerals, no more than a few dozen are abundant. Collectively, these few often occur with each other in the rocks of the earth's crust and are classified as the **rock-forming minerals.**

PHYSICAL PROPERTIES OF MINERALS

The physical properties of minerals are those that can be determined by observation or by performing some simple tests. The primary physical properties that are determined for all minerals include optical properties (in particular, luster and the ability to transmit light), hardness, color, streak, crystal form, cleavage or fracture, and specific gravity. Secondary (or "special") properties, including magnetism, taste, feel, striations, tenacity, and the reaction with dilute hydrochloric acid, are also useful in identifying certain minerals.

Optical Properties. Of the many optical properties of minerals, two, luster and the ability to transmit light, are frequently determined for hand specimens.

Luster describes the manner in which light is reflected from the surface of a mineral. Any mineral that shines with a metal-like appearance has a **metallic luster**. Those minerals that do not have a metallic luster are termed **nonmetallic** and may have one of a variety of lusters which include vitreous (glassy), pearly (like a pearl), or earthy (dull, like soil or concrete). In general, many minerals with metallic luster produce a dark gray, black, or other distinctively colored powder when they are rubbed on a hard porcelain plate (this property, called *streak*, will be investigated later in the exercise).

Observe the mineral photographs shown in Figures 1.2 through 1.13. The minerals illustrated in Figures 1.5 and 1.6 have definite metallic lusters. The mineral in Figure 1.9 has a nonmetallic, vitreous (glassy) luster.

TABLE 1.1 Mineral Uses

Mineral	Use
Chalcopyrite	Mined for copper
Feldspar	Ceramics and porcelain
Fluorite	Used in steel manufacturing
Galena	Mined for lead
Graphite	Pencil "lead," lubricant
Gypsum	Drywall, plaster of paris, wallboard
Halite	Table salt, road salt. source of sodium and chlorine
Hematite	Mined for iron
Limonite	Mined for iron
Magnetite	Mined for iron
Pyrite	Mined for sulfur and iron
Quartz	In the pure form, for making glass
Sphalerite	Mined for zinc
Talc	Used in ceramics, paint, talcum powder

|← ————————— 9 cm ————————— →|

FIGURE 1.2 Fluorite (left), halite (center), and calcite (right) exhibit smooth cleavage planes that are produced when the mineral is broken.

FIGURE 1.3 Sphalerite. An ore of zinc with six directions of cleavage and a white to yellow-brown streak.

FIGURE 1.5 Galena. An ore of lead with a high specific gravity, metallic luster, and three directions of cleavage at 90°, producing cubic crystals.

FIGURE 1.4 Limonite. An ore of iron, with an earthy luster and "rusty" appearance. When broken, the mineral fractures.

FIGURE 1.6 Pyrite. A brassy-yellow mineral with a metallic luster that is commonly known as "fool's gold." Although the mineral often occurs as cubic crystals with striations (illustrated), it fractures when broken (not illustrated).

FIGURE 1.7 Hematite. An ore of iron that has both a metallic (right) and nonmetallic (left) form. Regardless of luster or color, the streak of the mineral is characteristically red-brown.

FIGURE 1.9 Muscovite mica. Biotite mica (not shown) is similar in appearance, except it is black. Both of these rock–forming minerals have a nonmetallic, glassy luster and one direction of cleavage.

FIGURE 1.8 Two varieties of the mineral quartz . Rose quartz (left) and smoky quartz (right) each have a hardness of seven and fracture when broken.

FIGURE 1.10 Hornblende. A generally green to black, rock–forming, amphibole mineral that often occurs in prismatic crystals (left) and has two directions of cleavage at 56° and 124° (right).

FIGURE 1.13 Plagioclase feldspar, variety labradorite. It has two directions of cleavage with striations commonly visible on one cleavage surface.

FIGURE 1.11 Augite. A dark green to black, rock-forming, pyroxene mineral with short, stout crystals (illustrated). It has two directions of cleavage at 87° and 93° (not illustrated).

FIGURE 1.12 Potassium feldspar, variety microcline. The potassium feldspars are common rock-forming minerals with various colors (commonly white, cream, or pink) and two directions of cleavage at nearly 90°.

Some minerals, such as hematite (Figure 1.7), occur in both metallic and nonmetallic varieties.

The ability of a mineral to transmit light can be described as either **opaque**, when no light is transmitted (e.g. Figure 1.3); **translucent**, when light but not an image is transmitted; or, **transparent**, when an image is visible through the mineral (e.g. Figure 1.1). In general, most minerals with a metallic luster are opaque, while vitreous minerals are either translucent or transparent.

Examine the mineral specimens provided by your instructor and answer the following questions.

1. How many of your specimens can be grouped into each of the following luster types?

   Metallic: _____   Nonmetallic–glassy: _____

   Nonmetallic–earthy: _____

2. How many of your specimens are transparent and how many are opaque?

   Transparent: _____   Opaque: _____

Hardness. **Hardness,** one of the most useful diagnostic properties of a mineral, is a measure of the resistance of a mineral to abrasion or scratching. It is a relative property in that a harder substance will scratch, or cut into, a softer one.

In order to establish a common system for determining hardness, Friedrich Mohs (1773–1839), a German mineralogist, developed a reference scale of mineral hardness. The Mohs scale of hardness (Figure 1.14), widely used today by geologists and engineers, utilizes

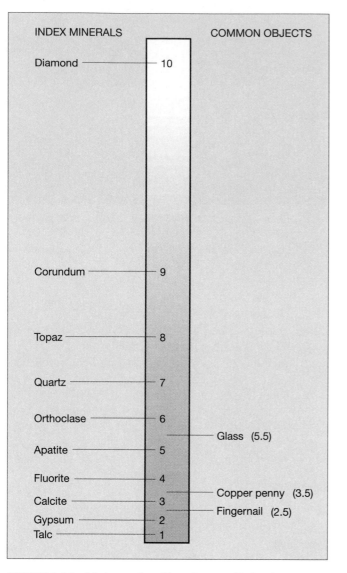

INDEX MINERALS   COMMON OBJECTS

Diamond — 10

Corundum — 9

Topaz — 8

Quartz — 7

Orthoclase — 6

Glass (5.5)

Apatite — 5

Fluorite — 4

Calcite — 3   Copper penny (3.5)

Fingernail (2.5)

Gypsum — 2

Talc — 1

FIGURE 1.14 Mohs scale of hardness with the hardness of some common objects.

ten index minerals as a reference set to determine the hardness of other minerals. The hardness value of 1 is assigned to the softest mineral in the set, talc, and 10 is assigned to the hardest mineral, diamond. Higher-numbered minerals will scratch lower-numbered minerals. For example, quartz, with a hardness of 7, will scratch calcite, which has a hardness of 3. It should be remembered that Mohs scale is a *relative ranking* and does *not* imply that mineral number 2, gypsum, is twice as hard as mineral number 1, talc.

Most people do not have a set of Mohs reference minerals available. However, by knowing the hardness of some common objects, such as those listed on Mohs scale in Figure 1.14, a hardness value can be assigned to a mineral. For example, a mineral that has a hardness greater than 5.5 will scratch glass. Table 1.2 can serve as a guide for determining the hardness of a mineral.

TABLE 1.2 Hardness guide

Hardness	Description
Less than 2.5	A mineral that can be scratched by your fingernail (hardness = 2.5).
2.5 to 5.5	A mineral that cannot be scratched by your fingernail (hardness = 2.5), and cannot scratch glass (hardness = 5.5).
Greater than 5.5	A mineral that scratches glass (hardness = 5.5).

3. Test the hardness of several of the mineral specimens provided by your instructor by rubbing any two together to determine which are hard (the minerals that do the scratching) and which are soft (the minerals that are scratched). Doing this will give you an indication of what is meant by the term "relative hardness" of minerals.

4. Use the hardness guide in Table 1.2 to find an example of a mineral supplied by your instructor that falls in each of the three categories.

Color.   **Color,** although an obvious feature of minerals, may also be misleading. For example, slight impurities in a mineral may result in one sample of the mineral having one color while a different sample of the same mineral may have an entirely different color. *Thus, color is one of the least reliable physical properties.*

5. Observe the minerals in Figure 1.8. Both of the minerals shown are varieties of the same mineral, quartz. What is the reason for the variety of colors that quartz exhibits?

_____

_____

6. Examine the mineral specimens supplied by your instructor and see if any appear to be the same mineral but with variable colors.

Streak.   The **streak** of a mineral is the color of the fine powder of a mineral obtained by rubbing a corner across a piece of unglazed porcelain—called a *streak plate*. Whereas the color of a mineral may vary from sample to sample, its streak usually does not and is therefore the more reliable property (see Figure 1.7). In many cases the color of a mineral's streak may not be the same as the color of the mineral. (*Note:* Minerals that have about the same hardness as, or are harder than, a streak plate (about 7 on Mohs scale of hardness), may not powder or produce a streak.)

---

7. Select three of the mineral specimens provided by your instructor. Do they exhibit a streak? If so, is the streak the same color as the mineral specimen? List your observations for each specimen in the following space.

	Color of Specimen	**Streak**
Specimen 1:	_____	_____
Specimen 2:	_____	_____
Specimen 3:	_____	_____

Crystal Form. **Crystal form** is the external appearance or shape of a mineral that results from the internal, orderly arrangement of atoms (Figure 1.15). Most inorganic substances consist of crystals. The flat external surfaces on a crystal are called *crystal faces*. A mineral that forms without space restrictions will exhibit well-formed crystal faces. However, most of the time, minerals must compete for space and the result is a dense intergrown mass in which crystals do not exhibit their crystal form, especially to the unaided eye.

One of the most useful instruments for measuring the angle between crystal faces on large crystals is the **contact goniometer** (Figure 1.16).

---

8. The mineral shown in Figure 1.1 has a well–developed crystal form with six faces that intersect at about 120° and come to a point. Several varieties of the same mineral are shown in Figure 1.8. Why do those in Figure 1.8 not exhibit crystal form?

_____

_____

9. Select one of the photographed minerals, other than Figure 1.1, that exhibits its crystal form and describe its shape.

Figure _____ : _____

10. Observe the various size crystals of the mineral quartz on display in the lab. Use the contact go-

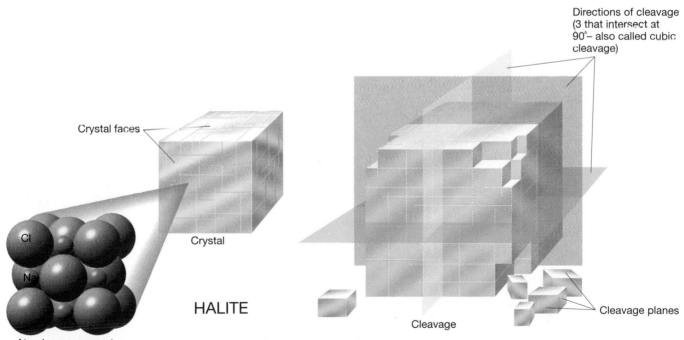

FIGURE 1.15 Atomic arrangement, crystal form, and cleavage of the mineral halite. Halite (NaCl) contains sodium and chlorine atoms arranged in a one-to-one ratio forming a cube. The internal, orderly arrangement of atoms produces the external cubic crystal form of the mineral. Planes of weak bonding between atoms in the internal crystalline structure are responsible for halite's cubic cleavage.

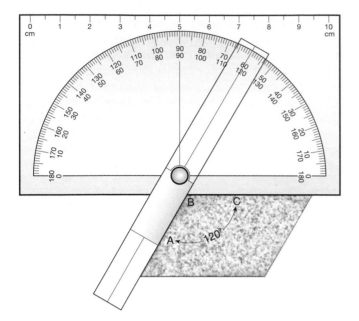

niometer to measure the angle between similar, adjacent crystal faces on several crystals. Then, write a statement relating the angle between adjacent crystal faces to the size of the crystal.

_____

_____

FIGURE 1.16 Contact goniometer. To use the instrument, hold the straight edge of the protractor in contact with one crystal face and the edge of the celluloid strip in contact with the other face. The angle is read where the fine line on the celluloid strip overlaps the degrees on the protractor. For example, the angle between the adjacent crystal faces on the mineral illustrated (angle ABC) is 120°.

Cleavage and Fracture. **Cleavage** is the tendency of some minerals to break along regular planes of weak bonding between atoms in the internal crystalline structure (see Figure 1.15). When broken, minerals that exhibit cleavage produce smooth, flat surfaces, called **cleavage planes.**

Cleavage is described by noting the number of **directions of cleavage,** which is the number of different sets of planes that form the surfaces of a mineral crystal when it cleaves, and the angle(s) at which the directions of cleavage meet (see Figure 1.15). Each cleavage plane of a mineral crystal that has a different orientation is counted as a different direction of cleavage. When two or more cleavage planes are parallel or line up with each other, they are counted only once, as one direction of cleavage. Minerals may have one, two, three, four, or six directions of cleavage. Figure 1.17 illustrates some of the common cleavage directions of minerals.

FIGURE 1.17 Common cleavage directions of minerals. (After AGI/NAGT, Laboratory Manual in Physical Geology, 2nd ed., New York: Macmillan Publishing Company, 1990.)

Number of Cleavage Directions	Shape	Sketch
0 No cleavage, only fracture	Irregular masses	
1	Flat sheets	
2 at 90°	Elongated form with rectangle cross section (prism)	
2 not at 90°	Elongated form with parallelogram cross section (prism)	
3 at 90°	Cube	
3 not at 90°	Rhombohedron	
4	Octahedron	
6	Dodecahedron	

Observe the minerals shown in Figures 1.5 and 1.15. These minerals have broken with regularity and exhibit cleavage. The crystals produced are in the form of a cube. Although there are six planes of cleavage surrounding each crystal, each exhibits only three directions of cleavage: top and bottom form one parallel set of planes—hence the first direction of cleavage; the two sides are a second parallel set—a second direction of cleavage; and the front and back form the third direction of cleavage. The cleavage of both minerals is described as three directions of cleavage that intersect at 90° (also commonly called *cubic cleavage*).

Cleavage and crystal form are *not* the same. Some mineral crystals cleave while others do not. Cleavage is determined by the bonds that hold atoms together, while crystal form results from the internal, orderly arrangement of the atoms. The best way to determine whether or not a mineral cleaves is to break it and carefully examine the results.

Minerals that do not exhibit cleavage when broken are said to **fracture** (see Figure 1.4). Fracturing can be irregular, splintery, or conchoidal (smooth curved surfaces resembling broken glass). Some minerals may cleave in one or two directions and exhibit fracturing where not cleaved (see Figure 1.12).

▬▬▬

11. The mineral shown in Figure 1.9 has one direction of cleavage. Describe the appearance of a mineral that exhibits this type of cleavage.

_____

12. Observe the photograph of calcite, the mineral on the right in Figure 1.2. Several smooth, flat planes result when the mineral is broken.
    a. How many planes of cleavage are present on the specimen?

    _____ planes of cleavage

    b. How many directions of cleavage are present on the specimen?

    _____ directions of cleavage

    c. The cleavage directions meet at (90° angles, angles other than 90°). Circle your answer.
13. Select one mineral specimen supplied by your instructor that exhibits cleavage. Describe its cleavage by completing the following statement.

    _____ directions of cleavage at _____ degrees

Specific Gravity.
Specific gravity is a number which represents the ratio of the weight of a mineral to an equal volume of water. For example, the mineral quartz, Figures 1.1 and 1.8, has a specific gravity of 2.65; this means it weighs 2.65 times more than an equal volume of water. The mineral galena, Figure 1.5, with a specific gravity of 7.4, feels heavy when held in your hand. With a little practice, you can estimate the specific gravity of a mineral by hefting it in your hand. The average specific gravity of minerals is about 2.7, but some metallic minerals have a specific gravity two or three times greater than the average. (*Note:* Exercise Twenty-two, "The Metric System, Measurements, and Scientific Inquiry," contains a simple experiment for estimating the specific gravity of a solid.)

▬▬▬

14. Find a mineral specimen supplied by your instructor that exhibits a high specific gravity by giving each mineral a heft in your hand.

Other Properties of Minerals.
Luster, the ability to transmit light, hardness, color, streak, crystal form, cleavage or fracture, and specific gravity are the most basic and common physical properties used to identify minerals. However, other special properties can also be used to identify certain minerals. These other properties include

Magnetism Magnetism is characteristic of minerals, such as magnetite, that have a high iron content and are attracted by a magnet. A variety of magnetite called *lodestone* is itself actually magnetic and will pick up paper clips.

Taste The mineral halite (Figure 1.2, center) has a "salty" taste. (*CAUTION*: DO NOT TASTE ANY MINERALS OR OTHER MATERIALS WITHOUT KNOWING IT IS *ABSOLUTELY* SAFE TO DO SO.)

Feel The mineral talc often feels "soapy" while the mineral graphite has a "greasy" feel.

Striations Striations are closely-spaced, fine lines on the crystal faces of some minerals. They resemble the surface of a phonograph record but are straight (see Figure 1.6). In addition to pyrite, certain plagioclase feldspar minerals often exhibit striations on one cleavage surface (see Figure 1.13).

Tenacity Tenacity is the manner in which a substance resists breaking. Terms like *flexible* (a thin piece of plastic) and *brittle* (glass) are used to describe this property.

Reaction to Dilute Hydrochloric Acid A very small drop of dilute hydrochloric acid, when placed on a freshly exposed surface of some minerals, will cause them to "fizz" (effervesce) as the gas carbon dioxide is released. The test is often used to identify a group of minerals called the *carbonate minerals*. The mineral calcite, Figure 1.2 (right), (chemical name: calcium carbonate) is the most common carbonate mineral and is frequently found in rocks. (*CAUTION:* Hydrochloric acid can discolor, decompose, and disintegrate mineral and rock samples. USE THE ACID ONLY AFTER YOU HAVE RECEIVED SPECIFIC INSTRUCTIONS ON ITS USE FROM YOUR INSTRUCTOR. NEVER TASTE MINERALS THAT HAVE HAD ACID PLACED ON THEM.)

15. Following the directions given by your instructor, examine the mineral specimens to determine if any exhibit one or more of the special properties listed above.

IDENTIFICATION OF MINERALS

Having investigated the physical properties of minerals, you are now prepared to proceed with the identification of the minerals supplied by your instructor.

To identify a mineral, you must first determine, using available tools, as many of its physical properties as you can. Next, knowing the properties of the mineral, you proceed to a mineral identification key, which often functions like an outline, to narrow down the choices and arrive at a specific name. *As you complete the exercise, remember that the goal is to learn the procedure for identifying minerals through observation and not simply to put a name on them.*

Arrange your mineral specimens by placing them on a numbered sheet of paper. Locate the mineral identification chart, Figure 1.18, and write the numbers of your mineral specimens, in order, under the column labeled "Specimen Number."

Using a Mineral Identification Key.
Figure 1.19 is a mineral identification key that uses the property of luster as the primary division of minerals into two groups, those with metallic lusters and those with nonmetallic lusters. Color (either dark or light colored) is used as a secondary division for the nonmetallic minerals. Examine the mineral identification key closely to see how it is arranged.

16. Use the mineral identification key (Figure 1.19). What would be the name of the mineral with these properties: nonmetallic luster, light colored, softer than a fingernail, produces small, thin plates or sheets when scratched by a fingernail, white color, and a "soapy" feel?

Mineral name: _____ (Check your answer with your instructor before proceeding.)

17. Complete the mineral identification chart, Figure 1.18, by listing the properties of each of the mineral specimens supplied by your instructor. Use the mineral identification key, Figure 1.19, to determine the name of each of the minerals.

18. Use Table 1.1, Mineral Uses, to determine which of the minerals you identify have an economic use. List their use in the column to the right of their name on the mineral identification chart, Figure 1.18.

19. The mineral photographs, Figure 1.2 (right, calcite) and Figures 1.8 through 1.13, show some of the most common rock-forming minerals. If your mineral specimens include examples of these minerals, indicate that they are rock-forming in the column to the right of their name on your mineral identification chart.

REVIEW

Having completed the exercise, you should know the following:

1. A mineral is a natural combination of atoms that has characteristic physical properties.
2. How to determine and use the physical properties of a mineral for identification with the aid of a mineral classification key.
3. Some minerals have economic uses while others are called rock-forming minerals.

MINERAL IDENTIFICATION CHART

Specimen Number	Luster	Hardness	Color	Streak	Fracture	Cleavage (number of directions and angle of intersection)	Other Properties	Name	Economic Use or Rock-forming

FIGURE 1.18   Mineral identification chart.

Hardness	Streak	Other Diagnostic Properties	Name (Chemical Composition)
Harder than glass	Black streak	Black; magnetic; hardness = 6; specific gravity = 5.2; often granular	Magnetite (Fe_3O_4)
	Greenish-black streak	Brass yellow; hardness = 6; specific gravity = 5.2; generally an aggregate of cubic crystals	Pyrite (FeS_2)— fool's gold
	Red-brown streak	Gray or reddish brown; hardness = 5–6; specific gravity = 5; platy appearance	Hematite (Fe_2O_3)
Softer than glass	Greenish-black streak	Golden yellow; hardness = 4; specific gravity = 4.2; massive	Chalcopyrite ($CuFeS_2$)
	Gray-black streak	Silvery gray; hardness = 2.5; specific gravity = 7.6 (very heavy); good cubic cleavage	Galena (PbS)
	Yellow-brown streak	Yellow brown to dark brown; hardness variable (1–6); specific gravity = 3.5–4; often found in rounded masses; earthy appearance	Limonite ($Fe_2O_3 \cdot H_2O$)
	Gray-black streak	Black to bronze; tarnishes to purples and greens; hardness = 3; specific gravity = 5; massive	Bornite (Cu_5FeS_4)
Softer than your fingernail	Dark gray streak	Silvery gray; hardness = 1 (very soft); specific gravity = 2.2; massive to platy; writes on paper (pencil lead); feels greasy	Graphite (C)

FIGURE 1.19 Mineral identification key.

Group II
Nonmetallic luster (dark colored).

Hardness	Cleavage	Other Diagnostic Properties	Name (Chemical Composition)
Harder than glass	Cleavage present	Black to greenish black; hardness = 5–6; specific gravity = 3.4; fair cleavage, two directions at nearly 90 degrees	Augite (Ca, Mg, Fe, Al silicate)
		Black to greenish black; hardness = 5–6; specific gravity = 3.2; fair cleavage, two directions at nearly 60 degrees and 120 degrees	Hornblende (Ca, Na, Mg, Fe, OH, Al silicate)
		Red to reddish brown; hardness = 6.5–7.5; conchoidal fracture; glassy luster	Garnet (Fe, Mg, Ca, Al silicate)
	Cleavage not prominent	Gray to brown; hardness = 9; specific gravity = 4; hexagonal crystals common	Corundum (Al_2O_3)
		Dark brown to black; hardness = 7; conchoidal fracture; glassy luster	Smoky quartz (SiO_2)
		Olive green; hardness = 6.5–7; small glassy grains	Olivine $(Mg, Fe)_2SiO_4$
Softer than glass	Cleavage present	Yellow brown to black; hardness = 4; good cleavage in six directions, light yellow streak that has the smell of sulfur	Sphalerite (ZnS)
		Dark brown to black; hardness = 2.5–3, excellent cleavage in one direction; elastic in thin sheets; black mica	Biotite (K, Mg, Fe, OH, Al silicate)
	Cleavage absent	Generally tarnished to brown or green; hardness = 2.5; specific gravity = 9; massive	Native copper (Cu)
Softer than your fingernail	Cleavage not prominent	Reddish brown; hardness = 1–5; specific gravity = 4–5; red streak; earthy appearance	Hematite (Fe_2O_3)
		Yellow brown; hardness = 1–3; specific gravity = 3.5; earthy appearance; powders easily	Limonite ($Fe_2O_3 \cdot H_2O$)

FIGURE 1.19 *(continued)*

Hardness	Cleavage	Other Diagnostic Properties	Name (Chemical Composition)
Harder than glass	Cleavage present	Flesh colored or white to gray; hardness = 6; specific gravity=2.6; two directions of cleavage at nearly right angles	Potassium feldspar $(KAlSi_3O_8)$ Plagioclase feldspar $(NaAlSi_3O_8$ to $CaAl_2Si_2O_8)$
	Cleavage absent	Any color; hardness = 7; specific gravity = 2.65; conchoidal fracture; glassy appearance; varieties: milky, rose, smoky, amethyst (violet)	Quartz (SiO_2)
Softer than glass	Cleavage present	White, yellowish to colorless; hardness = 3; three directions of cleavage at 75 degrees (rhombohedral); effervesces in HCl; often transparent	Calcite $(CaCO_3)$
		White to colorless; hardness = 2.5; three directions of cleavage at 90 degrees (cubic); salty taste	Halite $(NaCl)$
		Yellow, purple, green, colorless; hardness = 4; white streak; translucent to transparent; four directions of cleavage	Fluorite (CaF_2)
Softer than your fingernail	Cleavage present	Colorless; hardness = 2–2.5; transparent and elastic in thin sheets; excellent cleavage in one direction; light mica	Muscovite (K, OH, Al silicate)
		White to transparent, hardness = 2; when in sheets, is flexible but not elastic; varieties: selenite (transparent, three directions of cleavage); satin spar (fibrous, silky luster); alabaster (aggregate of small crystals)	Gypsum $(CaSO_4 \cdot 2H_2O)$
	Cleavage not prominent	White, pink, green; hardness = 1–2; forms in thin plates; soapy feel; pearly luster	Talc (Mg silicate)
		Yellow; hardness = 1–2.5	Sulfur (S)
		White; hardness = 2; smooth feel; earthy odor; when moistened, has typical clay texture	Kaolinite (Hydrous Al silicate)
		Green; hardness = 2.5; fibrous; variety of serpentine	Asbestos (Mg, Al silicate)
		Pale to dark reddish brown; hardness = 1–3; dull luster; earthy; often contains spheroidal-shaped particles; not a true mineral	Bauxite (Hydrous Al oxide)

FIGURE 1.19 *(continued)*

The Study of Minerals

SUMMARY/REPORT PAGE

Date Due: _____

Name: _____

Date: _____

Class: _____

After you have finished Exercise One, complete the following questions. You may have to refer to the exercise for assistance or to locate specific answers. Be prepared to submit this summary/report to your instructor at the designated time.

1. Describe the procedure for identifying a mineral and arriving at its name.

_____

_____

_____

2. Name the physical property of a mineral that is described by each of the following statements.

Physical Property

Breaks along smooth planes: _____

Scratches glass: _____

Shines like a metal: _____

A red-colored powder on unglazed porcelain:

_____

3. Describe the shape of a mineral that has three directions of cleavage that intersect at 90°.

_____

4. Name two minerals you identified that have good cleavage. Describe the cleavage of each mineral.

Mineral	Cleavage
_____ :	_____
_____ :	_____

5. Select one mineral you identified and list its name and physical properties.

_____ : _____

_____

_____

6. Name one mineral that you identified that has an economic use.

Mineral	Mined for
_____ :	_____

7. List the name and hardness of two minerals you identified.

Mineral	Hardness
_____ :	_____
_____ :	_____

8. How many directions of cleavage do the feldspar minerals, potassium feldspar and plagioclase feldspar, have?

_____ directions of cleavage

9. What was your conclusion concerning the angles between similar crystal faces on different size crystals of the same mineral?

_____

10. List the name(s) of the minerals you identified that had a special property such as magnetism or feel. Write the special property that you observed next to the name of the mineral.

Mineral	Special Property
_____ :	_____
_____ :	_____
_____ :	_____

EXERCISE TWO

Common Rocks

To an earth scientist, rocks represent much more than usable substances. They are the materials of the earth; understanding their origin and how they change allows us to begin to understand the earth and its processes. It is often said that "the history of the earth is written in the rocks"—we just have to be smart enough to read the "words."

In this exercise, you will investigate some of the common rocks that are found on and near the surface of the earth. The criteria used to classify a rock as being of either igneous, sedimentary, or metamorphic origin are examined, as well as the procedure for identifying rocks within each of these three families.

OBJECTIVES

After you have completed this exercise, you should be able to

1. Examine a rock and determine if it is an igneous, sedimentary, or metamorphic rock.
2. List and define the terms used to describe the textures of igneous, sedimentary, and metamorphic rocks.
3. Name the dominant mineral(s) found in the most common igneous, sedimentary, and metamorphic rocks.
4. Use a classification key to identify a rock.
5. Recognize and name some of the common rocks by sight.

TEXTBOOK REFERENCE

Chapter 1

MATERIALS

metric ruler

Materials Supplied by Your Instructor

igneous rocks
sedimentary rocks
metamorphic rocks
hand lens or binocular microscope

dilute hydrochloric acid
streak plate
glass plate
copper penny

TERMS

rock	weathering	composition
rock cycle	sediment	detrital material
igneous rock	lithification	chemical material
magma	metamorphic rock	foliation
sedimentary rock	texture	

INTRODUCTION

For thousands of years rocks have proven their usefulness because of their durability and beauty. Granite and limestone are widely used as building materials and decorative coverings. Slate is utilized for hard, enduring table surfaces. And everyone recognizes the beauty of a sculpture carved from a block of fine marble.

Most **rocks** are aggregates (mixtures) of minerals. However, there are some rocks that are composed essentially of one mineral found in large impure quantities. The rock limestone, consisting almost entirely of the mineral calcite, is a good example.

Rocks are classified into three types, based on the processes that formed them. One of the most useful devices for understanding rock types and the geologic

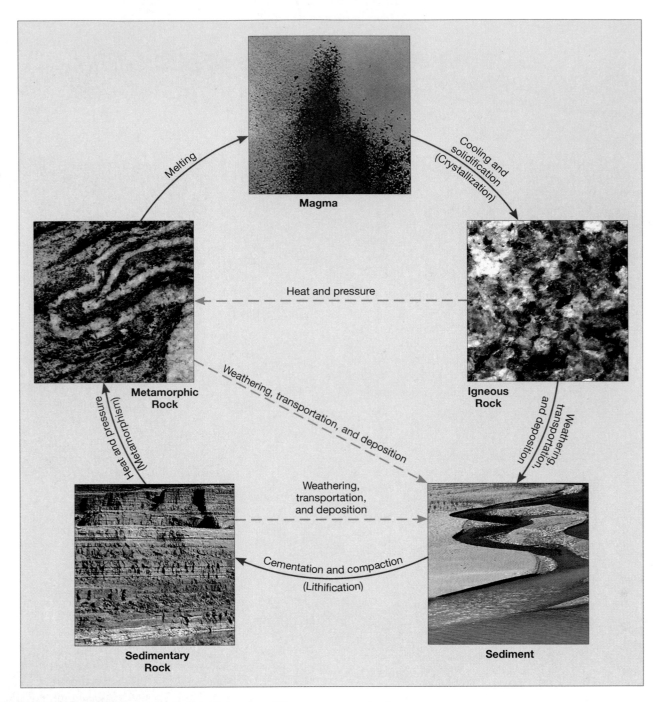

FIGURE 2.1 The rock cycle illustrating the role of the various geologic processes that act to transform one rock type into another. (Photos by J. D. Griggs, U.S.G.S. (top); E. J. Tarbuck (top right, bottom right); and Stephen Trimble (top left, bottom left))

processes that transform one rock type into another is the **rock cycle.** The cycle, shown in Figure 2.1, illustrates the various earth materials and uses arrows to indicate chemical and physical processes. As you examine the rock cycle and read the following definitions, notice the references to the origin of each rock type.

The three types of rock are igneous, sedimentary, and metamorphic.

Igneous Igneous rocks (Figures 2.2–2.7) are the solidified products of once-molten material called **magma.** The distinguishing feature of most igneous rocks is an interlocking arrangement of mineral crystals that forms as the molten material cools and crystals grow. *Intrusive* igneous rocks form below the surface of the earth, while those that form at the surface from lava are termed *extrusive.*

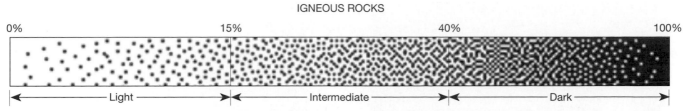

0%   15%   40%   100%

|← Light →|← Intermediate →|← Dark →|

Color index used for estimating the percent of dark minerals in an igneous rock.

|← 6 cm →|

FIGURE 2.2 Granite, a common coarse-grained, intrusive igneous rock composed primarily of quartz and potassium feldspar mineral crystals. (Photo by E. J. Tarbuck)

|← 6 cm →|

FIGURE 2.4 Diorite, a coarse-grained, intrusive igneous rock composed primarily of plagioclase feldspar and amphibole minerals.

|← 5 cm →|

FIGURE 2.3 Rhyolite, a fine-grained, extrusive igneous rock composed of light-colored feldspars and quartz.

|← 7 cm →|

FIGURE 2.5 Basalt, a fine-grained igneous rock composed primarily of pyroxene and plagioclase feldspar minerals that forms on, or near, the earth's surface.

FIGURE 2.6 Andesite porphry, an igneous rock with a porphyritic texture. The larger crystals are termed *phenocrysts* and the smaller, surrounding crystals, the *groundmass* (or *matrix*). (Photo by E. J. Tarbuck)

FIGURE 2.7 Obsidian, an igneous rock with a glassy texture that forms when lava is cooled very suddenly and no mineral crystals form.

Sedimentary These rocks (Figures 2.8–2.13) form at or near the earth's surface from the accumulated products of **weathering**, called **sediment.** These products may be solid particles or material that was formerly dissolved and then precipitated by either inorganic or organic processes. The process of **lithification** trans-

forms the sediment into hard rock. Since sedimentary rocks form at, or very near, the earth's surface, they often contain organic matter and/or fossils. The layering (or bedding) that develops as sediment is sorted by and settles out from a transporting material (usually water or air) helps make sedimentary rocks recognizable.

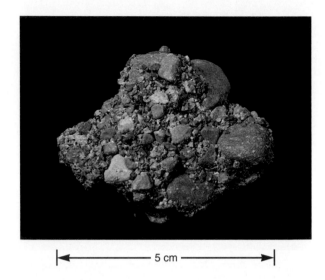

FIGURE 2.8 Conglomerate, a detrital sedimentary rock composed of large rounded grains. (Photo by E. J. Tarbuck)

FIGURE 2.9 Sandstone, a common detrital sedimentary rock composed primarily of quartz grains. (Photo by E. J. Tarbuck)

FIGURE 2.12 Rock salt, a chemical sedimentary rock formed as water evaporates leaving formerly dissolved material to precipitate.

← 7 cm →

FIGURE 2.10 Shale, a detrital sedimentary rock composed of very fine grains, primarily of the mineral clay.

← 18 cm →

FIGURE 2.11 Fossiliferous limestone, a biochemical sedimentary rock containing shells or shell fragments composed of the mineral calcite that have been cemented together with fine, crystalline calcite.

← 6 cm →

FIGURE 2.13 Bituminous coal, a sedimentary rock composed of altered plant remains.

Metamorphic These rocks (Figures 2.14–2.19) form below the earth's surface where high temperatures, pressures, and/or chemical fluids change preexisting rocks without melting them.

Minerals are identified by using their physical and chemical properties. However, rock types and the names of individual rocks are determined by describing their textures and compositions. The key to success in rock identification lies in learning to accurately determine and describe texture and composition.

Texture refers to the shape, arrangement, and size of mineral grains in a rock. The shape and arrangement of mineral grains help determine the type (igneous, sedimentary, or metamorphic) of rock. Mineral grain size is often used to separate rocks within a particular type. Each rock type uses different terms to describe its textures.

Composition refers to the minerals that are found in a rock. Often the larger mineral grains can be identified by sight or by using their physical properties. In some cases, small mineral grains may require the use of a hand lens or microscope. Occasionally, very small grains cannot be identified with the normal magnification of a microscope. Practice and increased familiarity with the minerals will make this assessment easier.

|← ———————— 3 cm ———————— →|

FIGURE 2.15 Phyllite, a foliated metamorphic rock with barely visible grains and silky appearance.

|← ———————— 8 cm ———————— →|

FIGURE 2.14 Slate, a fine-grained, foliated metamorphic rock.

|← ———————— 6 cm ———————— →|

FIGURE 2.16 Schist, a foliated metamorphic rock with visible grains. The specimen illustrated is called mica schist. (Photo by E. J. Tarbuck)

FIGURE 2.17   Gneiss, a foliated-banded metamorphic rock that often forms during intensive metamorphism. The light bands often consist of quartz and feldspar minerals, while the dark bands are usually platy or linear minerals such as mica or amphibole. (Photo by E. J. Tarbuck)

|← 5 cm →|

|← 6 cm →|

FIGURE 2.18   Pink marble, a nonfoliated metamorphic rock that forms from the metamorphism of the sedimentary rock limestone and is composed of the mineral calcite. (Photo by E. J. Tarbuck)

|← 5 cm →|

FIGURE 2.19   Anthracite coal, often called hard coal, forms from the metamorphism of bituminous coal.

COMPARING IGNEOUS, SEDIMENTARY, AND METAMORPHIC ROCKS

One of the first steps in the identification of rocks is to determine the rock type. Each of the three rock types has a somewhat unique appearance that helps to distinguish one type from the other.

Examine the specimens of the three rock types supplied by your instructor, as well as the photographs of the rocks in Figures 2.2–2.19. Then answer the following questions.

1. Which two of the three rock types appear to be made primarily of intergrown crystals?

   _____ rocks and _____ rocks

2. Which one of the two rock types you listed in question 1 has the mineral crystals aligned or arranged so that they point in the same direction in a linear, linelike manner?

   _____

3. Which one of the two rock types you listed in question 1 has the mineral crystals in most of the rocks arranged in a dense interlocking mass with no alignment?

   _____

4. Of the three rock types, (igneous, sedimentary, meta-morphic) rocks often contain haphazardly arranged pieces or fragments, rather than crystals. Circle your answer.

IGNEOUS ROCK IDENTIFICATION

Igneous rocks form from the cooling and crystallization of magma. The interlocking network of mineral crystals that develops as the molten material cools gives most igneous rocks their distinctive crystalline appearance.

Textures of Igneous Rocks. The rate of cooling of the magma determines the size of the interlocking crystals found in igneous rocks. The slower the cooling rate, the larger the mineral crystals. The five principal textures of igneous rocks are

> **Coarse Grained** (or *phaneritic*) The majority of mineral crystals are of a uniform size and large enough to be identifiable without a microscope. This texture occurs when magma cools slowly inside the earth.

> **Fine Grained** (or *aphanitic*) Very small crystals, which are generally not identifiable without strong magnification, develop when molten material cools quickly on, or very near, the surface of the earth.

> **Porphyritic** Two very contrasting sizes of crystals are caused by magma having two different rates of cooling. The larger crystals are termed *phenocrysts;* and the smaller, surrounding crystals are termed *groundmass* (or *matrix*).

> **Glassy** No mineral crystals develop because of very rapid cooling. This lack of crystals causes the rock to have a glassy appearance. In some cases, rapidly escaping gases may produce a frothy appearance similar to spun glass.

> **Fragmental** The rock contains broken, angular fragments of rocky materials produced during an explosive volcanic eruption.

Examine the igneous rock photographs in Figures 2.2–2.7. Then answer the following questions.

5. The igneous rock illustrated in Figure 2.2 is made of large mineral crystals that are all about the same size. The rock formed from magma that cooled (slowly, rapidly) (inside, on the surface of) the earth. Circle your answers.

6. The rock shown in Figure 2.5 is made of mineral crystals that are all small and not identifiable without a microscope. The rock formed from magma

that cooled (slowly, rapidly) (inside, on/near the surface of) the earth. Circle your answers.

7. The igneous rock in Figure 2.6 has a porphyritic texture. The large crystals are called _____, and the surrounding, smaller crystals are called _____.

8. The rocks in Figures 2.2 and 2.3 have nearly the same mineral composition. What fact about the mineral crystals in the rocks makes their appearances so different? What caused this difference?

_____

_____

Select a coarse-grained rock from the igneous rock specimens supplied by your instructor and examine the mineral crystals closely using a hand lens or microscope.

9. Sketch a diagram showing the arrangement of the mineral crystals in the igneous rock specimen you examined in the space provided below. Indicate the scale of your sketch by writing the appropriate length within the ( ) provided on the bar scale.

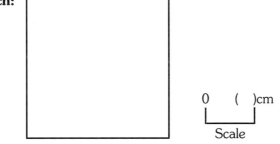

Sketch:

0   (   )cm
Scale

Composition of Igneous Rocks. The specific mineral composition of an igneous rock is ultimately determined by the chemical composition of the magma from which it crystallized. However, the minerals found in igneous rocks can be arranged into three groups. Each group can be identified by observing the proportion of dark-colored minerals compared to light-colored minerals. The three groups are

> **Felsic** (or *granitic*)—composed mainly of the light-colored minerals quartz and potassium feldspars. Dark–colored minerals account for less than 15% of the minerals in rocks found in this group.

> **Intermediate** (or *andesitic*)—a mixture of both light-colored and dark-colored minerals. Dark minerals comprise about 15% to 40% of these rocks.

> **Mafic** (or *basaltic*)—dark-colored minerals such as pyroxene and olivine account for over 40% of the composition of these rocks.

10. Use the color guide included on the same page as the igneous rock photographs to estimate the percent of dark minerals contained in the igneous rock in Figure 2.4. The rock's color is (light, intermediate, dark). Circle your answer.
11. The rocks shown in Figures 2.3 and 2.5 have the same texture. What fact about the mineral crystals makes their appearances so different?

Using an Igneous Rock Identification Key. The name of an igneous rock can be found by first determining its texture and color (an indication of mineral composition), identifying visible mineral grains, and then using an igneous rock identification key such as the one shown in Figure 2.20 to determine the name.

For example, the igneous rock shown in Figure 2.2 has a coarse-grained texture and is light colored (quartz and potassium feldspar dominant). Intersecting the light-colored column with the coarse-grained row on the igneous rock identification key, Figure 2.20, determines that the name of the rock is "granite."

12. Place each of the igneous rocks supplied by your instructor on a numbered piece of paper. Then complete the igneous rock identification chart, Figure 2.21, for each rock. Use the igneous rock identification key, Figure 2.20, to determine each specimen's name.

SEDIMENTARY ROCK IDENTIFICATION

Sedimentary rocks, Figures 2.8–2.13, form from the accumulated products of weathering called *sediment*. Sedimentary rocks can be made of either, or a combination of, detrital or chemical material.

Detrital material consists of mineral grains or rock fragments derived from the process of mechanical weathering that are transported and deposited as solid

IGNEOUS ROCK IDENTIFICATION KEY

		0% 15% 40% 100%		
Color Index and Graphic Illustration				
COLOR		Light	Medium	Dark
CHEMICAL COMPOSITION		Felsic	Intermediate	Mafic
DOMINANT MINERALS		Quartz / Potassium feldspar	Amphibole / Plagioclase feldspar	Pyroxene / Plagioclase feldspar
T E X T U R E	Coarse Grained[1]	**Granite**	**Diorite**	**Gabbro**
	Fine Grained[2]	**Rhyolite**	**Andesite**	**Basalt**[4]
	Porphyritic[3]	"Porphyritic" precedes any of the above names whenever there are appreciable phenocrysts		
	Glassy	**Obsidian** (compact glass) / **Pumice** (frothy glass)		
	Fragmental	**Tuff** (fragments less than 2 mm) / **Volcanic Breccia** (fragments greater than 2 mm)		

[1] Also called *phaneritic*. Crystals generally 1-10 mm (1 cm). The term *pegmatite* is added to the rock name when crystals are greater than 1 cm; e.g. *granite-pegmatite*.
[2] Also called *aphanitic*. Crystals generally less than 1 mm.
[3] For example, a granite with phenocrysts is called *porphyritic granite*.
[4] Basalt with a cinder-like appearance that develops from gas bubbles trapped in cooling lava is called *scoria*.

FIGURE 2.20 Igneous rock identification key. Color, with associated mineral composition, is shown along the top axis. Each rock in a column has the color and composition indicated at the top of the column. Texture is shown along the left side of the key. Each rock in a row has the texture indicated for that row.
To determine the name of a rock, intersect the appropriate column (color & mineral composition) with the appropriate row (texture) and read the name at the place of intersection.

IGNEOUS ROCK IDENTIFICATION CHART

Specimen Number	Texture	Color (light-intermediate-dark)	Dominant Minerals	Rock Name

FIGURE 2.21 Igneous rock identification chart.

particles. Rocks formed in this manner are called *detrital sedimentary rocks*. The mineral pieces that make up a detrital sedimentary rock are called *grains* (or *fragments* if they are pieces of rock). The identification of a detrital sedimentary rock is determined primarily by the size of the grains or fragments. Mineral composition of the rock is a secondary concern.

Chemical material was previously dissolved in water and later precipitated by either inorganic or organic processes. Rocks formed in this manner are called *chemical sedimentary rocks*. If the material is the result of the life processes of water-dwelling organisms—for example, the formation of a shell—it is said to be of biochemical origin. Mineral composition is the primary consideration in the identification of chemical sedimentary rocks.

Sedimentary rocks come in many varieties that have formed in many different ways. For the purpose of examination, this investigation divides the sedimentary rocks into the two groups, *detrital* and *chemical,* based upon the type of material found in the rock.

Examining Sedimentary Rocks. Examine the sedimentary rock specimens supplied by your instructor. Separate those that are made of pieces or fragments of mineral and/or rock material. They are the detrital sedimentary rocks. Do *not* include any rocks that have

abundant shells or shell fragments. You may find the photographs of the detrital sedimentary rocks in Figures 2.8–2.10 helpful. The remaining sedimentary rocks, those with shells or shell fragments and those that consist of crystals, are the chemical rocks.

Pick up each detrital rock specimen and rub your finger over it to feel the size of the grains or fragments.

▬▬▬

13. How many of your detrital specimens feel rough like sand? How many feel smooth like mud or clay?

_____ specimens feel rough and _____ feel smooth

Use a hand lens or microscope to examine the grains or fragments of several coarse detrital rock specimens. Notice that they are not crystals.

▬▬▬

14. Sketch the magnified pieces and surrounding material, called *cement* (or *matrix*), of a coarse detrital rock in the space provided below. Indicate the scale of your sketch by writing the appropriate length within the ( ) provided on the bar scale.

Sketch:

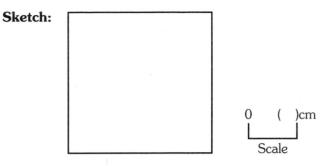

0 (   )cm

Scale

a. Observe the material surrounding the grains or fragments in the rock specimen closely with a hand lens or microscope. The material is (coarse, fine). Circle your answer.

b. Write a brief description of the detrital rock specimen you have examined.

_____

_____

Two of the minerals that often comprise the grains of detrital sedimentary rocks are quartz, a hard (hardness = 7) mineral with a glassy luster, and clay, a soft, fine mineral that consists of microscopic platy particles. The difference in appearance and hardness of quartz and clay is helpful in distinguishing them.

15. How many of your detrital specimens are made of quartz, and how many appear to be made of clay?

_____ specimens have quartz grains and

_____ have clay grains

As a result of their method of formation, many chemical sedimentary rocks are fine-to-coarse crystalline while others consist of shells or shell fragments.

16. How many of your chemical sedimentary rocks are crystalline, and how many contain abundant shells or shell fragments?

_____ specimens are crystalline and

_____ contain shells or shell fragments.

The rock limestone, Figure 2.11, is the most abundant chemical sedimentary rock. It has several origins and many different varieties; however, one thing that all limestones have in common is that they are made of the calcium carbonate mineral called *calcite*. Calcite can precipitate directly from the sea to form limestone or can be used by marine organisms to make shells. After the organisms die, the shells become sediment and eventually the sedimentary rock limestone.

Calcite is a mineral that reacts with dilute hydrochloric acid and effervesces as carbon dioxide gas is released. Most limestones react readily when a small drop of acid is placed on them, thus providing a good test for identifying the rock. Many limestones also contain fragments of sea shells, which also aid in their identification.

17. Follow the directions of your instructor to test the specified sedimentary rock(s) with the dilute hydrochloric acid provided and observe the results. (*Note:* Several detrital sedimentary rocks have calcite surrounding their grains or fragments (calcite cement) that will effervesce with acid and give a *false* test for limestone. Observe the acid reaction closely.)

Using a Sedimentary Rock Identification Key. The sedimentary rock identification key in Figure 2.22 divides the sedimentary rocks into detrital and chemical types. Notice that the primary subdivisions for the detrital rocks are based upon grain size, whereas composition is used to subdivide the chemical rocks.

18. Place each of the sedimentary rocks supplied by your instructor on a numbered piece of paper. Then complete the sedimentary rock identification chart, Figure 2.23, for each rock. Use the sedimentary rock identification key, Figure 2.22, to determine each specimen's name.

Sedimentary Rocks and Environments. Sedimentary rocks are extremely important in the study of earth history. Particle size and the materials from which they are made often suggest something about the place, or environment, in which the rock formed. The fossils that often are found in a sedimentary rock also provide information about the rock's history.

Re-examine the sedimentary rocks and think of them as representing a "place" on the earth where the sediment was deposited. For example, Figure 2.9 is the rock sandstone which formed from sand.

19. Where on the earth do you find sand, the primary material of sandstone, being deposited today?

_____

SEDIMENTARY ROCK IDENTIFICATION KEY

DETRITAL ROCKS

Texture (grain size)	Composition	Rock Name
Coarse (over 2 mm) with large grains	Rounded fragments of quartz and/or chert	**Conglomerate**
	Angular fragments of quartz and/or chert	**Breccia**
Medium (1/16 to 2 mm) feels "sandy"	Quartz usually dominates (If abundant feldspar is present the rock is called **Arkose**)	**Sandstone**
Fine (1/16 to 1/256 mm)	Quartz and clay	**Siltstone**
Very fine (less than 1/256 mm)	Quartz and clay	**Shale**

CHEMICAL ROCKS

Composition	Texture (grain size)	Rock Name	
Calcite, $CaCO_3$ (will effervesce)	Fine to coarse crystalline	**Crystalline Limestone**	
	Visible shells and shell fragments loosely cemented	**Coquina**	**B i o c h e m i c a l** **L i m e s t o n e**
	Various size shells and shell fragments cemented with calcite cement	**Fossiliferous Limestone**	
	Microscopic shells and clay	**Chalk**	
Dolomite $CaMg(CO_3)_2$ (will effervesce if powdered)	Fine to coarse crystalline	**Dolostone**	
Quartz, SiO_2	Very fine crystalline	**Chert (light colored) Flint (dark colored)**	
Gypsum $CaSO_4 \cdot 2H_2O$	Fine to coarse crystalline	**Rock Gypsum**	
Halite, NaCl	Fine to coarse crystalline	**Rock Salt**	
Altered plant fragments	Various size fragments	**Bituminous Coal**	

FIGURE 2.22 Sedimentary rock identification key. Sedimentary rocks are divided into two groups, detrital and chemical, depending upon the type of material that composes them. Detrital rocks are further subdivided by the size of their grains, while the subdivision of the chemical rocks is determined by composition.

SEDIMENTARY ROCK IDENTIFICATION CHART

Specimen Number	Detrital or Chemical	Texture (grain size)	Composition	Rock Name

FIGURE 2.23 Sedimentary rock identification chart.

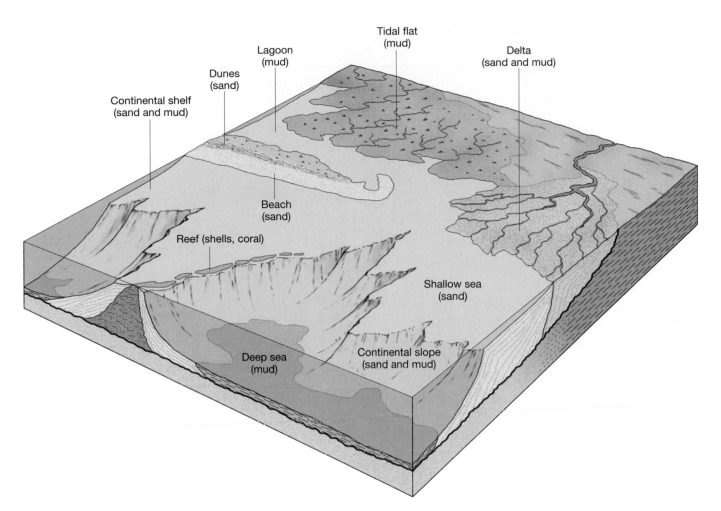

FIGURE 2.24 Generalized illustration of sedimentary environments. Although many environments exist on both the land and in the sea, only some of those that are found along a coast and in the ocean are illustrated.

Figure 2.24 shows a few generalized environments (places) where sediment accumulates. Often, an environment is characterized by the type of sediment and life forms associated with it.

---

20. Use Figure 2.24 to name the environment(s) where, in the past, the sediment for the following sedimentary rocks may have been deposited.

Original Sediment		Environment(s)
Sandstone:	(sand)	_____
Shale:	(mud)	_____
Limestone:	(coral, shells)	_____

METAMORPHIC ROCK IDENTIFICATION

Metamorphic rocks were previously igneous, sedimentary, or other metamorphic rocks that were changed by any combination of heat, pressure, and/or chemical fluids during the process of **metamorphism.** They are most often located beneath sedimentary rocks on the continents and in the cores of mountains.

During metamorphism new minerals may form, and/or existing minerals can grow larger as metamorphism becomes more intense. Frequently, mineral crystals that are elongated (like hornblende) or have a sheet structure (like the micas—biotite and muscovite) become oriented perpendicular to compressional forces. The resulting parallel, linear alignment of mineral crystals perpendicular to compressional forces (stress) is called **foliation** (Figure 2.25). Foliation is unique to many metamorphic rocks and gives them a layered or banded appearance.

Metamorphic rocks are divided into two groups based on texture—foliated and nonfoliated (see Figure 2.25). These textural divisions provide the basis for the identification of metamorphic rocks.

Foliated Metamorphic Rocks. The mineral crystals in foliated metamorphic rocks are either elongated or

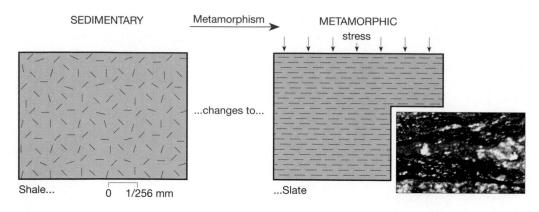

Shale... 0 1/256 mm ...Slate

A. Foliated texture

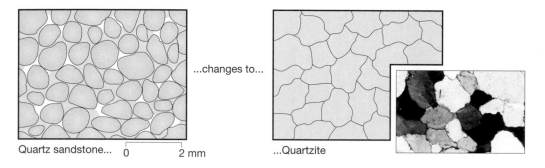

Quartz sandstone... 0 2 mm ...Quartzite

B. Nonfoliated texture

FIGURE 2.25 Metamorphic rock textures. **A.** A foliated texture results when elongated mineral crystals are aligned in a parallel, linear arrangement perpendicular to the compressional force (stress). The layered or banded appearance that results is evident in the photomicrograph (enlarged photograph taken through a microscope) of slate. **B.** When mineral crystals are not elongated, the result is a nonfoliated texture consisting of interlocking mineral grains.

have a sheet structure and are arranged in a parallel or "layered" manner. *During metamorphism, increased heat and pressure can cause the mineral crystals to become larger and the foliation more obvious.* The metamorphic rocks in Figures 2.14–2.17 exhibit foliated textures.

21. From the rocks illustrated in Figures 2.14 and 2.16, the (slate, schist) resulted from more intensive heat and pressure. Circle your answer.
22. From the metamorphic rocks in Figures 2.15 and 2.17, the (phyllite, gneiss) shows the minerals separated into light and dark bands. Circle your answer. The foliated-banded texture of the rock that you have selected often results from the most intensive heat and pressure during metamorphism.

Select several of the foliated metamorphic rock specimens supplied by your instructor that have large

crystals and examine them with a hand lens or microscope.

23. Sketch the appearance of the magnified crystals of one foliated metamorphic rock in the space provided below. Indicate the scale of your sketch by writing the appropriate length within the ( ) provided on the bar scale.

Sketch:

0   (   )cm

Scale

Nonfoliated Metamorphic Rocks. Nonfoliated metamorphic rocks are most often identified by determining their mineral composition. The minerals that comprise them, most often calcite or quartz, are neither elongated nor sheet structured and therefore cannot be aligned (see Figure 2.25). Hence, no foliation develops during metamorphism.

──────

24. Examine the nonfoliated metamorphic rocks supplied by your instructor to determine if any are composed of calcite or quartz. Hardness and the reaction to dilute hydrochloric acid often provide a clue. FOLLOW THE DIRECTIONS OF YOUR INSTRUCTOR WHEN USING ACID TO TEST FOR CALCITE.

Using a Metamorphic Rock Identification Key. A metamorphic rock identification key is presented in Figure 2.26. To use the key, first determine a rock's texture, foliated or nonfoliated, and then proceed to further subdivisions to arrive at a name. The names of the coarse foliated rocks are often modified with the mineral composition placed in front of the name, e.g. "mica schist."

──────

25. Place each of the metamorphic rocks supplied by your instructor on a numbered piece of paper. Then complete the metamorphic rock identification chart, Figure 2.27, for each rock. Use the metamorphic rock identification key, Figure 2.26, to determine each specimen's name.

METAMORPHIC ROCK IDENTIFICATION KEY

Texture	Grain Size	Rock Name		Comments
Foliated	Very fine (not visible)	Slate	Increasing metamorphism →	Often shiny, hard, smooth
	Fine (barely visible)	Phyllite		"Silky" appearance
	Medium to coarse (visible)	Schist		Various types based on mineral content, e.g. biotite schist
	Banded (with segregation) medium to coarse	Gneiss		Color banding due to segregation of minerals into layers
Nonfoliated	Crystalline (fine to coarse)	Marble		Interlocking calcite or dolomite grains, forms from limestone or dolostone
	Fused quartz grains (fine to coarse)	Quartzite		Interlocking quartz grains, forms from quartz sandstone
	Fine	Anthracite Coal		Bright, hard coal, forms from bituminous coal

FIGURE 2.26 Metamorphic rock identification key. Metamorphic rocks are divided into the two textural groups, foliated and nonfoliated. Foliated rocks are further subdivided based upon the size of the mineral grains.

METAMORPHIC ROCK IDENTIFICATION CHART

Specimen Number	Foliated or Nonfoliated	Grain Size	Composition (if identifiable)	Rock Name

FIGURE 2.27 Metamorphic rock identification chart.

REVIEW

Having completed the exercise, you should know the following:

1. Most rocks are combinations of minerals.
2. There are three rock types, and each type has unique physical characteristics that can be used for recognition.
3. Rocks form from a variety of processes and one rock type can be changed into any other.
4. Rocks are identified by using texture and mineral composition.
5. How to use texture and mineral composition to identify a rock with the aid of a rock identification key.
6. How rocks can be used to help understand the history of the earth.

Common Rocks

SUMMARY/REPORT PAGE

Date Due: _____

Name: _____

Date: _____

Class: _____

After you have finished Exercise Two, complete the following questions. You may have to refer to the exercise for assistance or to locate specific answers. Be prepared to submit this summary/report to your instructor at the designated time.

1. Write a brief definition of each of the three rock types.

   Igneous rocks: _____

   _____

   Sedimentary rocks: _____

   _____

   Metamorphic rocks: _____

   _____

2. All three rock types can contain crystals. What unique factor about the arrangement of mineral crystals occurs in many metamorphic rocks?

   _____

3. Describe the procedure you would follow to determine the name of a specific igneous rock.

   _____

   _____

   _____

4. Describe the basic difference between detrital and chemical sedimentary rocks.

   _____

   _____

5. List the texture and mineral *composition* of each of the following rocks.

	Texture	Mineral Composition
Granite:	_____	_____
Marble:	_____	_____
Sandstone:	_____	_____

6. What is one possible environment for the origin of the sedimentary rock sandstone?

   _____

7. Of the three rock types, which one is most likely to contain fossils? Explain the reason for your choice.

   _____

   _____

8. What factor determines the size of the crystals in igneous rocks?

   _____

9. What is a good chemical test to determine the primary mineral in limestone?

   _____

10. What factor determines the size of crystals in metamorphic rocks?

    _____

11. If the sedimentary rock limestone is subjected to metamorphism, what metamorphic rock will likely form?

    _____

12. Describe the processes and changes that an igneous rock will undergo as it is changed first to a sedimentary rock, which then becomes a metamorphic rock.

_____

_____

_____

Introduction to Aerial Photographs and Topographic Maps

Aerial photographs and topographic maps are important research tools that provide insight into the various processes that shape the surface of the land. The ability of an earth scientist to effectively interpret and use these tools is essential to identifying and understanding earth features.

OBJECTIVES

After you have completed this exercise, you should be able to

1. Use a stereoscope to view a stereogram, a pair of aerial photographs.
2. Explain what a topographic map is and how it can be used to study landforms.
3. Use map scales to determine distances.
4. Determine the latitude and longitude of a place from a topographic map.
5. Use the Public Land Survey system to locate features.
6. Explain how contour lines are drawn and be able to use contours to determine elevation, relief, and slope of the land.
7. Construct a simple contour map.
8. Construct a topographic profile.

TEXTBOOK REFERENCE

Appendix D

MATERIALS

ruler

Materials Supplied by Your Instructor

stereoscope
topographic map
United States and world wall maps

TERMS

cartographer	graphic scale	section
topographic map	Public Land	contour line
stereoscope	Survey	contour interval
stereogram	base line	index contour
datum	principal meridian	bench mark
quadrangle	township	slope
magnetic declination	range	relief
map scale	congressional	topographic
fractional scale	township	profile

INTRODUCTION

Aerial photographs and maps are indispensable methods for reducing vast amounts of data to a scale that can be easily managed. Vertical aerial photographs have long been used by earth scientists to study earth features in great detail, and since the early 1970's, satellite imagery has added a new dimension to our understanding of surface features and their origin. In addition to aerial photographs, an infinite variety of maps limited only by the imagination of the map maker, called a **cartographer,** is possible. However, one kind of map, the **topographic map,** is most useful when investigating the many kinds of landforms that exist on the earth's surface.

AERIAL PHOTOGRAPHS

Aerial photographs are useful for geological, environmental, agricultural, and related studies. Photographs of the same feature, when taken sequentially and overlapped, can be viewed in three dimensions through a viewer called a **stereoscope.**

To view a stereoscopic aerial photograph, called a **stereogram,** the stereoscope is placed directly over the line separating any two photos of the same feature (Figure 3.1). As you look through a stereoscope, it may have to be moved around slightly until the image appears in three dimensions. The observed heights will be vertically exaggerated and the difference in heights you see through the stereoscope will not be the same as the actual difference in heights on the land.

To provide some practice viewing stereograms, obtain a stereoscope from your instructor, unfold it, and center it over the line that separates the two aerial photographs of the volcanic cone in Figure 3.2. As you view the photographs, adjust the stereoscope until the cone appears in three dimensions. You may have to be patient until your eyes focus.

Use the stereogram in Figure 3.2 to answer questions 1–5.

1. Identify and label the crater at the summit of the volcano.
2. Outline and label the lava flow at the base of the volcano.
3. What is the white, curved feature that extends from the base of the cone to its summit?

_____

4. Mark the highest point on the volcano with an "X."
5. Assume the summit of the volcanic cone is 1500 feet above the surrounding land. While viewing the cone

through the stereoscope, draw lines around the volcano at approximately 400 foot intervals above the local surface.

TOPOGRAPHIC MAPS

Topography means "the shape of the land." Each topographic map shows, to scale, the width, length, and variable height of the land above a **datum** or reference plane—generally average sea level. The maps, which are also referred to as **quadrangles,** are two-dimensional representations of the three-dimensional surface of the earth. Their primary value to the earth scientist is for determining locations, landform types, elevations, and other physical data.

Topographic maps have been produced by the United States Geological Survey since the late 1800s. Today, a vast area of the United States has been accurately portrayed on these commercially available maps. To facilitate their use, the maps follow a similar format. In addition to standard colors and symbols, each contains information about where the area mapped is located, date when the mapping was done or revised, scale, north arrow, and names of adjoining quadrangle maps.

To help understand the basics of topographic maps, obtain a copy of a topographic map from your instructor and examine it. You will use this map to answer specific groups of questions that follow. PLEASE DO NOT WRITE OR MARK ON THE MAP.

General Map Information. Every topographic map contains useful information printed in its margin. Locate and record the following information for your map.

Each topographic map is assigned a name for reference. The name of a topographic map is located in the upper right corner of the map.

6. What is the name of your map?

   Map name: _____

Notice the small reference map and compass arrow in the lower margin of the map.

7. In what part of the state (North, Southwest, etc.) is the area covered by your map located?

_____

The names of adjoining maps are given along the four margins and four corners of the map.

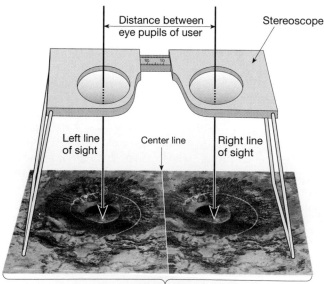

FIGURE 3.1 Aligning a stereoscope to view a stereogram.

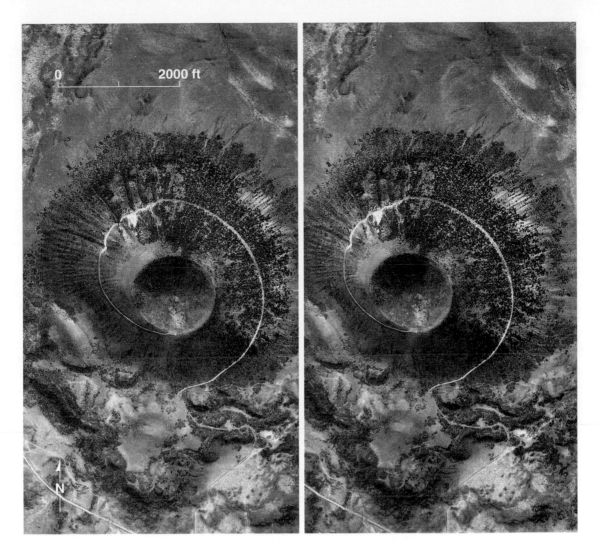

FIGURE 3.2 Stereogram of Mt. Capulin, a volcanic cinder cone located in northeastern New Mexico. The stereogram is composed from two overlapping aerial photographs taken from an altitude of approximately 16,000 feet. To view the three-dimensional image of the volcano, center a stereoscope over the line that separates the two photographs. Then, while looking through the stereoscope, adjust the stereoscope until the image appears in three dimensions. (Courtesy of U.S. Geological Survey)

8. What is the name of the map that adjoins the northeast corner of your map?

   Adjoining map: _____

   Information about when the area was surveyed and the map published is provided in the margin of the map.

9. When was the area surveyed? When was the map published? If the map has been revised, when was the revision completed?

   Surveyed: _____ Published: _____ Revised: _____

Since the geographic North Pole and magnetic North Pole of the earth do not coincide, the north arrow on a topographic map often shows the difference between true north (TN) and magnetic north (MN), the direction a compass would point, for the area represented. The difference in degrees is called the **magnetic declination.**

10. What is the magnetic declination of the area shown on your map?

    Magnetic declination: _____

Map Colors and Symbols. Each symbol and color used on a U.S. Geological Survey topographic map has a meaning. Refer to Appendix D of the text or the inside cover of this manual and carefully examine the standard U.S. Geological Survey topographic map symbols. Using the standard map symbols as a guide, locate examples of various types of roads, buildings, and streams on the topographic map supplied by your instructor.

11. In general, what color(s) are used for the following types of features?

Highways and roads: _____

Buildings: _____

Urban areas: _____

Wooded areas: _____

Water features: _____

Map Scale. Many people have built or seen scale model airplanes or cars that are miniature representations of the actual objects. Maps are similar in that they are "scale models" of the earth's surface. Each map will have a **map scale** that expresses the relation between distance on the map to the true distance on the earth's surface. Different map scales depict an area on the earth with more or less detail (Figure 3.3). On a topographic map, scale is usually indicated in the lower margin and is expressed in two ways.

Fractional scale (e.g. 1/24,000 or 1:24,000) means that a distance of 1 unit on the map represents a distance of 24,000 of the *same* units on the surface of the earth. For example, one inch on the map equals 24,000 inches on the earth, or one centimeter on the map equals 24,000 centimeters on the earth. Maps with small fractional scales (fractions with large numbers in the denominator, e.g., 1/250,000) cover large areas. Those with large fractional scales (fractions with small numbers in the denominator, e.g., 1/1,000) cover small areas.

Graphic, or **bar, scale** is a bar that is divided into segments that show the relation between distance on the map to actual distance on the earth (Figure 3.4). Scales showing miles, feet, and kilometers are generally included. The left side of the bar is often divided into fractions to allow for more accurate measurement of distance. The graphic scale is more useful than the fractional scale for measuring distances between points. Graphic scales can be used to make your own "map ruler" for measuring distances on the map.

FIGURE 3.3 Portions of two topographic maps of the same area showing the effect that different map scales have on the detail illustrated.

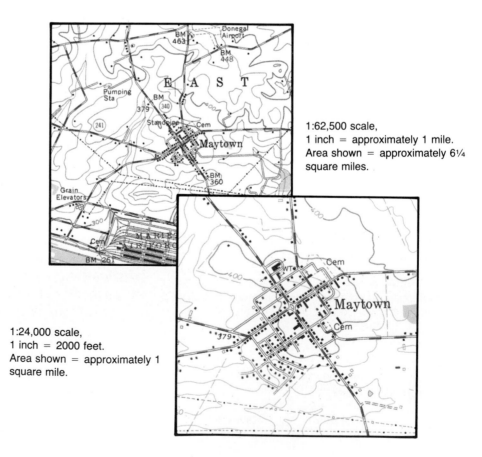

1:62,500 scale,
1 inch = approximately 1 mile.
Area shown = approximately 6¼ square miles.

1:24,000 scale,
1 inch = 2000 feet.
Area shown = approximately 1 square mile.

FIGURE 3.4 Typical graphic scale.

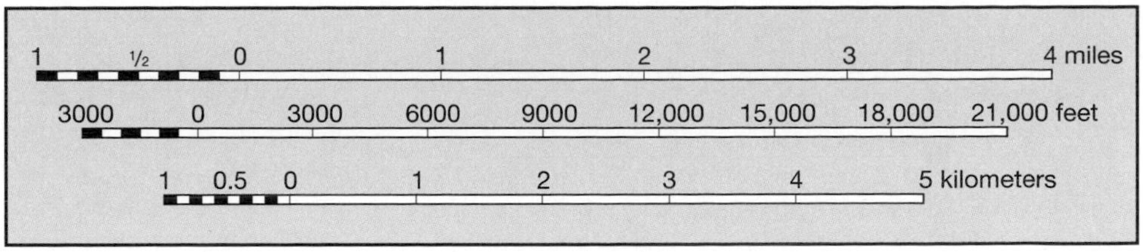

12. Examine your topographic map as well as the large wall maps in the laboratory and write out the fractional scale for each in the following space. Then answer questions 12a and 12b.

Topographic map:＿＿＿＿＿＿:＿＿＿＿＿＿

Wall map of the United States

(if available): ＿＿＿＿＿＿:＿＿＿＿＿＿

World map

(if available): ＿＿＿＿＿＿:＿＿＿＿＿＿

    a. Which of the three maps has the smallest scale (largest denominator in the fractional scale)?

    ＿＿＿＿＿＿＿＿＿＿＿＿＿＿＿＿＿

    b. Which of the three maps covers more square miles?

    ＿＿＿＿＿＿＿＿＿＿＿＿＿＿＿＿＿

13. Appendix D of the text includes a table that illustrates what one inch on a topographic map represents on the earth at various map scales. Locate the table in the text and use it to convert the following scales.

Scale	1 Inch on the Map Represents
1:24,000	＿＿＿＿＿＿ feet on the earth
1:63,360	＿＿＿＿＿＿ mile on the earth
1:250,000	＿＿＿＿＿＿ miles on the earth

14. Use the graphic scale provided on your topographic map to construct a "map ruler" in miles and measure the following distances that are represented on the map.

Width of the map along the south edge =

＿＿＿＿＿＿＿ miles

Length of the map along the east edge =

＿＿＿＿＿＿＿ miles

15. How many square miles are represented on your topographic map? (HINT: the area of a rec-

tangle is calculated using the formula, Area = width × length)

    Map area equals ＿＿＿＿＿＿＿ square miles

Location. One of the most useful functions of a topographic map is determining the precise location of a feature on the earth's surface. The two most often used methods for designating location are 1) latitude and longitude to determine the location of a point and 2) the **Public Land Survey** system (PLS) to define an area. Because topographic maps are very accurate, both methods of location can be used to provide information helpful to engineers, surveyors, realtors, and others.

Latitude and Longitude. Topographic maps are bounded by parallels of latitude on the north and south, and meridians of longitude on the east and west. The latitudes and longitudes covered by the quadrangles are printed at the four corners of the map in degrees (°), minutes ('), and seconds (") and are indicated at intervals along the margins. Maps that cover 15 minutes of latitude and 15 minutes of longitude are called *15-minute series topographic maps.* A 7 1/2-minute series topographic map covers 7 1/2 minutes of latitude and 7 1/2 minutes of longitude. (*Note:* There are 60 minutes of arc in one degree and 60 seconds of arc in one minute of arc. Therefore, 1/2 minute is the same as 30 seconds.) A more complete examination of latitude and longitude can be found in Exercise Twenty-one, "Location and Distance on the Earth."

    Use the topographic map supplied by your instructor to answer questions 16–22. PLEASE DO NOT WRITE OR MARK ON THE MAPS.

16. What are the latitudes of the southern edge and northern edge of the map to the nearest 1/2 minute of latitude?

Latitude of southern edge: ＿＿＿＿＿＿＿

Latitude of northern edge: ＿＿＿＿＿＿＿

17. How many total minutes of latitude does the map cover?

＿＿＿＿＿＿＿ minutes of latitude

18. What are the longitudes of the eastern edge and western edge of the map to the nearest 1/2 minute of longitude?

Longitude of eastern edge: _____

Longitude of western edge: _____

19. How many total minutes of longitude does the map cover?

_____ minutes of longitude

20. The map is a _____ -minute series topographic map because it covers _____ minutes of latitude and _____ minutes of longitude.

21. The total minutes of latitude and total minutes of longitude covered by the map are equal. Why is the appearance of the map rectangular rather than square?

_____

_____

22. Your instructor will supply you with the names of two features (school, church, etc.) located on the map. Write the name of each feature, as well as its latitude and longitude to the nearest minute, in the following spaces.

Feature name: _____

Latitude: _____ Longitude: _____

Feature name:

Latitude: _____ Longitude: _____

Public Land Survey. The Public Land Survey (PLS) provides a precise method for identifying the location of land in most states by establishing a grid system that systematically subdivides the land area (Figure 3.5). The Public Land Survey begins at an initial point (generally there are one or more initial points for each state). An east-west line, called a **base line**, and a north-south line, called a **principal meridian**, extend through the initial point and provide the basis of the grid (see Figure 3.5).

Horizontal lines at six mile intervals that parallel the base line establish east-west tracts, called **townships**. Each township is numbered north and south from the base line. The first horizontal six mile wide tract north of the base line is designated Township One North (T1N), the second T2N, etc. Vertical lines at six mile intervals that parallel the principal meridian define north-south tracts, called **ranges**. Each range is numbered east and west of the principal meridian. The first vertical six mile wide tract west of the principal meridian is designated Range One West (R1W), the second R2W, etc. *On a topographic map, the townships and ranges covered by the map are printed in red along the margins.*

The intersection of a township and a range defines a six mile by six mile rectangle, called a **congressional township**, which may or may not coincide with a civil township. Each congressional township is identified by referring to its township and range numbers. For example, in Figure 3.5A, the shaded congressional township would be identified as T1N, R4W.

Each congressional township is divided into 36 one-mile square parcels of land, called **sections**, with each section containing 640 acres. Sections are numbered beginning with number one in the northeast corner of the congressional township and ending with number 36 in the southeast (Figure 3.5B). The shaded section of land in Figure 3.5B would be designated as Section 11, T1N, R4W. *On a topographic map, the sections are outlined and their numbers are printed in red.*

For more detailed descriptions, sections may be subdivided into halves, quarters, or quarters of a quarter (see Figure 3.5C). Each of these subdivisions are identified by their compass position. For example, the forty acre area designated with the letter X in Figure 3.5C would be described as the SW1/4 (southwest 1/4), of the SE 1/4 (southeast 1/4) of Section 11. Hence, the complete locational description of the area marked with the letter X would be: SW1/4, SE1/4, Sec. 11, T1N, R4W. *By convention, in the description the smallest subdivision is given first and the township number precedes the range number.*

Figure 3.6 illustrates a hypothetical area that has been surveyed using the Public Land Survey system. Figure 3.6A is a township and range diagram, 3.6B represents a congressional township within the township and range system, and Figure 3.6C is a section of the congressional township. Use Figures 3.6A–3.6C to complete questions 23–25.

23. Use the PLS system to label the townships, ranges, and sections in Figure 3.6A and 3.6B with their proper designation.

24. Follow each of the letters, A through D, through Figure 3.6 and write the PLS location description of each in the following space. As an example, A has already been done.

A: NW 1/4, SW 1/4, Sec. 8 , T 3N , R 4W

B: ___1/4, ___1/4, Sec. ___, T___, R___

C: ___1/4, ___1/4, Sec. ___, T___, R___

D: ___1/4, ___1/4, Sec. ___, T___, R___

25. Locate each of the areas described below on Figure 3.6 by placing the appropriate letter in the proper places in Figure 3.6A–3.6C.

E: SW1/4, SW1/4, Sec. 5, T5N, R3E
F: SE1/4, NE1/4, Sec. 34, T4S, R7W

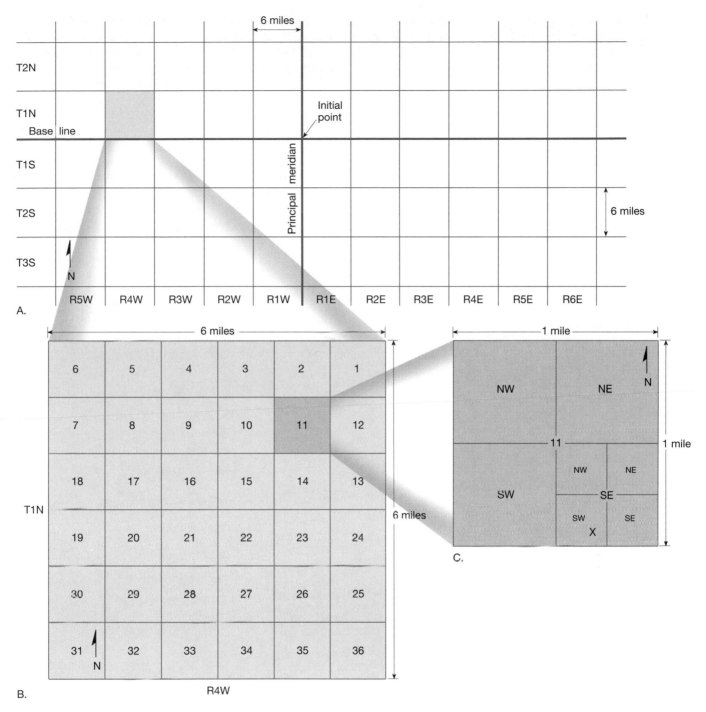

FIGURE 3.5 The Public Land Survey system (PLS).

Use the topographic map supplied by your instructor to answer questions 26–29. PLEASE DO NOT WRITE OR MARK ON THE MAPS.

————

26. List the townships and ranges represented on the map.

Townships: _____

Ranges: _____

27. To reach the principal meridian that was used to survey the land represented on the map, people living within the area would have to travel (eastward, westward). To reach the base line they would travel to the (north, south). Circle your answers.

28. What is the section, township, and range at each of the following locations on the map?

Exact center of the map:

Sec. _____, T _____, R _____

Extreme NE corner of the map:

Sec. _____, T _____, R _____

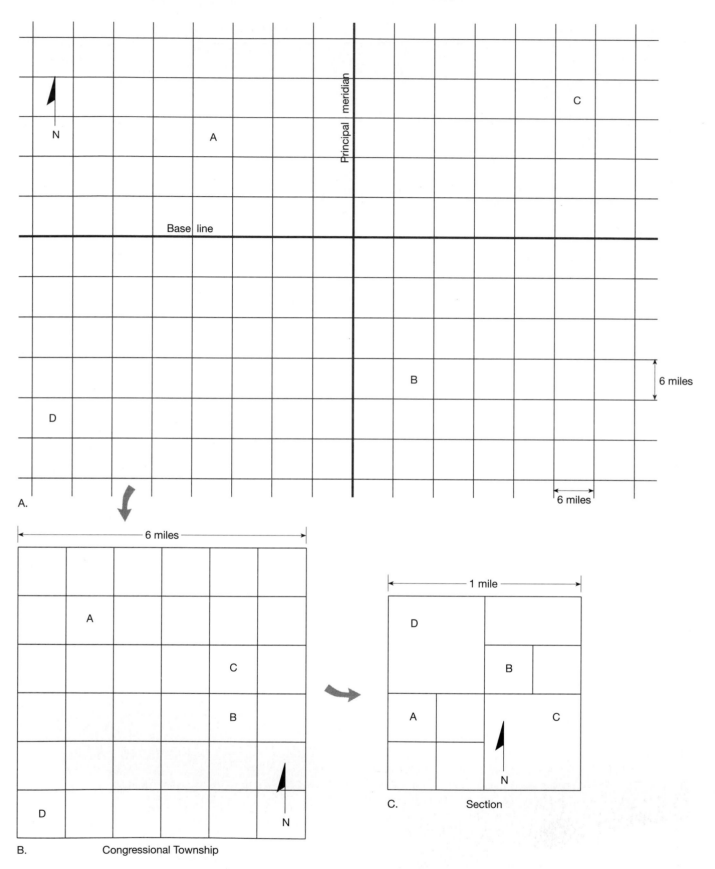

FIGURE 3.6 Hypothetical Public Land Survey system showing the locations of various points.

FIGURE 3.7 Schematic illustration showing how contour lines are determined when topographic maps are constructed. **A** is the ocean surface. **B** is an imaginary plane 20 feet above the ocean that intersects the land. **C** is an imaginary plane 40 feet above the ocean that intersects the land. **D** is an imaginary plane 60 feet above the ocean that intersects the land. **E** is the topographic map that results when the contour lines that mark where the imaginary planes intersect the surface are drawn on a map.

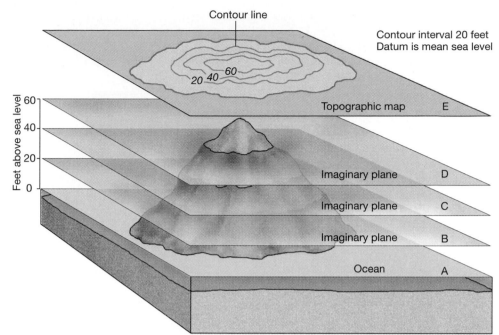

29. Your instructor will supply you with the name of a feature (school, church, etc.) located on the map. Using the Public Land Survey system, write the feature's complete location to the nearest 1/4 of a 1/4 section in the following space.

    Feature name: _____

    Location: _____

Contour Lines. Depicting the height or elevation of the land, thereby showing the shape of landforms, is what makes a topographic map unique. A **contour line** is a line drawn on a topographic map that connects all points that have equal elevations above or below a datum or reference plane on the earth's surface (Figure 3.7). The reference plane from which elevations are measured for most topographic maps is mean (average) sea level. The datum for a map is usually indicated in the lower, central margin of the map with the phrase, "Datum is . . ."

Contour lines must conform to certain guidelines. Figure 3.8 presents some of the general rules that apply to contour lines.

The **contour interval** (CI) is the vertical difference in elevation between adjacent contour lines. All contour lines are multiples of the contour interval. For example, for a contour interval of 20 feet, the lines may read 420', 440', 460', etc. Most maps use the smallest contour interval possible to provide the greatest detail for the surface that is being mapped. The contour interval of a topographic map is usually indicated in the lower central margin of the map with the phrase, "Contour interval . . ." *The contour interval should always be known before using a topographic map.*

To help determine the elevation of the contour lines, on most topographic maps every fifth contour line, called an **index contour,** is printed as a bold line and the elevation of the line is indicated. Reference points of elevation, called **bench marks** (BM), are also often present on the map and can be used to establish elevations.

Contour lines that are close together indicate a steep **slope** (vertical change in elevation per horizontal distance, usually expressed in feet/mile or meters/kilome-

---

GENERAL RULES FOR CONTOUR LINES

1. A contour line connects points of equal elevation.

2. A contour line never branches or splits.

3. Steep slopes are shown by closely spaced contours.

4. Contour lines never cross, except to show an overhanging cliff. (To show an overhanging cliff, the hidden contours are dashed. Contour lines can also merge to form a single line along a vertical cliff.)

5. Hills are represented by a concentric series of closed contour lines.

6. A concentric series of closed contours with hachure marks on the downhill side represents a closed depression.

7. When contour lines cross streams or dry stream channels, they form a "V" that points upstream.

FIGURE 3.8 Some general rules for contour lines.

ter), while widely spaced lines show a gradual slope. Consequently, the "shading" which results from closely spaced contour lines allows for the recognition of such features as hills, valleys, ridges, etc.

Relief is defined as the difference in elevation between two points on a map. *Total relief* is the difference between the highest and lowest points on a map. *Local relief* refers to the difference in elevation between two specified points, for example, a hill and nearby valley.

Examining Contour Lines. Figure 3.9 shows a contour map of volcanic cones along with a stereoscopic contour map of the same features. The large volcano illustrated is very similar to the one you observed in the stereogram, Figure 3.2.

30. Examine the stereoscopic contour map in Figure 3.9 by centering your stereoscope over the center line and observing the three-dimensional image. Before continuing the exercise, compare the stereoscopic image closely with the contour map. Examine the features noted in the caption on both

FIGURE 3.9 Contour map and stereoscopic contour map of the same volcanic cones. The two maps show a large, steep-sided cinder cone (**A**) with a well-developed crater (**C**) at its summit and lava flows (**D**) at its base. Ridges (**E**) and depressions (**F**) are evident on the lava flows, while water erosion has carved gullies (**B**) on the sides of the cones. (From Horace MacMahan, Jr., *Stereogram Book of Contours*, p. 18. Copyright (c) 1972, Hubbard Scientific Company. Reprinted by permission of American Educational Products—Hubbard Scientific)

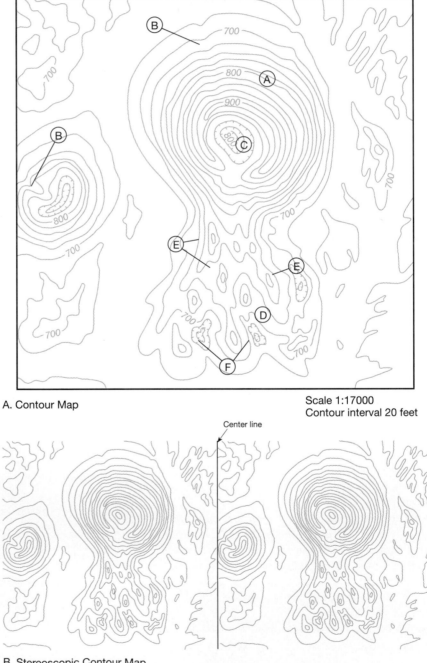

A. Contour Map

Scale 1:17000
Contour interval 20 feet

B. Stereoscopic Contour Map

maps, paying particular attention to the use of hachure marks on contours to show a depression.

Figure 3.10 shows both a perspective view and contour map of a hypothetical area situated along an ocean coast. The elevations in feet of several contour lines and points are identified on the map for reference. Datum for the map is mean sea level. Use Figure 3.10 to answer questions 31–36.

━━━━

31. What is the contour interval that has been used on the map?

Contour interval: _____ feet

32. Indicate the two areas on the map that have the steepest slopes by writing the word "steep" on the map. What characteristic of the contour lines shows that the slopes are steep?

_____

33. Notice what happens to the contour lines as they cross a stream. The "peak" formed by a contour line as it crosses a stream points (upstream, downstream). Circle your answer.

34. What are the elevations of the points designated with the following letters?

Point A: _____ feet

Point B: _____ feet

Point C: _____ feet

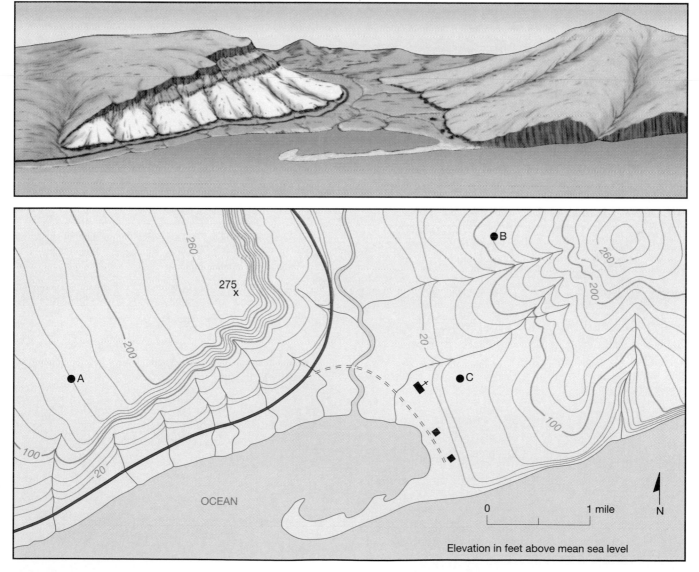

FIGURE 3.10 Perspective and map view of a hypothetical coastal area. (After U.S. Geological Survey)

35. The approximate elevation of the church is (12, 22, 32) feet. Circle your answer.

36. What is the total relief shown on the map?

Highest elevation (_____ft) − lowest elevation

(_____ft) = total relief (_____ft)

Use the topographic map supplied by your instructor to answer questions 37–41. PLEASE DO NOT WRITE OR MARK ON THE MAPS.

▬▬▬▬

37. What is the datum that has been used for determining the elevations on the map?

Datum: _____

38. What is the contour interval of the map?

Contour interval: _____ feet

39. What are the lowest and highest elevations found on the map?

Lowest elevation: _____ feet

Highest elevation: _____ feet

40. What is the elevation of the exact center of the map?

Elevation: _____ feet

41. After examining the contour lines, etc., write a brief description of the slope of the land represented by the map.

_____

_____

Using Contour Lines and Stereograms. Figure 3.11 is a portion of the Cheakamus River, East, British Columbia, Canada topographic map printed in black and white rather than the normal colors. Questions 42–54 refer to Figure 3.11.

▬▬▬▬

42. What is the fractional scale of the map?

Scale: _____

43. One inch on the map equals _____ inches on the surface of the earth.

44. The straight line distance between the tops of Panorama Ridge and Mount Price, as measured with the graphic scale, is _____ miles,

or _____ meters.

45. The approximate width of Garibaldi Lake along a line from Mount Price to Panorama Ridge to the nearest quarter mile is _____ miles?

46. What is the datum that has been used to determine the elevations on the Cheakamus River topographic map?

Datum: _____

47. The contour interval that has been used on the map is _____ feet.

48. The elevation at the top of Panorama Ridge, the highest point on the map, is _____ feet.

The lowest point on the map, about 3,000 ft, is near the northwest corner.

▬▬▬▬

49. What is the total relief found on the map?

Total relief: _____ feet

50. The steepest slope on the map is located near _____ ridge.

51. The elevation at the top of Mount Price is _____ feet.

52. The elevation of the shoreline of Garibaldi Lake is _____ feet.

53. What is the local relief between the summit of Mount Price and the shore of Garibaldi Lake, directly east of the mountain?

There are _____ feet of local relief from the summit of Mount Price to the shore of Garibaldi Lake.

Slope is vertical change in elevation (relief) per horizontal distance. The distance from the top of Mount Price directly eastward to the shore of Garibaldi Lake is approximately one mile.

▬▬▬▬

54. What is the slope in feet per mile of the eastern side of Mount Price from the summit to Garibaldi Lake?

$$\text{Slope} = \frac{\text{vertical change in elevation in feet}}{\text{horizontal distance in mile(s)}} = \underline{\hspace{1cm}} \text{ feet/mile}$$

Figure 3.12 is a stereogram composed of two overlapping photographs of the Cheakamus River, British Columbia region. The region shown on the stereogram includes much of the same area that is represented on the topographic map, Figure 3.11.

Compare the stereogram, Figure 3.12, to the topographic map of Cheakamus River, Figure 3.11. See if you can locate the same features on *both* the map and stereogram. Then using the two figures, complete questions 55–58.

55. Label Garibaldi Lake, Driftwood Bay, Table Bay, Battleship Islands, and Mount Price on one of the aerial photographs.
56. Outline the area covered by the aerial photographs on the topographic map, Figure 3.11.

Locate Lesser Garibaldi Lake on the topographic map, Figure 3.11. Identify the lake on the stereogram and view it with the stereoscope.

57. The elevation of Lesser Garibaldi Lake is (higher, lower) than Garibaldi Lake. Circle your answer.
58. Draw an arrow on one of the aerial photographs that shows the direction of flow of the river that connects Garibaldi Lake and Lesser Garibaldi Lake.

   a. The river is flowing (eastward, westward). Circle your answer.
   b. What is the approximate slope of the river between the two lakes?

   Slope: _____

Constructing a Contour Map.
Originally, contour maps were constructed by first surveying an area and establishing the elevations of several points in the field.

The surveyor then sketched contour lines on the map by estimating their location between the points of known elevation. Today, topographic maps are made from stereoscopic aerial photographs that are computer processed to determine elevations and contours.

59. To help understand the process of drawing a contour map, using a pencil, complete the contour map shown in Figure 3.13. The points illustrated are of known elevation. The 100 foot contour line has been drawn to provide a reference. Using a 20-ft contour interval, draw a contour line for each 20-ft change in elevation below and above 100 ft—for example, 80 ft, 60 ft, and 120 ft, etc. You will have to estimate the elevations between the points. Label each of the lines with the proper elevation.
60. In Figure 3.13, in general, the land slopes toward the (north, south). Circle your answer.
61. After examining the contour lines and elevations in Figure 3.13, show the directions that the intermittent streams are flowing by drawing arrows on the map.
62. What is the average slope of the intermittent stream on the west side of the map you drew in Figure 3.13?

   Slope: _____ feet/mile

Drawing a Topographic Profile.
Topographic maps, like most other maps, depict the earth viewed from above. Often a topographic profile or "side-view" will provide a more useful representation of the elevations and slopes of an area. To change an overhead or map view into a profile, follow the steps illustrated in Figure 3.14.

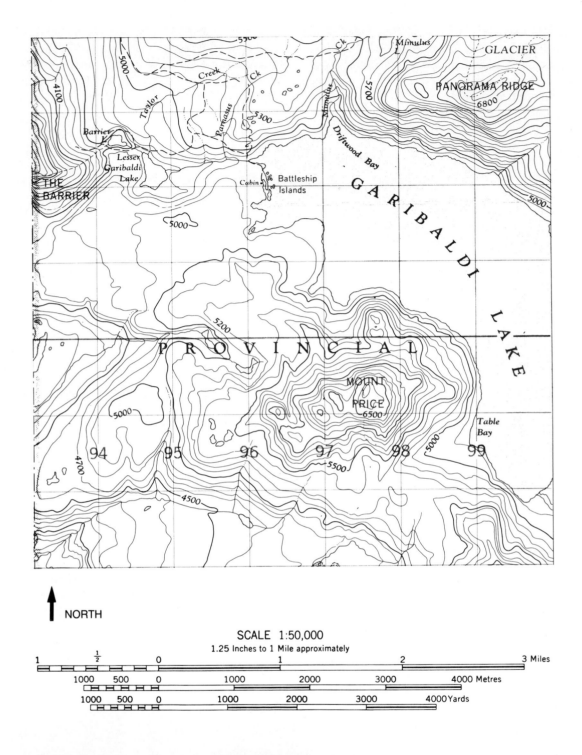

↑ NORTH

SCALE 1:50,000
1.25 Inches to 1 Mile approximately

1   ½   0   1   2   3 Miles

1000   500   0   1000   2000   3000   4000 Metres

1000   500   0   1000   2000   3000   4000 Yards

CONTOUR INTERVAL 100 FEET
DATUM IS MEAN SEA LEVEL

FIGURE 3.11 Portion of the Cheakamus River, East, British Columbia, topographic map. (Reproduced courtesy of Surveys and Resource Mapping Branch, Ministry of Environment, Government of British Columbia, Canada)

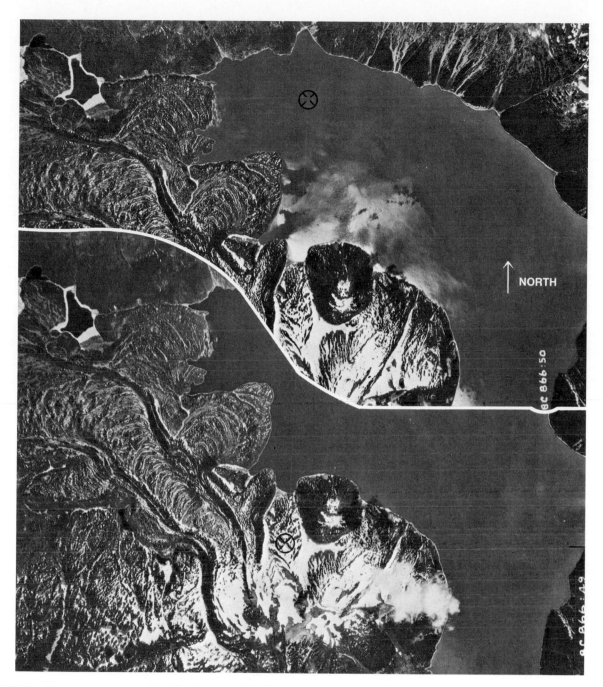

NORTH

FIGURE 3.12 Stereogram composed of two aerial photographs of the Cheakamus River, British Columbia region. (Reproduced courtesy of Surveys and Resource Mapping Branch, Ministry of Environment, Government of British Columbia, Canada)

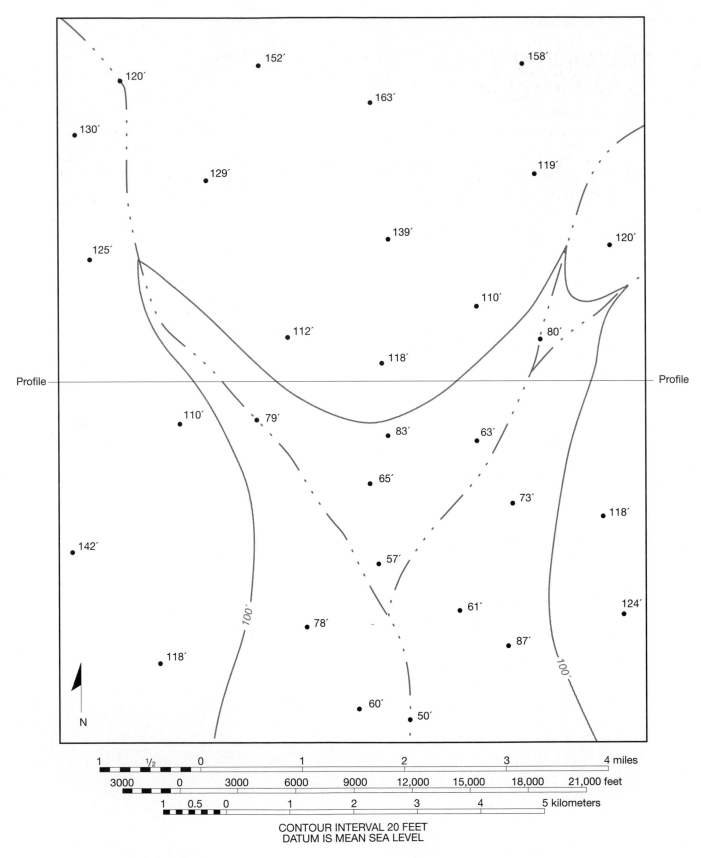

Profile ——————————————————————— Profile

CONTOUR INTERVAL 20 FEET
DATUM IS MEAN SEA LEVEL

FIGURE 3.13 Points of elevation with 100-ft contour line drawn.

FIGURE 3.14 Construction of a topographic profile. Step 1. On the topographic map, draw a line along which the profile is to be constructed. Label the line A–A'.

As shown in Step 2, lay a piece of paper along line A–A'. Mark each place where a contour line intersects the edge of the paper and note the elevation of the contour line by each mark.

In Step 3, on a separate piece of paper, draw a horizontal line slightly longer than your profile line, A-A'. Select a vertical scale for your profile that begins slightly below the lowest elevation along the profile and extends slightly beyond the highest elevation. Mark this scale off on either side of the horizontal line. Lay the marked paper edge (from Step 2) along the horizontal line. Wherever you have marked a contour line on the edge of the paper, place a dot directly above the mark at an elevation on the vertical scale equal to that of the contour line. Connect the dots on the profile with a smooth line to see the finished product. (Note: Since you have more or less arbitrarily selected the vertical scale for the profile, *the finished profile may be somewhat vertically exaggerated and not the same as you would see it from the ground.*)

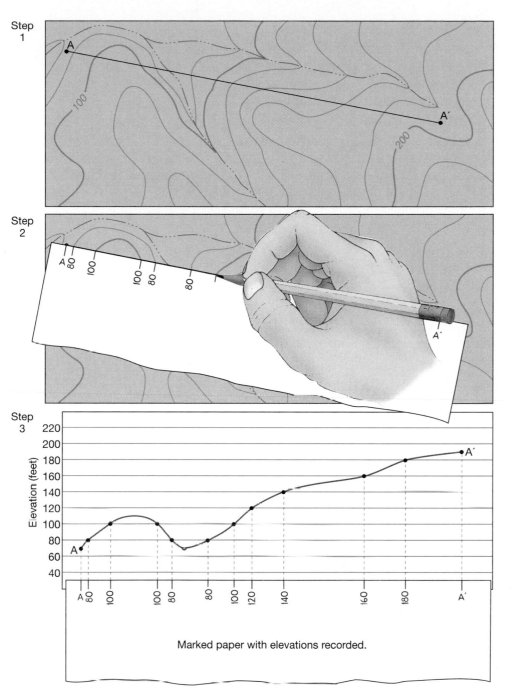

Marked paper with elevations recorded.

63. Use the horizontal line and vertical scale in Diagram 3.1 to construct a west-east profile along the profile line indicated on the contour map you have drawn in Figure 3.13. Follow the guidelines for preparing a topographic profile in Figure 3.14.

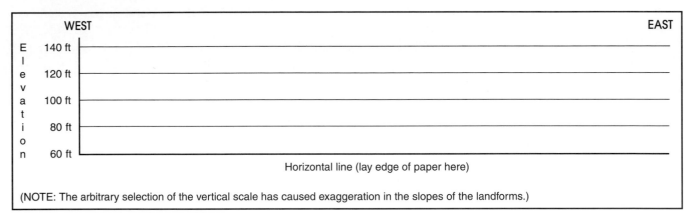

DIAGRAM 3.1

REVIEW

Having completed the exercise, you should know the following:

1. Aerial photographs are useful tools for studying landforms.
2. Reference information such as map name, publication date, revision date, direction, magnetic declination, and adjoining maps is shown in the margins of topographic maps.
3. Standard colors used on topographic maps are red for major highways, brown for contour lines, blue for water features, green for areas of vegetation, and black for buildings, secondary roads, and trails.
4. The scale of a map shows the relation between map distance and actual distance on the earth. Maps that have small fractional scales cover large areas, while maps with large fractional scales cover small areas.
5. The latitude and longitude of a feature can be determined from a topographic map.
6. Topographic maps come in several series determined by the minutes of latitude and longitude covered by the map.
7. The Public Land Survey is a grid system that locates features with reference to a township, range, section, and fractional part of a section.
8. Topographic maps use contour lines to show the elevation of the land and shapes of landforms. Contour lines connect points of equal elevation and conform to certain rules.
9. Topographic maps use various contour intervals and scales depending on the detail desired.
10. Topographic maps can be used to determine the relief and slope of the land.
11. A topographic profile shows an exaggerated ground-level view of the shape of the land surface.

Introduction to Aerial Photographs and Topographic Maps

SUMMARY/REPORT PAGE

Date Due: _____

Name: _____

Date: _____

Class: _____

After you have finished Exercise Three, complete the following questions. You may have to refer to the exercise for assistance or to locate specific answers. Be prepared to submit this summary/report to your instructor at the designated time.

1. Define the following:

   Stereogram: _____

   _____

   Topographic map: _____

   _____

   Contour line: _____

   _____

   Contour interval: _____

   _____

   Public Land Survey system: _____

   _____

2. What is the latitude and longitude to the nearest minute of the *exact center* of the topographic map supplied by your instructor?

   Latitude: _____ Longitude: _____

3. The topographic map supplied by your instructor is a _____ -minute series topographic map, which means that it:

   _____

4. What are the numbers of the townships and ranges covered by the topographic map supplied by your instructor?

   Townships: _____

   Ranges: _____

5. Use the Public Land Survey system to give the name and location of the feature your instructor requested that you locate on your topographic map in question 29.

   Name of feature: _____
   Location:

   _____ 1/4, _____ 1/4, Sec _____ ,T _____ , R _____

6. What was your calculated slope for the eastern side of Mount Price illustrated in Figure 3.11?

   Slope _____ feet/mile

WEST   EAST

DIAGRAM 3.2

7. Use Figure 3.11 to determine the distance between Mount Price and Lesser Garibaldi Lake as accurately as possible in miles and kilometers.

Miles: _____ miles   Kilometers: _____ km

8. In Diagram 3.2, sketch a copy of the west-east topographic profile you constructed in question 63. Label the appropriate elevations on the vertical axis of your sketch.

9. Are the following statements true or false?

T F   a. A contour line never splits into two contour lines.

T F   b. Contour lines that are far apart indicate a steep slope.

T F   c. The Public Land Survey system uses only one initial point to survey all the land in the United States.

T F   d. A stereoscope exaggerates the relief of the land in a stereogram.

T F   e. To properly use a stereoscope, you must look at the same feature with both eyes on a single photograph.

EXERCISE FOUR

Shaping the Earth's Surface
Running Water and Groundwater

The study of the processes that modify the earth's surface is of major significance to the earth scientist. By understanding those processes and the features they produce, scientists gain insights into the geologic history of an area and make predictions concerning its future development.

Some of the agents that are responsible for modifying the earth's surface are running water, groundwater, glacial ice, wind, and volcanic activity. Each produces a unique landscape with characteristic features that can be recognized on topographic maps. Exercises Four and Five examine several of these agents, the variety of landforms associated with them, and some of the consequences of human interaction with these natural systems.

OBJECTIVES

After you have completed this exercise, you should be able to

1. Sketch, label, and discuss the complete hydrologic cycle.
2. Explain the relation between infiltration and runoff that occurs during a rainfall.
3. Discuss the effect that urbanization has on the runoff and infiltration of an area.
4. Identify on a topographic map the following features that are associated with rivers and valleys; rapids, meanders, floodplain, oxbow lakes, and back swamps.
5. Explain the occurrence, fluctuation, use, and misuse of groundwater supplies.
6. Identify on a topographic map the following features associated with karst landscapes: sinkholes, disappearing streams, and solution valleys.

TEXTBOOK REFERENCE

Chapter 3 and Appendix D

MATERIALS

calculator
ruler

Materials Supplied by Your Instructor

graduated measuring cylinder (100 ml)
beaker (100 ml)
small funnel
cotton
coarse sand, fine sand, soil
stereoscope

TERMS

hydrologic cycle	permeability	aquifer
infiltration	hydrograph	karst topography
groundwater	discharge	disappearing stream
runoff	base level	solution valley
erosion	meander belt	sinkhole
evaporation	zone of saturation	cave
transpiration	water table	cavern
porosity	zone of aeration	

INTRODUCTION

The earth's water is constantly being exchanged between its surface and atmosphere. The **hydrologic cycle,** illustrated in Figure 4.1, describes this continuous movement of water from the oceans to the atmosphere, from the atmosphere to the land, and from the land back to the sea.

A portion of the precipitation that falls on land will soak into the ground via **infiltration** and become

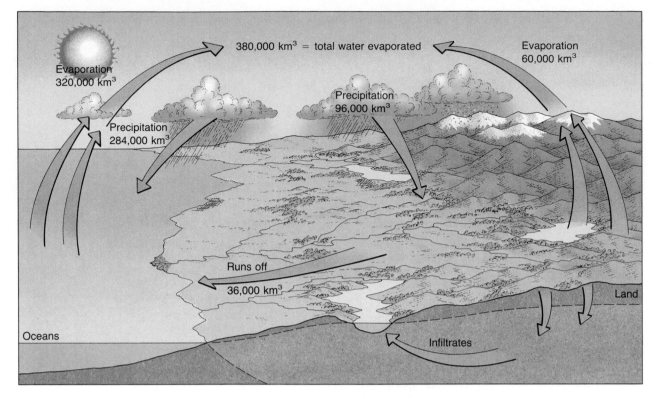

FIGURE 4.1 The earth's water balance, a quantitative view of the hydrologic cycle.

groundwater. If the rate of rainfall is greater than the earth's ability to absorb it, the additional water flows over the surface and becomes **runoff.** Runoff initially flows in broad sheets; however, it soon becomes confined and is channeled to form streams and rivers. **Erosion** by both groundwater and runoff wears down the land and modifies the shape of the earth's surface. Eventually runoff and groundwater from the continents return to the sea or the atmosphere, continuing the endless cycle.

EXAMINING THE HYDROLOGIC CYCLE

Figure 4.1 illustrates the earth's water balance, a quantitative view of the hydrologic cycle. Although the figure correctly implies a uniform exchange of water between the earth's atmosphere and surface on a worldwide basis, factors such as climate, soil type, vegetation, and urbanization often produce local variations.

Use Figure 4.1 and your text as a reference to answer questions 1–6.

1. On a worldwide basis, more water is evaporated into the atmosphere from the (oceans, land). Circle your answer.

2. Approximately what percent of the total water evaporated into the atmosphere comes from the oceans?

$$\text{Percent from oceans} = \frac{\text{ocean evaporation}}{\text{total evaporation}} \times 100$$
$$= \underline{\hspace{2cm}} \%$$

Notice in the figure that more water evaporates from the oceans than is returned directly to them by precipitation.

———

3. Since sea level is not dropping, what is the other source of water for the oceans in addition to precipitation?

_____

Over most of the earth, the quantity of precipitation that falls on the land must eventually be accounted for by the sum total of **evaporation, transpiration,** runoff, and infiltration.

———

4. Define each of the following four variables.

Evaporation: _____

Transpiration: _____

Runoff: _____

Infiltration: _____

At high elevations or high latitudes, some of the water that falls on the land does not immediately soak in, run off, evaporate, or transpire.

---

5. On the land at high elevations or high latitudes, where is water being temporarily stored?

_____

6. On a worldwide basis, about (35, 55, 75) percent of the precipitation that falls on the land becomes runoff. Circle your answer.

Infiltration and Runoff. During a rainfall most of the water that reaches the land surface will infiltrate or run off. The balance between infiltration and runoff is influenced by factors such as the **porosity** and **permeability** of the surface material, slope of the land, intensity of the rainfall, type of vegetation, as well as the amount of vegetation. After infiltration saturates the land and the ground contains all the water it can hold, runoff will begin to occur on the surface.

---

7. Describe the difference between the two terms *porosity* and *permeability*. Use the glossary of your text as a reference.

_____

_____

Permeability Experiment. To gain a better understanding of how the permeability of various earth materials affects the flow of groundwater, examine the equipment setup in Figure 4.2 and conduct the following experiment by completing each of the indicated steps.

Step 1. Obtain the following equipment and materials from your instructor:

    graduated measuring cylinder (1)

    beaker (1)

    small funnel (1)

    piece of cotton

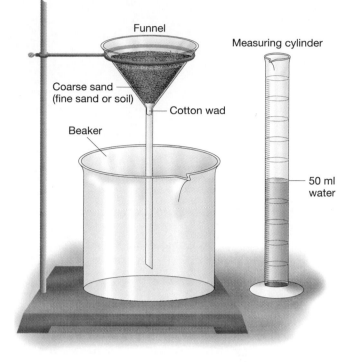

FIGURE 4.2 Equipment setup for permeability experiment.

    samples of coarse sand, fine sand, and soil (enough of each to fill the funnel approximately two-thirds full)

Step 2. Place a small wad of cotton in the neck of the funnel.

Step 3. Fill the funnel above the cotton about two-thirds full with coarse sand.

Step 4. With the bottom of the funnel placed in the beaker, measure the length of time that it takes for 50 ml of water to drain through the funnel filled with coarse sand. Record the time in the data table, Table 4.1.

Step 5. Using the measuring cylinder, measure the amount (in milliliters) of water that has drained into the beaker and record the measurement in the data table.

Step 6. Empty and clean the measuring cylinder, funnel, and beaker.

Step 7. Repeat the experiment two additional times, using fine sand and then soil. Record the results of each experiment at the appropriate place in the data table, Table 4.1. (*Note:* In each case, fill the funnel with the material to the same level that was used for the coarse sand.)

Step 8. Clean the glassware and return it along with any unused sand and soil to your instructor.

TABLE 4.1 Data Table For Permeability Experiment

	Length of time to drain 50 ml of water through funnel	Milliliters of water drained into beaker
Coarse sand	seconds	ml
Fine sand	seconds	ml
Soil	seconds	ml

8. Questions 8a–8c refer to the permeability experiment.
   a. Of the three materials you tested, the (coarse sand, fine sand, soil) has the greatest permeability. Circle your answer.
   b. Suggest a reason why different amounts of water were recovered in the beaker for each material that was tested.

   _____

   c. Write a brief statement summarizing the results of your permeability experiment.

   _____

   _____

   _____

9. What will be the effect of each of the following conditions on the relation between infiltration and runoff?

   Highly permeable surface material: _____

   _____

   Steep slope: _____

   _____

   Gentle rainfall: _____

   _____

   Dense ground vegetation: _____

   _____

10. What will be the relation between infiltration and runoff in a region with a moderate slope that has a permeable surface material covered with sparse vegetation?

   _____

   _____

Infiltration and Runoff in Urban Areas. In urban areas much of the land surface has been covered with buildings, concrete, and asphalt. The consequence of covering large areas with impervious materials is to alter the relation between runoff and infiltration of the region.

Figure 4.3 shows two hypothetical **hydrographs** (plots of stream flow, or runoff, over time) for an area before and after urbanization. The amount of precipitation the area receives is the same after urbanization as before. Runoff is evaluated by measuring the stream **discharge**, which is the volume of water flowing past a given point per unit of time, usually measured in cubic feet per second. Use Figure 4.3 to answer questions 11–14.

11. As illustrated in Figure 4.3, urbanization (increases, decreases) the peak, or maximum, stream flow. Circle your answer.
12. What is the effect that urbanization has on the lag time between the time of the rainfall and the time of peak stream discharge?

   _____

13. Total runoff occurs over a (longer, shorter) period of time in an area that has been urbanized. Circle your answer.
14. Based on what you have learned from the hydrographs, explain why urban areas often experience flash-flooding during intense rainfalls?

   _____

   _____

RUNNING WATER

Of all the agents that shape the earth's surface, running water is the most important. Rivers and streams are responsible for producing a vast array of erosional and depositional landforms in both humid and arid regions. As illustrated in Figure 4.4, many of these features are associated with the *headwaters* of a river, while others typically are found near the *mouth*.

An important factor that governs the flow of a river is its **base level.** Base level is the lowest point to which a river or stream may erode. The ultimate base level is

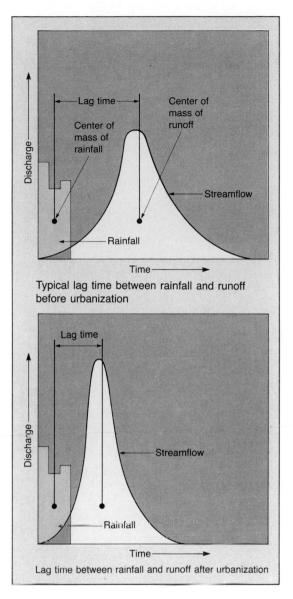

Typical lag time between rainfall and runoff before urbanization

Lag time between rainfall and runoff after urbanization

FIGURE 4.3 The effect of urbanization on stream flow. (After L. B. Leopold, U.S. Geological Survey Circular 559, 1968)

sea level. However, lakes, resistant rocks, and main rivers often act as temporary, or local, base levels that control the erosional and depositional activities of a river for a period of time.

Often the *head*, or source area, of a river is well above base level. At the headwaters, rivers typically have steep slopes and downcutting prevails. As the river deepens its valley it may encounter rocks that are resistant to erosion and form *rapids* and *waterfalls*. In arid areas rivers often erode narrow valleys with nearly vertical walls. In humid regions the effect of mass wast-

ing and slope erosion caused by heavy rainfall produce typical V-shaped valleys (Figure 4.4A).

In humid regions downstream from the headwaters, the gradient or slope of a river decreases while its discharge increases because of the additional water being added by tributaries. As the level of the channel begins to approach base level, the river's energy is directed from side to side and the channel begins to follow a sinuous path, or *meanders*. Lateral erosion by the meandering river widens the valley floor and a *floodplain* begins to form (Figures 4.4B and 4.4C).

Near the mouth of a river where the channel is nearly at base level, maximum discharge occurs and meandering often becomes very pronounced. Widespread lateral erosion by the meandering river produces a floodplain that is often several times wider than the river's *meander belt*. Features such as *oxbow lakes, natural levees, back swamps* or *marshes,* and *yazoo tributaries* commonly develop on broad floodplains (Figure 4.4D).

Questions 15 to 25 refer to the Portage, Montana, topographic map, Figure 4.5, and stereogram of a portion of the same region, Figure 4.6. On the map, notice the rapids indicated by A and the steep-sided valley walls of the Missouri River indicated by B.

---

15. Compare the aerial photograph to the map. Then, on the topographic map, outline the area that is shown in the photo.

16. Use the map to determine the approximate total *relief* (vertical distance between the lowest and highest points of the area represented).

Highest elevation (_____ ft) − lowest elevation

(_____ ft) = total relief (_____ ft)

17. On Diagram 4.1, *sketch* a north-south topographic profile through the center of the map along a line from north of Blackfeet Gulch to south of the Missouri River. Indicate the appropriate elevations on the vertical axis of the profile. Label Blackfeet Gulch and the Missouri River on the profile. (*Note:* Exercise Three contains a detailed explanation for constructing topographic profiles.)

18. Label the upland areas between stream valleys on the topographic profile in Diagram 4.1 with the word "upland."

19. The upland areas are (broad and flat, narrow ridges). Circle your answer.

20. Approximately what percent of the area shown on the map is stream valley and what percent upland?

Stream valley: _____ %

Upland: _____ %

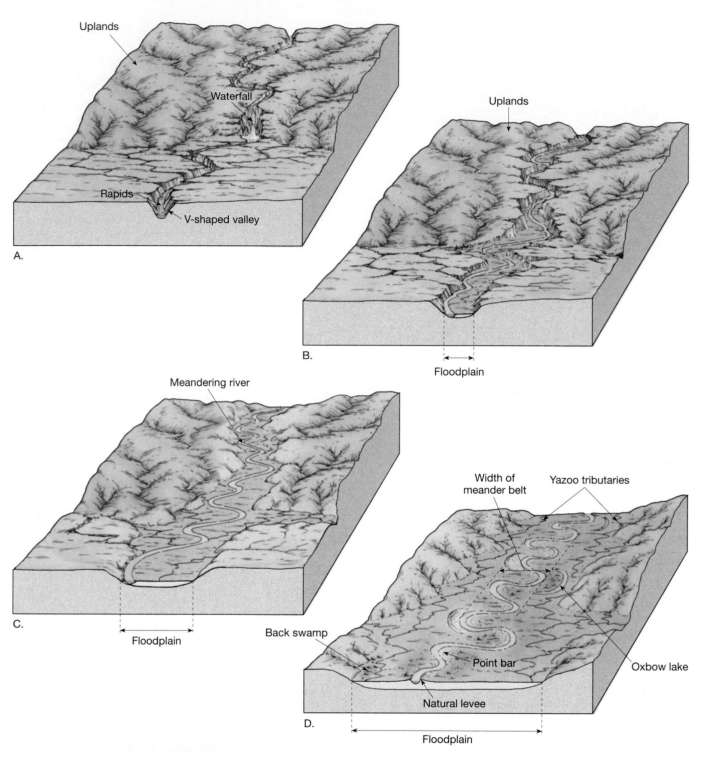

FIGURE 4.4 Common features of valleys. **A.** Near the headwaters. **B.** and **C.** In the middle. **D.** At the mouth. (After Ward's Natural Science Establishment, Inc., Rochester, New York)

NORTH   SOUTH

DIAGRAM 4.1

21. Approximately how deep would the Missouri River have to erode to reach ultimate base level?

_____ feet

22. It appears that the Missouri River and its tributaries are, for the most part, actively (eroding, depositing) in the area. Circle your answer.

23. With increasing time as the tributaries erode and lengthen their courses near the headwaters, what will happen to the upland areas?

_____

Notice the dams located along the Missouri River at C.

24. What effect have the dams had on the width of the river up-river from their locations?

_____

25. Assuming that climate, base level, and other factors remain unchanged, how might the area look millions of years from now?

_____

_____

Questions 26 to 31 refer to the Angelica, New York topographic map, Figure 4.7.

26. What is the approximate total relief shown on the map?

_____ feet of total relief

27. Draw an arrow on the map indicating the direction that the main river, the Genesee, is flowing.

(HINT: Use the elevations of the contour lines on the floodplain to determine your answer.)

28. What is the approximate *gradient* (the slope of a river; generally measured in feet per mile) of the Genesee River?

Average gradient = _____ feet/mile

29. The Genesee River (follows a straight course, meanders from valley wall to valley wall). Circle your answer.

30. Most of the areas separating the valleys on the Angelica map are (very broad and flat, relatively narrow ridges). Circle your answer.

31. Assume that erosion continues in the region without interruption. How might the appearance of the area change over a span of millions of years?

_____

_____

Questions 32–39 refer to the Campti, Louisiana, topographic map, Figure 4.8, and stereogram of the same area, Figure 4.9. On the map, A indicates the width of the floodplain of the Red River and the dashed lines, B, mark the two sides of the meander belt of the river.

32. Approximately what percent of the map area is floodplain?

Floodplain = _____ % of the map area

33. In Diagram 4.2, sketch a north-south topographic profile along a line from the south edge of the City of Campti to south of Bayou Pierre. Indicate the appropriate elevations on the vertical axis of the profile. Label the floodplain area and Bayou Pierre on the sketch.

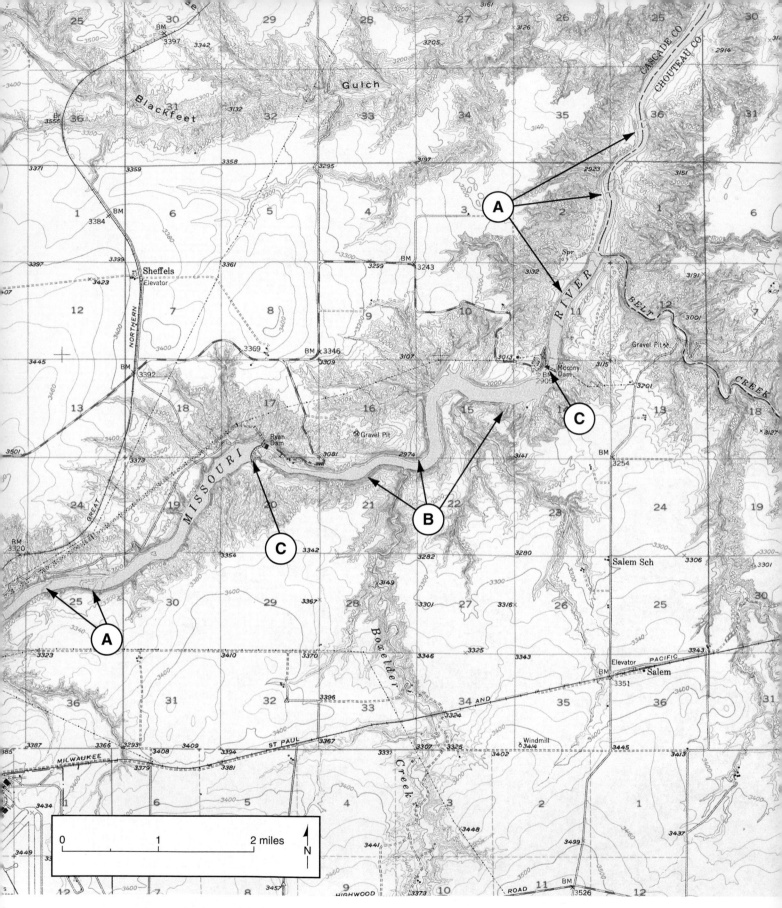

FIGURE 4.5 Portion of the Portage, Montana, topographic map. (Map source: United States Department of the Interior, Geological Survey)

SCALE 1:62500
CONTOUR INTERVAL 20 FEET
DATUM IS MEAN SEA LEVEL

MONTANA

QUADRANGLE LOCATION

62

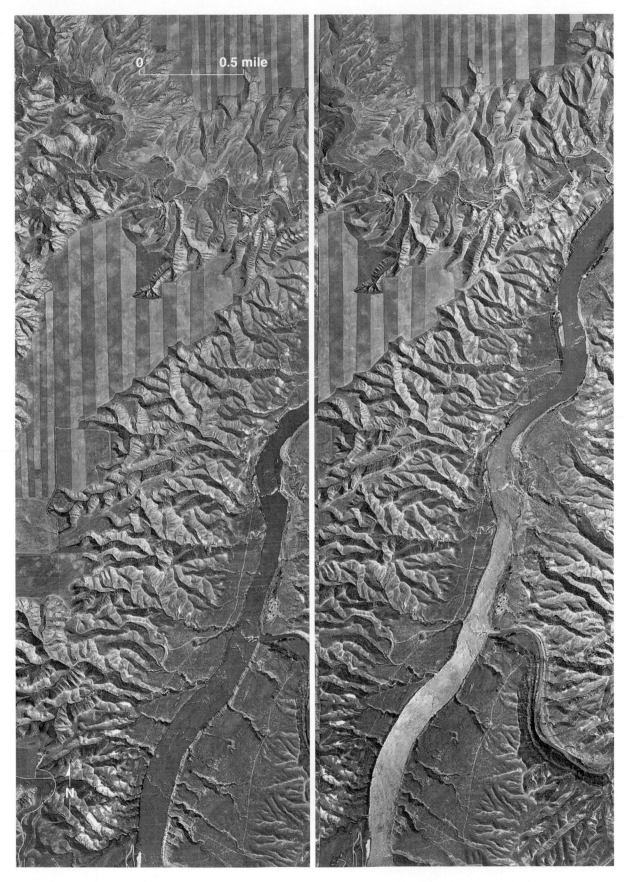

FIGURE 4.6 Stereogram of the Missouri River, vicinity of Portage, Montana.
(Courtesy of U.S. Geological Survey)

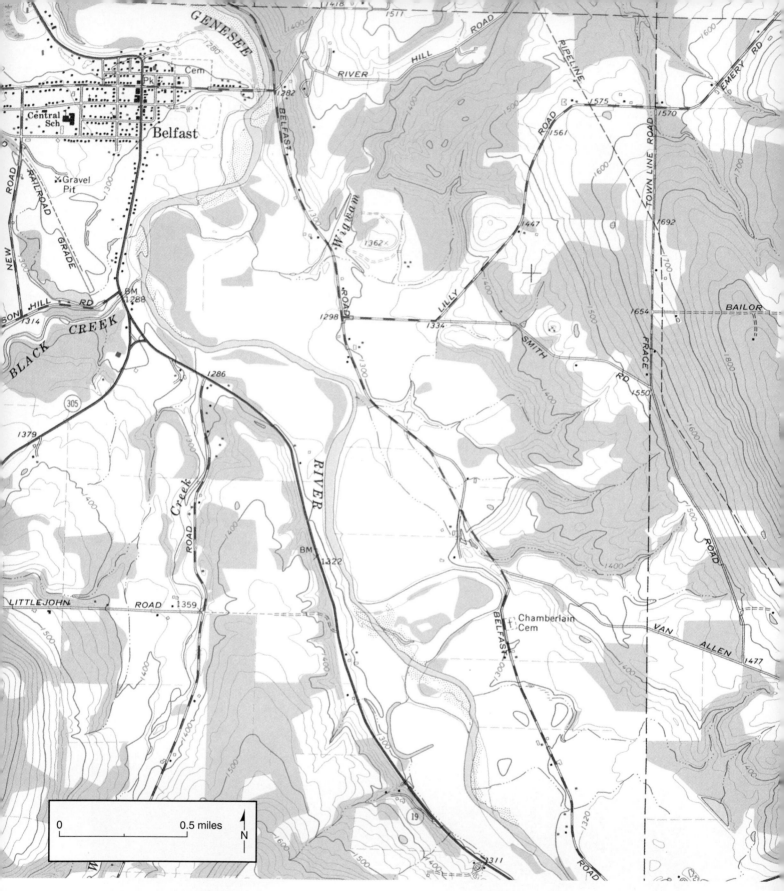

FIGURE 4.7 Portion of the Angelica, New York, topographic map. (Map source: United States Department of the Interior, Geological Survey)

SCALE 1:24000
CONTOUR INTERVAL 20 FEET
NATIONAL GEODETIC VERTICAL DATUM
OF 1929

QUADRANGLE LOCATION

34. Approximately how many feet is the floodplain above ultimate base level?

_____ feet above ultimate base level

35. Using your text and Figure 4.4 as references, identify the type of feature found at each of the following lettered positions on the map. Also, write a brief statement describing how each feature forms.

Letter C (in particular, _Old River_): _____

_____

Letter D: _____

_____

Letter E: _____

_____

Letter F: _____

_____

36. Identify and label examples of a point bar, cutbank, and an oxbow lake on the stereogram.

37. Write a statement that compares the width of the meander belt of the Red River to the width of its floodplain.

_____

_____

38. (Downcutting, Lateral erosion) is the dominant activity of the Red River. Circle your answer.

39. Assuming that erosion by the Red River continues without interruption, what will eventually happen to the width of its floodplain?

_____

_____

Answer questions 40–42 by comparing the Portage, Angelica, and Campti topographic maps.

40. On which of the three maps is the gradient of the main river the steepest?

_____

41. Which of the three areas has the greatest total relief (vertical distance between the lowest and highest elevations)?

_____

42. Choosing from the three topographic maps, write the name of the map that is best described by each of the following statements.

Primarily floodplain: _____

River valleys separated by broad, relatively flat upland areas: _____

Most of the area consists of steep slopes:

_____

Greatest number of streams and tributaries:

_____

Poorly drained lowland area with marshes and swamps: _____

Active downcutting by rivers and streams:

_____

Surface nearest to base level: _____

GROUNDWATER

As a resource, groundwater supplies much of our water needs for consumption, irrigation, and industry. On the other hand, as a hazard, groundwater can damage building foundations and aid the movement of materials during landslides and mudflows. In many areas, overuse and contamination of this valuable resource threaten the supply. One of the most serious problems faced by many localities is land subsidence caused by groundwater withdrawal.

Water Beneath the Surface. Groundwater is water that has soaked into the earth's surface and occupies all the pore spaces in the soil and bedrock in a zone called the **zone of saturation.** The upper surface of this saturated zone is called the **water table.** Above the water table in the **zone of aeration,** the pore spaces of the earth materials are unsaturated and mainly filled with air.

Figure 4.10 illustrates a profile through the subsurface of a hypothetical area. Use the figure and your text to answer questions 43–50.

43. Label the zone of saturation, zone of aeration, and water table on Figure 4.10.

44. Describe the shape of the water table in relation to the shape of the land surface.

FIGURE 4.8 Portion of the Campti, Louisiana, topographic map. (Map source: United States Department of the Interior, Geological Survey)

SCALE 1:62500
CONTOUR INTERVAL 20 FEET
DATUM IS MEAN SEA LEVEL

LOUISIANA
QUADRANGLE LOCATION

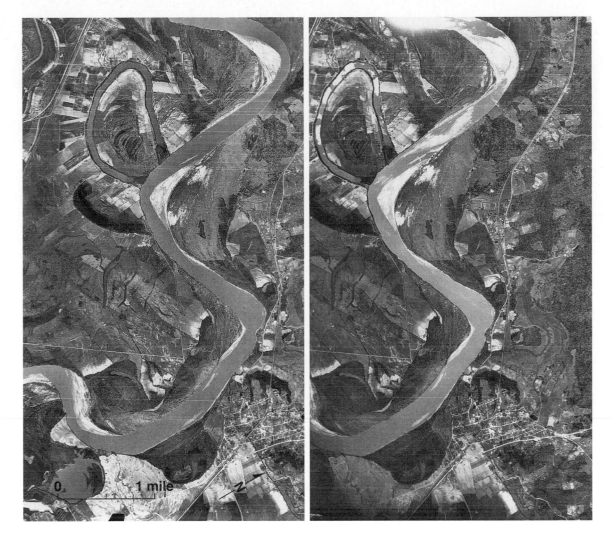

FIGURE 4.9 Stereogram of the Campti, Louisiana, area. (Courtesy of U.S. Geological Survey)

NORTH   SOUTH

Campti

DIAGRAM 4.2

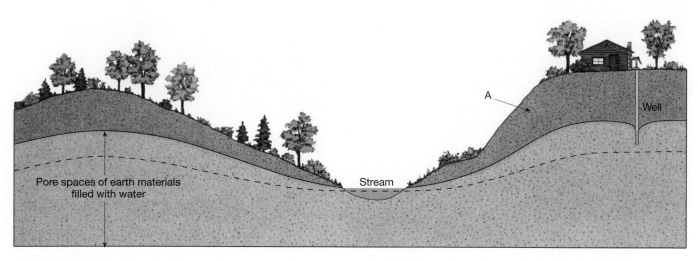

FIGURE 4.10 Subsurface of the earth showing saturated and unsaturated materials.

Pore spaces of earth materials filled with water

Stream

A

Well

---

45. What is the relation of the surface of the water in the stream to the water table?

46. What is the lowered surface in the water table around the well called? What has caused the lowering of the surface of the water table around the well? What will make it larger or smaller?

47. At point A, sketch a small, impermeable pocket of clay that intersects the valley wall.

48. Describe what will happen to water that infiltrates to the depth of the clay pocket at point A.

The dashed line in Figure 4.10 represents the level of the water table during the dry season when infiltration is no longer replenishing the groundwater.

49. What is the consequence of the lower elevation of the water table during the dry season on the operation of the well? How might the problem have been avoided?

50. What are two main sources of pollutants that can contaminate groundwater supplies?

The Problem of Ground Subsidence. As the demand for freshwater increases, surface subsidence caused by the withdrawal of groundwater from **aquifers** presents a serious problem for many areas. Several major urban areas such as Las Vegas, Houston-Galveston, Mexico City, and the Central Valley of California are experiencing subsidence caused by over-pumping wells. In Mexico City alone, compaction of the subsurface material resulting from the reduction of fluid pressure as the water table is lowered has caused as much as seven meters of subsidence. Fortunately, in many areas an increased reliance on surface water and replenishing the groundwater supply has slowed the trend.

A classic example of land subsidence caused from groundwater withdrawal is in the Santa Clara Valley, which borders the southern part of San Francisco Bay in California. The graph presented in Figure 4.11 illustrates the relation between ground subsidence in the valley and the level of water in a well in the same area. Questions 51–55 refer to Figure 4.11.

51. What is the general relation between the ground subsidence and level of water in the well illustrated on the graph?

52. What was the total ground subsidence and total drop in the level of water in the well during the period shown on the graph?

FIGURE 4.11 Ground subsidence and water level in a well in the Santa Clara Valley, California. (Data courtesy of U.S. Geological Survey)

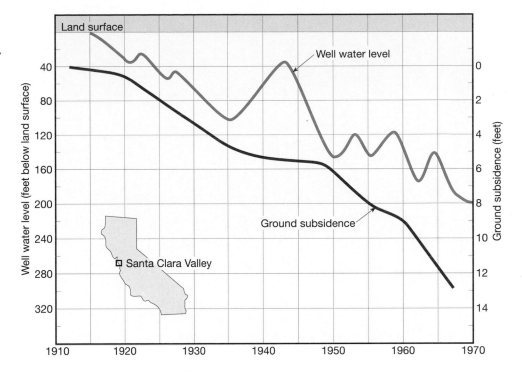

Total ground subsidence = _____ feet

Total drop in well level = _____ feet

53. During the period shown on the graph, on an average, about (1 foot, 5 feet, 10 feet) of land subsidence occurred with each 20-foot decrease in the level of water in the well. Circle your answer.

54. The ground subsidence that took place during the twenty years before 1950 was (less, greater) than the subsidence that took place between 1950 and 1970. Circle your answer.

55. Notice that minimal subsidence took place between 1935 and 1950. After referring to the well water level during the same period of time, suggest a possible reason for the reduced rate of subsidence between 1935 and 1950?

_____

Examining a Karst Landscape. Landscapes that are dominated by features that form from groundwater dissolving the underlying rock are said to exhibit **karst topography.** On the surface, karst topography is characterized by irregular terrain, springs, **disappearing streams, solution valleys,** and depressions called **sinkholes** (Figure 4.12). Beneath the surface, dissolution of soluble rock may result in **caves** and **caverns.**

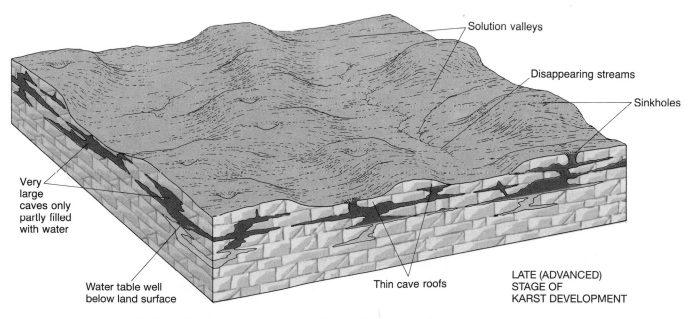

FIGURE 4.12 Generalized features of an advanced stage of karst topography.

FIGURE 4.13 Portion of the Mammoth Cave, Kentucky, topographic map. (Map source: United States Department of the Interior, Geological Survey)

SCALE 1:62500
CONTOUR INTERVAL 20 FEET
DATUM IS MEAN SEA LEVEL

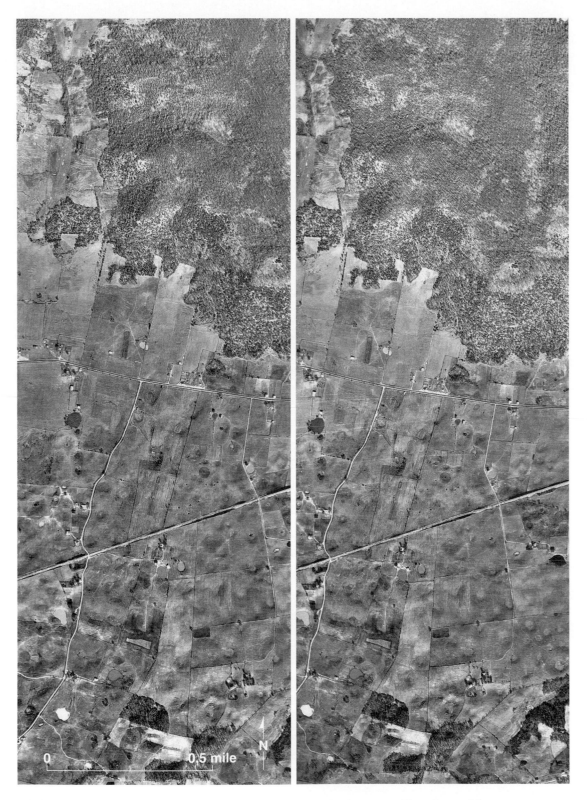

FIGURE 4.14 Stereogram of the Mammoth Cave, Kentucky, area. (Courtesy of the U.S. Geological Survey)

One of the classic karst regions in the United States is the Mammoth Cave, Kentucky, area. Locate and examine the Mammoth Cave, Kentucky, topographic map, Figure 4.13. An insoluble sandstone layer is the surface rock that forms the upland area in the northern quarter of the map. Underneath the sandstone layer is a soluble limestone. Erosion has removed all the sandstone in the southern three-fourths of the map and exposed the limestone. On the limestone surface, numerous sinkholes, indicated by closed contour lines with hachures, are present, as well as several disappearing streams (A).

Use the Mammoth Cave topographic map, Figure 4.13, and stereogram of the same area, Figure 4.14, to answer questions 56–60.

---

56. On the topographic map, outline the area that is shown on the stereogram.

57. What does the absence of water in the majority of sinkholes indicate about the depth of the water table in the area?

_____

58. Examine both the stereogram and map. Then describe the difference in appearance between the northern quarter and southern three-fourths of the mapped area.

_____

_____

59. Describe what is happening to Gardner Creek in the area indicated with the letter B on the map.

_____

60. Use your text as a reference. List the two ways that sinkholes commonly form.

   a. _____

   b. _____

REVIEW

Having completed the exercise, you should know the following:

1. The hydrologic cycle illustrates the exchange of water between the oceans, atmosphere, and land.
2. Precipitation that falls over land may infiltrate the surface material and become groundwater, run off over the surface, evaporate, or be transpired by plants. At high elevations and high latitudes, some water may also become stored for a period of time in snowfields and glaciers.
3. The permeability of a material is related to its grain size.
4. In urban areas, peak stream flow is greater and occurs sooner than in rural areas.
5. In humid regions, rivers that are high above base level often have steep gradients and are actively downcutting toward base level. Waterfalls and rapids are common and the valleys often have a characteristic V-shape with little or no floodplain.
6. In humid regions, rivers near base level exhibit pronounced meandering and have floodplains several times wider than their meander belts. Natural levees, oxbow lakes, back swamps, and yazoo tributaries are common features in these valleys.
7. The water table is the upper surface of the zone of saturation. In the zone of saturation, the pore spaces of the earth materials are filled with water.
8. The shape of the water table generally conforms to the shape of the land surface and its depth fluctuates with the seasons. Lakes and streams generally occupy areas where the land surface is below the water table.
9. Excessive withdrawal of groundwater from an aquifer may cause compaction of earth materials beneath the surface and ground subsidence on the surface.
10. Karst topography is caused by mildly acidic groundwater dissolving soluble rock at or near the earth's surface. Sinkholes, disappearing streams, and caves are common features of karst landscapes.

EXERCISE FOUR

Shaping the Earth's Surface
Running Water and Groundwater

SUMMARY/REPORT PAGE

Date Due: _____

Name: _____

Date: _____

Class: _____

After you have finished Exercise Four, complete the following questions. You may have to refer to the exercise for assistance or to locate specific answers. Be prepared to submit this summary/report to your instructor at the designated time.

1. Write a statement that describes the movement of water through the hydrologic cycle, citing several of the processes that are involved.

_____

_____

_____

2. Assume you are assigned a project to determine the quantity of infiltration that takes place in an area. What are the variables you must measure or know before you can arrive at your answer?

_____

_____

3. Write a brief paragraph summarizing the results of your permeability experiment in question 8 of the exercise.

_____

_____

_____

4. Describe the effects that urbanization has on the stream flow of a region.

_____

_____

5. Refer to the proportion of water that either infiltrates or runs off. Why does a soil-covered hillside with sparse vegetation often experience severe soil erosion? What are some soil conservation methods that could be used to reduce the erosion?

_____

_____

_____

6. Define the following terms:

Base level: _____

_____

Meander: _____

_____

Water table: _____

_____

Permeability: _____

_____

Aquifer: _____

_____

Karst topography: _____

_____

Sinkhole: _____

_____

7. Name and describe three features you would expect to find on the floodplain of a widely meandering river near its mouth.

Feature   **Description**

_____   _____

_____   _____

_____   _____

8. Assume you have decided to drill a water well. What are at least two factors concerning the water table and zone of saturation that should be considered prior to drilling?

_____

_____

9. How might a rapidly growing urban area that relies on groundwater as a freshwater source avoid the problem of land subsidence from groundwater withdrawal?

_____

_____

10. Name and describe two features you would expect to find in a region with karst topography.

Feature   **Description**

_____   _____

_____   _____

EXERCISE FIVE

Shaping the Earth's Surface
Arid and Glacial Landscapes

The previous exercise explored the hydrologic cycle and the role of running water and groundwater in shaping the landscape in humid regions. However, when taken together, the dry regions of the world and those areas whose surfaces have been modified by glacial ice also comprise a significant portion of the earth's surface. Since desert or near-desert conditions and glaciated regions prevail over a large area of the earth, an understanding of the landforms and processes that shaped these regions is essential to the earth scientist.

In this exercise you will investigate some of the features produced by running water in desert regions. Also to be examined are the erosional effects and depositional features of alpine glaciers and the vast continental ice sheets that at one time covered almost 30 percent of the earth's land area.

OBJECTIVES

After you have completed this exercise, you should be able to

1. Locate the desert and steppe regions of North America.
2. Describe the evolution of the landforms that exist in the mountainous desert areas of the Basin and Range region of the western United States.
3. Describe the different types of glacial deposits and the features they compose.
4. Identify and explain the formation of the features commonly found in areas where the landforms are the result of deposition by continental ice sheets.
5. Describe the evolution and appearance of a glaciated mountainous area.
6. Identify and explain the formation of the features caused by alpine glaciation.

TEXTBOOK REFERENCE

Chapter 4 and Appendix D

MATERIALS

calculator   ruler

Materials Supplied by Your Instructor

stereoscope

TERMS

desert	bajada	till
steppe	playa lake	stratified drift
flash flood	inselberg	moraine
Basin and Range	pediment	Pleistocene epoch
fault-block	alpine glacier	arête
mountains	continental ice sheet	cirque
alluvial fan	drift	horn

DESERT LANDSCAPES

Introduction. Arid **(desert)** and semiarid **(steppe)** climates cover about 30 percent of the earth's land area (Figure 5.1). At first glance, many desert landscapes with their angular hills and steep canyon walls may appear to have been shaped by processes other than those that are responsible for landforms in regions with an abundance of water. However, as striking as the contrasts may be, running water is still the dominant agent responsible for most of the erosional work in deserts. Wind erosion, although more significant in dry areas than elsewhere, is only of secondary importance.

The distinct effects that running water has on humid and dry areas are the result of the same processes

75

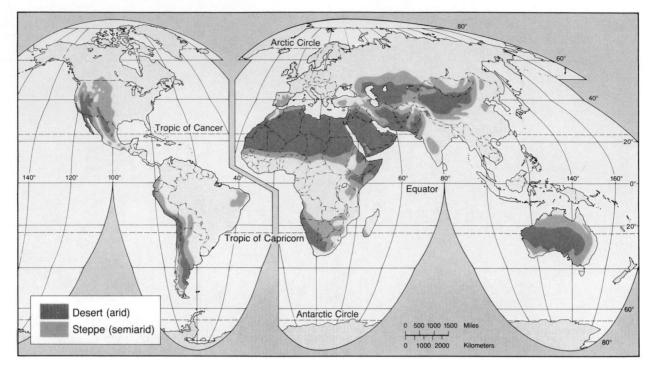

FIGURE 5.1   Global arid and semiarid climates.

operating under different climatic conditions. Precipitation in the dry climates is minimal, often sporadic, and frequently comes in the form of torrential downpours that last only a short time. Consequently, in desert areas **flash floods** occur and few streams or rivers reach the sea because the water often evaporates and/or infiltrates into the ground.

Evolution of a Desert Landscape.   Desert landscapes have developed in response to a variety of geologic processes. A classic region for studying the effects of running water in dry areas is the western United States. Throughout much of this **Basin and Range** region, which includes southeastern California, Nevada, western Utah, southern Oregon, southern Arizona and New Mexico, the erosion of mountain ranges and subsequent deposition of sediment in adjoining basins have produced a landscape characterized by several unique landforms (Figure 5.2).

In a large area of the Basin and Range region of the western United States **fault-block mountains** have formed as large blocks of the earth's crust have been forced upward (Figure 5.2A). The infrequent and intermittent precipitation in this desert region typically results in streams that carry their eroded material from the mountains into interior basins. **Alluvial fans** and **bajadas** often form as streams deposit sediment on the less steep slopes at the base of the mountains (Figure 5.2B). On rare occasions when streams flow across the alluvial fans, a shallow **playa lake** may develop near the center of a basin.

Continuing erosion in the mountains and deposition in the basins may eventually fill the basin and only isolated peaks, called **inselbergs,** surrounded by gently sloping sediment, remain. As the front of the mountain is worn back by erosion, a broad, sloping bedrock surface called a **pediment,** covered by a thin layer of sediment, often forms at its base (Figure 5.2C). In the final stages, even the inselbergs will disappear, and all that remains is a nearly flat, sediment-covered surface underlain by the erosional remnants of mountains.

Use Figure 5.1 and your text as a reference to answer questions 1 and 2.

1. Where are the desert and steppe regions of North America?

   Desert: _____

   Steppe: _____

2. Using an "X" to mark your selection(s), indicate which of the following statements are commonly held misconceptions concerning the world's dry lands?

   ____ The world's dry lands are always hot.

   ____ Desert landscapes are almost completely covered with sand dunes.

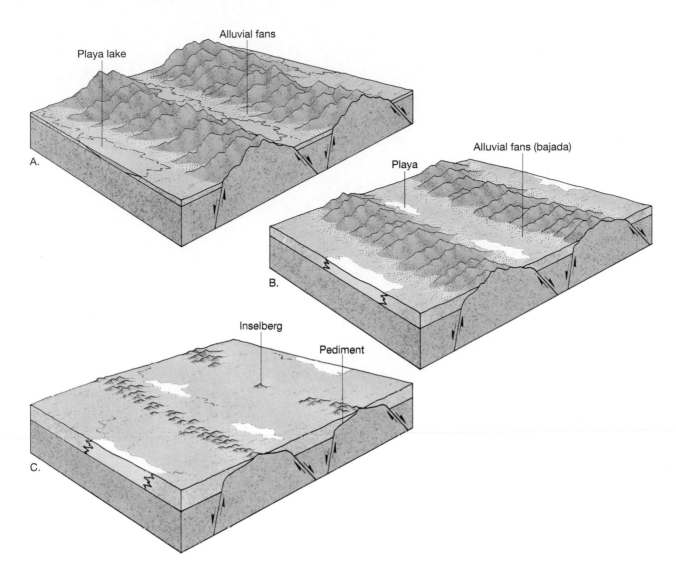

FIGURE 5.2 Stages of landscape evolution in a block-faulted, mountainous desert. **A.** Early stage; **B.** Middle stage; **C.** Late stage.

_____ The dry regions of the world encompass about 30% of the earth's land surface.

_____ Dry lands are all practically lifeless.

Figure 5.3 is a portion of the Antelope Peak, Arizona, topographic map that illustrates many of the features of the desert landscapes found in the western United States. Use the map and accompanying stereogram of the area (Figure 5.4) to answer questions 3–13. You may find the diagrams in Figure 5.2 helpful.

3. On the map, outline the area that is illustrated in the stereogram.

Use a stereoscope to examine the stereogram, Figure 5.4.

4. The vegetation in the area is (dense, sparse) and there are (few, many) dry stream courses. Circle your answers.
5. By examining the map, determine the total relief of the map area.

   Total relief = _____ feet
6. (Continuously flowing, Intermittent) streams dominate the area shown on the map. Circle your answer.
7. On the map, of the two lines, A or B, (A, B) follows the steepest slope. Circle your answer.
8. By drawing arrows on the map, indicate the directions that intermittent streams will flow as they leave the mountains.
9. Where on the map is the most likely place that surface water may accumulate? Label the area "possible lake."

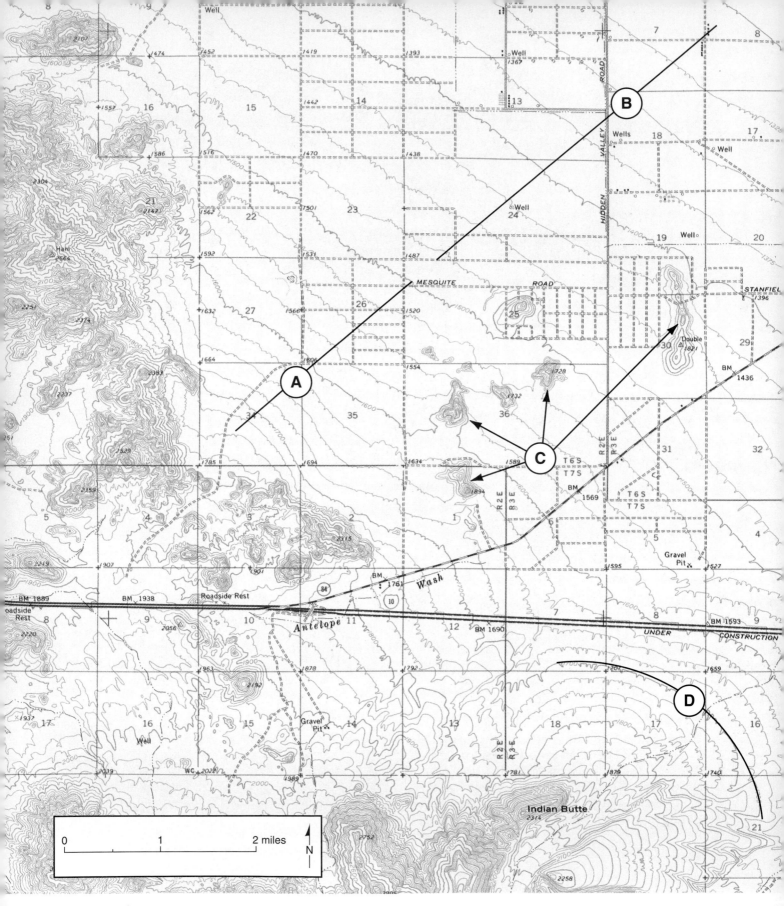

FIGURE 5.3 Portion of the Antelope Peak, Arizona, topographic map. (Map source: United States Department of the Interior, Geological Survey)

SCALE 1:62500
CONTOUR INTERVAL 25 FEET
DATUM IS MEAN SEA LEVEL

ARIZONA

QUADRANGLE LOCATION

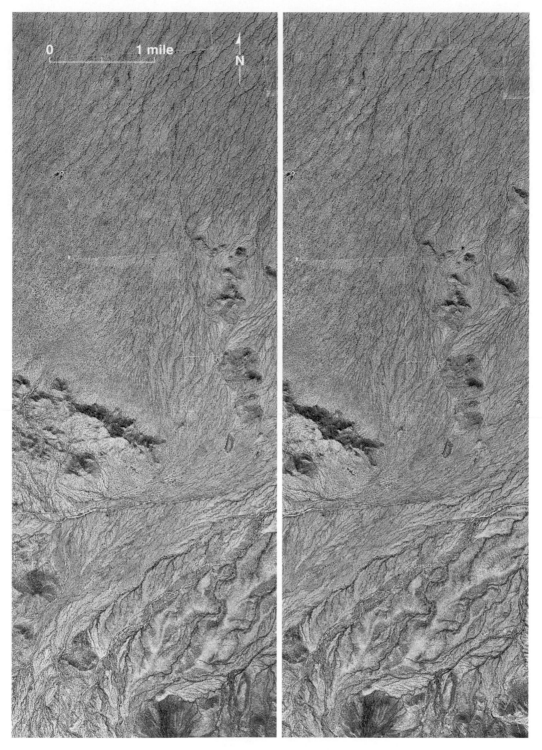

FIGURE 5.4 Stereogram of the Antelope Peak, Arizona, area. (Courtesy of the U.S. Geological Survey)

10. Identify the features indicated on the map by the following letters and briefly describe how each formed.

Letter C: _____

_____

Letter D: _____

_____

The area at A on the map is a bedrock surface covered by a thin layer of sediment.

11. The feature labeled A is called a _____ .

12. Briefly describe how the Antelope Peak area may have looked millions of years ago.

_____

_____

13. Assume that erosion continues in the area without interruption. How might the area look millions of years from now?

_____

_____

GLACIAL LANDSCAPES

Introduction. Slightly more than two percent of the world's water is in the form of glacial ice that covers nearly 10 percent of the earth's land area. However, to an earth scientist, glaciers represent more than storehouses of fresh water in the hydrologic cycle. Like the other agents that modify the earth's surface, glaciers are dynamic forces capable of eroding, transporting, and depositing sediment.

Literally thousands of glaciers exist on Earth today. They occur in regions where, over long periods of time, the yearly snowfall has exceeded the quantity lost by melting or evaporation. **Alpine,** or **valley, glaciers** form from snow and ice at high altitudes. At high latitudes, enormous **continental ice sheets** cover much of Greenland and Antarctica.

Glacial erosion and deposition leave an unmistakable imprint on the earth's surface. In regions once covered by continental ice sheets, glacially scoured surfaces and subdued terrain dominated by glacial deposits are the rule. By contrast, erosion by alpine glaciers in mountainous areas tends to accentuate the irregularity of the topography, often resulting in spectacular scenery characterized by sharp, angular features.

Glacial Deposits and Depositional Features. The general term **drift** applies to all sediments of glacial origin, no matter how, where, or in what form they were deposited. There are two types of glacial drift: (1) **till,** which is characteristically unsorted sediment deposited directly by the glacier, and (2) **stratified drift,** which is material that has been sorted and deposited by glacial meltwater. The most widespread depositional features of glaciers are **moraines,** which are ridges of till that form along the edges of glaciers and layers of till that accumulate on the ground as the ice melts and recedes. There are several types of moraines, some common only to alpine glaciers, as well as other kinds of glacial depositional features.

14. Use your text as a reference to name the type of moraine described by each of the following statements.

_____ moraines form at the terminus of a glacier.

_____ moraines form along the sides of a valley.

_____ moraines are end moraines that mark the farthest advance of a glacier.

_____ moraines form as the ice front periodically becomes stationary during retreat.

_____ moraines form as the glacier recedes and lays down a layer of till.

_____ moraines form when two valley glaciers coalesce to form a single ice stream.

Figure 5.5 illustrates a hypothetical area during and after glaciation. Use Figure 5.5 and your text to answer questions 15–17.

15. Draw a large arrow on Figure 5.5B that indicates the direction of glacial ice movement in the area. Label the arrow, "ice flow."

16. On Figure 5.5B, label an example of a terminal moraine, recessional moraine, and ground moraine.

17. After you review the appropriate section of the text, briefly describe each of the following glacial depositional features and select the letter on Figure 5.5B that indicates an example of each.

Drumlin: _____

_____ Letter: _____

Esker: _____

_____ Letter: _____

Kame: _____

_____ Letter: _____

Kettle: _____

_____ Letter: _____

Outwash plain: _____

_____ Letter: _____

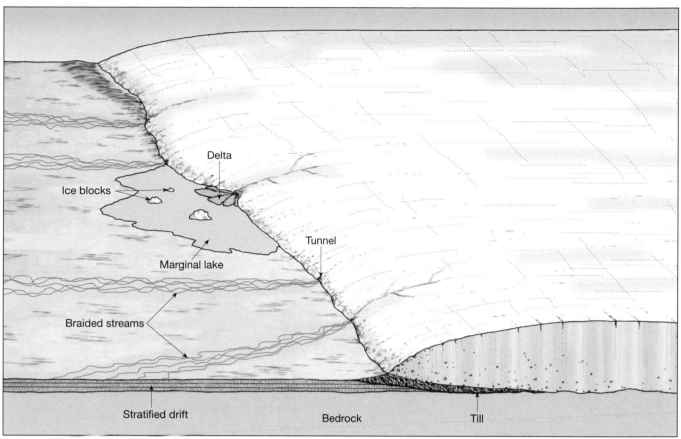

A.

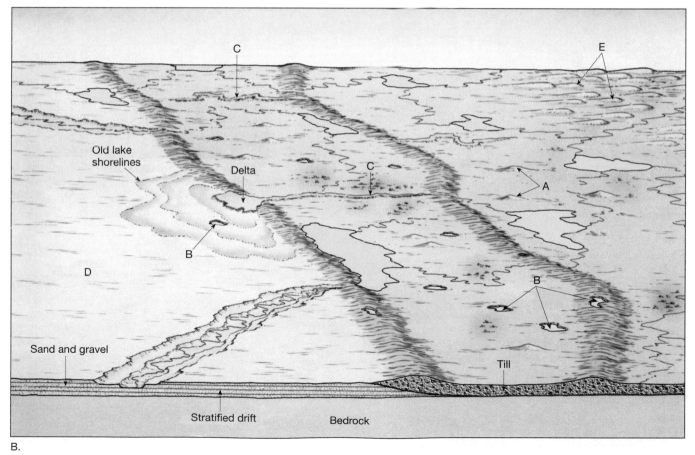

B.

FIGURE 5.5 Characteristic depositional features of glaciers. **A.** During glaciation and **B.** After glaciation.

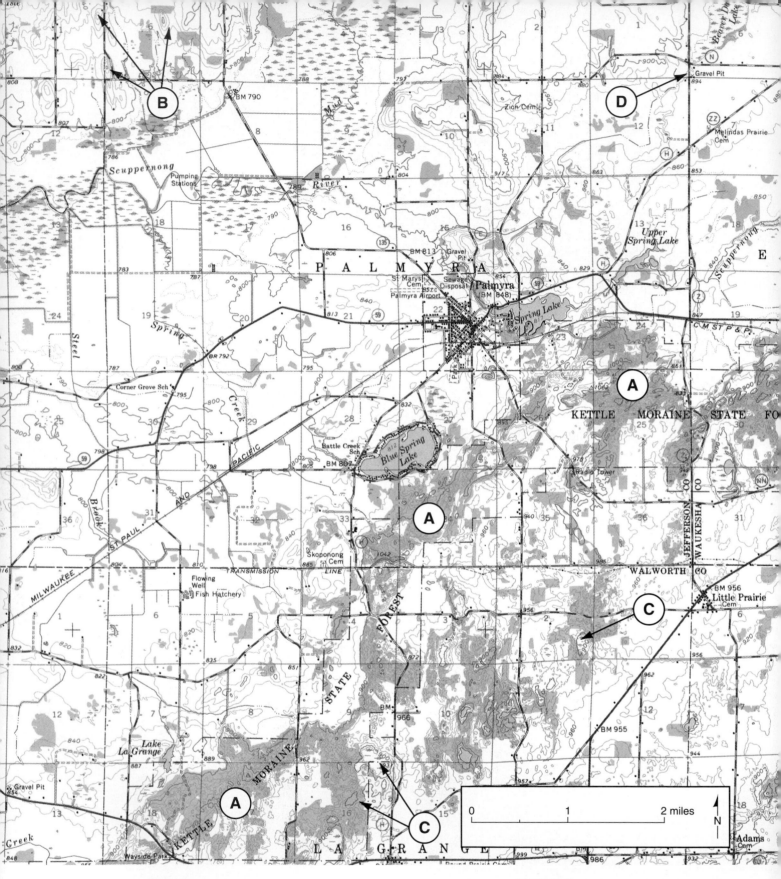

FIGURE 5.6 Portion of the Whitewater, Wisconsin, topographic map. (Map source: United States Department of the Interior, Geological Survey)

SCALE 1:62500
CONTOUR INTERVAL 20 FEET
DOTTED LINES REPRESENT 10-FOOT CONTOURS
DATUM IS MEAN SEA LEVEL

WISCONSIN

QUADRANGLE LOCATION

FIGURE 5.7 Stereogram of the Whitewater, Wisconsin, area. (Courtesy of the U.S. Geological Survey)

Features of Continental Ice Sheets. Continental ice sheets, as well as alpine glaciers, were considerably more extensive over the earth's surface than they are today during a division of earth history called the **Pleistocene epoch.** At one time, these thick sheets of ice covered all of Canada, portions of Alaska, and much of the northern United States. Today, the impact that these ice sheets had on the landscape is still very obvious.

While alpine glaciers change the shape of the land surface primarily by erosion, landforms produced by continental ice sheets, especially those that covered portions of the United States during the Pleistocene epoch, are essentially depositional in origin. Some of the most extensive areas of glacial deposition occurred in the north-central United States. Here, glacial drift covers the surface and landscapes are dominated by moraines, outwash plains, kettles and other depositional features.

Use your text as a reference to answer question 18.

18. By listing state abbreviations, indicate the geographic area of the continental United States that Pleistocene glaciers covered during their maximum extent.

_____

_____

Figure 5.6 is a portion of the Whitewater, Wisconsin, topographic map, which illustrates many of the depositional features that are typical of continental glaciation. Use the map and the accompanying stereogram of the area (Figure 5.7) to answer questions 19–31.

19. After examining the map and stereogram, draw a line on the map that outlines the area illustrated on the photograph.
20. The general topography of the land in the southeast corner of the region is (higher, lower) in elevation and (more, less) irregular than the land in the northwest. Circle your answers.
21. What features on the map indicate that portions of the area are poorly drained? Where are these features located?

_____

_____

Examine the elevations of the feature indicated with the letter A that, in general, coincides with Kettle Moraine State Forest. Compare the elevations to those found to the northwest and southeast of the feature.

NORTHWEST   SOUTHEAST

Scuppernong River   Little Prairie

DIAGRAM 5.1

22. In Diagram 5.1, sketch a northwest-southeast topographic profile along a line that extends from the Scuppernong River to the city of Little Prairie. Indicate the appropriate elevations on the vertical axis of the profile.

23. The area that coincides with Kettle Moraine State Forest is (higher, lower) in elevation than the land to the northwest and southeast. Circle your answer.

24. The feature labeled A on the map is a long ridge composed of till called a (kettle, moraine, drumlin). Circle your answer.

25. The streamlined, asymmetrical hills composed of till, labeled B, are what type of feature?

_____

26. How are the features labeled B used to determine the direction of ice flow in a glaciated area?

_____

27. Use the features labeled B as a guide to draw an arrow on the map that indicates the direction of ice flow in the region.

28. Where on the map is the likely location of the outwash plain? Identify and label the area "outwash plain."

_____

29. Identify and label the ground moraine area on the map.

30. What term is applied to the numerous almost circular depressions designated C on the map?

_____

31. What is the material that is being mined at letter D and north of Palmyra? What is the probable source of the material?

_____

Features of Alpine or Valley Glaciation. As they flow, alpine glaciers often exaggerate the already irregular topography of a region by eroding the mountain slopes and deepening the valleys. Figure 5.8 illustrates the changes that a formerly unglaciated mountainous area (Figure 5.8A) experiences as the result of alpine glaciation. Many of the landforms produced by glacial erosion, such as **arêtes, cirques,** and **horns** are identified in Figure 5.8C.

Questions 32–34 refer to Figure 5.8.

32. How has glaciation changed the shape and depth of the main valley?

_____

_____

Prior to glaciation, tributary streams were adjusted to the depth of the main valley.

33. What has been the consequence of glacial erosion on the gradients or slopes of tributary streams?

_____

34. Use your own words to describe how the appearance of the area has changed from what it was prior to glaciation.

_____

_____

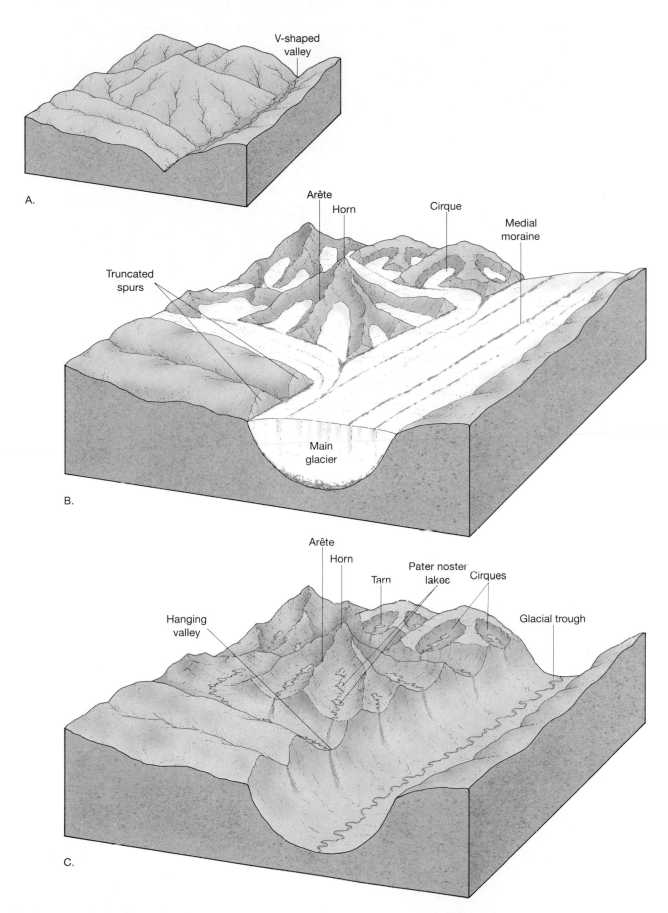

FIGURE 5.8 Landforms created by alpine glaciers. **A.** Landscape prior to glaciation; **B.** During glaciation; **C.** After glaciation.

FIGURE 5.9 Oblique aerial photograph of alpine glaciers and glacial features, Mont Blanc, France. (Photo by James E. Patterson)

Figure 5.9 is an oblique aerial photograph of an area experiencing alpine glacial erosion on Mont Blanc, France. Questions 35 and 36 refer to the figure.

▬▬▬

35. Draw arrows on the photograph that indicate the directions that the glaciers are flowing.
36. Give the name of the glacial feature described by each of the following statements as well as the letter on the photograph that labels an example of the feature. Use Figure 5.8C and your text as references.

a. Sinuous, sharp-edged ridge:

Name: _____

Letter of Example: _____

b. Hollowed-out, bowl-shaped depression that is the glacier's source and the area of snow accumulation and ice formation:

Name: _____

Letter of Example: _____

c. Moraine formed along the side of a valley:

Name: _____

Letter of Example: _____

d. Moraine formed when two valley glaciers coalesce to form a single ice stream:

Name: _____

Letter of Example: _____

FIGURE 5.10 Portion of the Holy Cross, Colorado, topographic map. (Map source: United States Department of the Interior, Geological Survey)

SCALE 1:62500
CONTOUR INTERVAL 50 FEET
DATUM IS MEAN SEA LEVEL

COLORADO

QUADRANGLE LOCATION

Figure 5.10 is a portion of the Holy Cross, Colorado, topographic map, a mountainous area that underwent alpine glaciation in the past. Questions 37–42 refer to the map.

▬▬▬▬

37. Following line A on the map, sketch a topographic profile of the valley of Lake Fork from Sugar Loaf Mtn. to Bear Lake on Diagram 5.2. Indicate the appropriate elevations along the vertical axis of the profile.

38. Describe the shape of the profile of the valley of

Lake Fork. The valley is called a glacial _____

_____ .

39. Identify the type of glacial feature indicated on the map at each of the following letters. Use Figure 5.8C as a reference.

Letter B: _____

Letter C: _____

40. Letter (D, E, F, G) on the map indicates a *tarn(s)*, a lake that forms in a cirque. Circle your answer.

The feature marked H on the map is composed of glacial till.

▬▬▬▬

41. What type of glacial feature is designated H? How did it form?

_____

42. What is the reason for the formation of Turquoise Lake?

_____

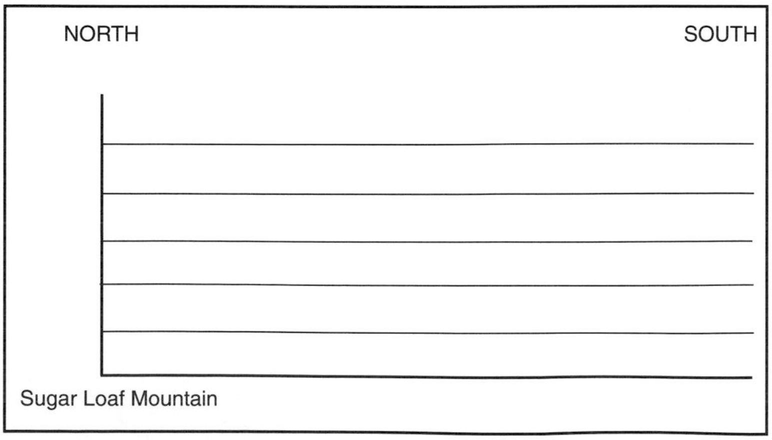

REVIEW

Having completed the exercise, you should know the following:

1. Running water is the erosional agent most responsible for shaping desert landscapes.
2. In the arid regions of the western United States, fault-block mountains often result in interior drainage. In these areas, sediment eroded from the mountains fills the basins and alluvial fans, bajadas, and playas are formed during the process. During the later stages of development, remnant mountain peaks, called inselbergs, often project through the sediment.
3. There are two types of glacial drift—till and stratified drift. Till is unsorted material deposited directly by the ice. Stratified drift is material that is sorted and deposited by glacial meltwater.
4. The two most significant types of glaciers are alpine or valley glaciers and continental ice sheets.
5. Many of the landscapes of the north-central and northeastern United States are characterized by glacial depositional features that were deposited by continental ice sheets during the Pleistocene epoch. Moraines, outwash plains, drumlins, kames, and eskers occur throughout these regions.
6. The effect of alpine glaciers is to accentuate and enhance the already irregular topography of mountains by erosion. Cirques, arêtes, horns, and U-shaped valleys are some of the features commonly found in glaciated mountains.

NORTH   SOUTH

Sugar Loaf Mountain

DIAGRAM 5.2

Shaping the Earth's Surface
Arid and Glacial Landscapes

SUMMARY/REPORT PAGE

Name: _____

Date Due: _____

Date: _____

Class: _____

After you have finished Exercise Five, complete the following questions. You may have to refer to the exercise for assistance or to locate specific answers. Be prepared to submit this summary/report to your instructor at the designated time.

1. What area of the United States is characterized by fault-block mountains with interior drainage into adjoining basins?

_____

2. Describe the sequence of geologic events that have produced the landforms in the Antelope Peak area of Arizona.

_____

_____

3. What type of feature is located at each of the following letters on the Antelope Peak, Arizona, topographic map, Figure 5.3?

Letter C: _____

Letter D: _____

4. Toward what direction does the pediment slope on the Antelope Peak, Arizona, topographic map?

_____

5. What is the reason for so many dry stream channels in the Antelope Peak, Arizona, area?

_____

6. If you were working in the field, explain how you might determine whether a glacial feature is a recessional moraine or an esker.

_____

_____

7. In the following space, sketch a map-view (the area viewed from above) of the Whitewater, Wisconsin, topographic map, Figure 5.6. Show and label the outwash plain, end moraine, area containing drumlins, and the area containing kettles and kettle lakes.

8. Assume you are hiking in the mountains. You suspect that the area was glaciated in the past. De-

scribe some of the features you would look for to confirm your suspicion.

_____

_____

9. What was your conclusion as to the reason for the formation of Turquoise Lake on the Holy Cross, Colorado, topographic map, Figure 5.10?

_____

10. Define each of the following terms.

Bajada: _____

_____

Glacial drift: _____

_____

End moraine: _____

_____

Horn: _____

_____

Pleistocene epoch: _____

_____

Inselberg: _____

_____

EXERCISE SIX

Determining Geologic Ages

The recognition of the vastness of geologic time and the ability to establish the sequence of geologic events that have occurred at various places at different times are among the great intellectual achievements of science. To accomplish the task of deciphering earth history, geologists have formulated several laws, principles, and doctrines that can be used to place geologic events in their proper sequence. Also, using the principles that govern the radioactive decay of certain elements, scientists are now able to determine the age of many earth materials with reasonable accuracy. In this exercise you will investigate some of the techniques and procedures used by earth scientists in their search to interpret the geologic history of the earth.

OBJECTIVES

After you have completed this exercise, you should be able to

1. List and explain each of the laws, principles, and doctrines that are used to determine the relative ages of geologic events.
2. Determine the sequence of geologic events that have occurred in an area by applying the techniques and procedures for relative dating.
3. Explain the methods of fossilization and how fossils are used to define the ages of rocks and correlate rock units.
4. Explain how the radioactive decay of certain elements can be used to determine the age of earth materials.
5. Apply the techniques of radiometric dating to determine the absolute age of a rock.
6. Describe the geologic time scale and list in proper order some of the major events that have taken place on the earth since its formation.

TEXTBOOK REFERENCE

Chapter 9

MATERIALS

ruler
calculator

Materials Supplied by Your Instructor

fossils and fossil questions (optional)

5 meter length of adding machine paper

meterstick or metric tape measure

TERMS

relative dating
uniformitarianism
original horizontality
superposition
inclusion

unconformity
cross-cutting
fossil
fossil succession

radiometric date
half-life
eon
era

INTRODUCTION

The history of geology, for the most part, can be described as a quest to comprehend the physical evolution of the earth. As a result of their investigations, geologists have become aware of certain fundamental concepts concerning the formation and character of rocks, fossils, processes at work on and within the earth, and the radioactive decay of certain atoms. Using this knowledge, geologists interpret earth history by determining the sequence of geologic events and/or age of earth materials.

RELATIVE DATING

Relative dating, the placing of geologic events in their proper sequence or order, does not tell how long ago something occurred, only that it preceded one event and followed another. Several logical doctrines, laws, and principles govern the techniques used to establish the relative age of an object or event.

Doctrine of Uniformitarianism.
First proposed by James Hutton in the late 1700s, this doctrine states that the physical, chemical, and biological laws that operate today have operated throughout the earth's history. Although geologic processes such as erosion, deposition, and volcanism are governed by these unchanging laws, their rates and intensities may vary. The doctrine is often summarized in the sentence, "The present is the key to the past."

Principle of Original Horizontality.
Sediment, when deposited, forms nearly horizontal layers. There-fore, if we observe beds of sedimentary rocks that are folded or inclined at a steep angle, the implication is that some deforming force took place after the sediment was deposited (see Figure 6.1).

Law of Superposition.
In any sequence of unde-formed sedimentary rocks (or surface deposited igneous rocks such as lava flows and layers of volcanic ash), the oldest rock is always at the bottom and the youngest is at the top. Therefore, each layer of rock represents an interval of time that is more recent than that of the underlying rocks.

Assume the playing cards shown in Figure 6.2 are layers of sedimentary rocks viewed from above. Using the figure, answer questions 1 and 2.

1. In the space provided in Figure 6.2, list the order first (oldest) to last (youngest), that the cards were laid down.

FIGURE 6.1 Uplifted sedimentary strata in the Canadian Rockies. (Photo by E. J. Tarbuck)

FIGURE 6.2 Sequence of playing cards illustrating the law of superposition.

Youngest (last) _____

_____

_____

_____

_____

_____

_____

_____

Oldest (first) _____

2. Were you able to place all of the cards in sequence? If not, which one(s) could not be "relative" dated and why?

_____

_____

Figure 6.3 illustrates a geologic cross section, a side view, of the rocks beneath the surface of a hypothetical region. Use Figure 6.3 to answer questions 3 and 4.

3. Of the two sequences of rocks, A–D and E–G, (A–D, E–G) was disturbed by crustal movements after its deposition. Circle your answer. What law or principle did you apply to arrive at your answer?

_____

4. Apply the law of superposition to determine the relative ages of the *undisturbed* sequence of sedi-

FIGURE 6.3 Geologic cross section of a hypothetical region showing igneous intrusive features (C and H) and sedimentary rocks.

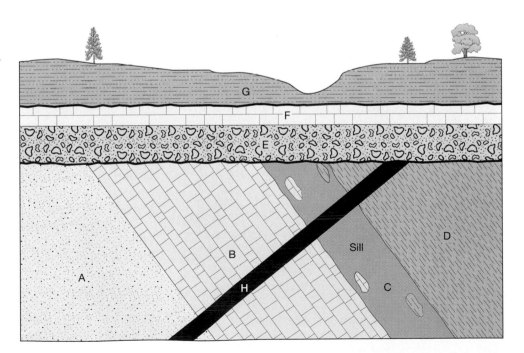

mentary rocks. List the letter of the oldest rock layer first.

Oldest _____ Youngest

Inclusions.

Inclusions are pieces of one rock unit that are contained within another unit, for example, pieces of granite embedded in a shale. The rock mass adjacent to the one containing the inclusions must have been there first in order to provide the rock fragments. Therefore, the rock containing the inclusions is the younger of the two.

Refer to Figure 6.4 to answer questions 5 and 6. The sedimentary layer B is a sandstone. Letter C is the sedimentary rock, shale.

5. Identify and label the inclusions in Figure 6.4.
6. Of the two rocks B and C, rock (B, C) is older. Circle your answer.

Unconformities.

As long as continuous sedimentation occurs at a particular place, there will be an uninterrupted record of the material and life forms. However, if the sedimentation process is suspended by an emergence of the area from below sea level, then no sediment will be deposited and an erosion surface will develop. The result is that no rock record will exist for a part of geologic time. Such a gap in the rock record is termed an **unconformity.** An unconformity is typically shown on a cross-sectional (side view) diagram by a wavy line (〰〰〰).

Review the discussion of unconformities in Chapter 9 of your text. Then answer question 7.

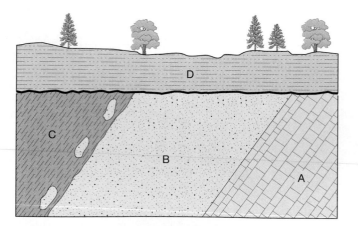

FIGURE 6.4 Geologic cross section showing sedimentary rocks.

7. Identify and label an example of an angular unconformity and a disconformity in Figure 6.3.

Principle of Cross-Cutting Relationships.

Whenever a fault or intrusive igneous rock cuts through an existing feature it is younger than the structure it cuts. For example, if a basalt dike cuts through a sandstone layer, the sandstone had to be there first and, therefore, is older than the dike.

Figure 6.5 is a geologic cross section showing sedimentary rocks (A, B, D, E, F, and G), an igneous intrusive feature called a *dike* (C), and a fault (H). Use Figure 6.5 to answer the questions 8–11.

8. The igneous intrusion C is (older, younger) than the sedimentary rocks B and D. Circle your answer.
9. Fault H is (older, younger) than the sedimentary beds A–E.
10. The relative age of fault H is (older, younger) than the sedimentary layer F.
11. Did the fault occur before or after the igneous intrusion? Explain how you arrived at your answer.

_____

_____

12. Refer to Figure 6.3. The igneous intrusion H is (older, younger) than rock layer E and (older, younger) than layer D. Circle your answers.
13. Refer to Figure 6.3. What evidence supports the conclusion that the igneous intrusive feature called a *sill,* C, is more recent than both of the rock layers B and D and older than the igneous intrusion H?

_____

_____

Fossils and the Principle of Fossil Succession.

Fossils are among the most important tools used to interpret earth history. They are used to define the ages of rocks, correlate one rock unit with another, and determine past environments on the earth.

The earth has been inhabited by different assemblages of plants and animals at different times. As rocks form, they often incorporate the preserved remains of these organisms as fossils. According to the principle of **fossil succession,** fossil organisms succeed each other in a definite and determinable order. Therefore, the time that a rock originated can frequently be determined by noting the kinds of fossils that are found within it.

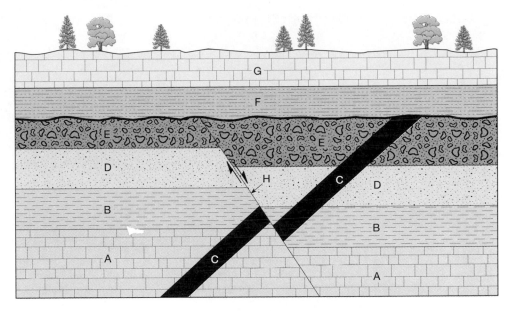

FIGURE 6.5 Geologic cross section of a hypothetical area showing an igneous intrusion (C), a fault (H), and sedimentary rocks.

Using your text and the materials supplied by your instructor, answer questions 14 and 15.

14. At the discretion of your instructor, there may be several stations with fossils and questions set up in the laboratory. Following the specific directions of your instructor, proceed to the stations.

15. What are the two conditions that favor the preservation of an organism as a fossil?

Condition 1: _____

Condition 2: _____

16. Refer to Figure 6.6. Select the photo, A, B, C, or D, that best illustrates each of the following methods of fossilization or fossil evidence.

Petrification The small internal cavities and pores of the original organism are filled with precipitated mineral matter. Photo: _____

Cast The space once occupied by a dissolved shell or other structure is subsequently filled with mineral matter. Photo: _____

Impression A replica of a former fossil left in fine-grained sediment after the fossilizing material, often carbon, is removed. Photo: _____

Indirect evidence Traces of prehistoric life, but not the organism itself. Photo: _____

Figure 6.7 shows a sequence of undeformed sedimentary rocks. Each layer of rock contains the fossils illustrated within it. The three rocks, Rocks 1, 2, and 3, illustrated below the layered sequence were found

nearby and each rock contained the fossils indicated. Answer question 17 using Figure 6.7.

17. Applying the principle of fossil succession, indicate the proper position of each of the three rocks relative to the rock layers by writing the words Rock 1, Rock 2, or Rock 3 at the appropriate position in the sequence.

Applying Relative Dating Techniques. Geologists often apply several of the techniques of relative dating when investigating the geologic history of an area.

Figure 6.8 is a geologic cross section of a hypothetical area. Letters K and L are igneous rocks. Letter M is a fault. All the remaining letters represent sedimentary rocks. Using Figure 6.8 to complete questions 18–24 will provide insight into how the relative geologic history of an area is determined.

18. Identify and label the unconformities indicated in the cross section.

19. Rock layer I is (older, younger) than layer J. Circle your answer. What law or principle have you applied to determine your answer?

_____

20. The fault is (older, younger) than rock layer I. Circle your answer. What law or principle have you applied to determine your answer?

_____

21. The igneous intrusion K is (older, younger) than layers A and B. Circle your answer. What two laws

A.

B.

C.

D.

FIGURE 6.6 Various types of fossilization. In photo **A.** the mineral quartz now occupies the internal spaces of what was once wood. **B.** is the replica of fish after the carbonized remains were removed. In photo **C.** mineral matter occupies the hollow space where a shell was once located. **D.** is a track left by a dinosaur in formerly soft sediment. (Photos by E. J. Tarbuck)

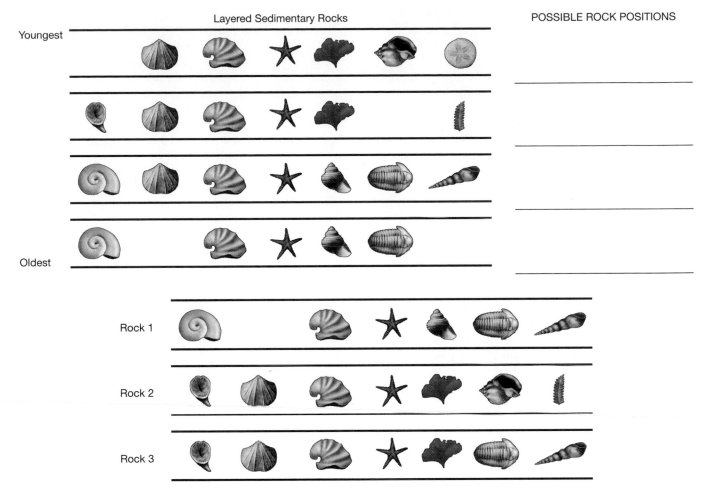

FIGURE 6.7 Layered sequence of sedimentary rocks with fossils and three separate rocks containing similar fossils.

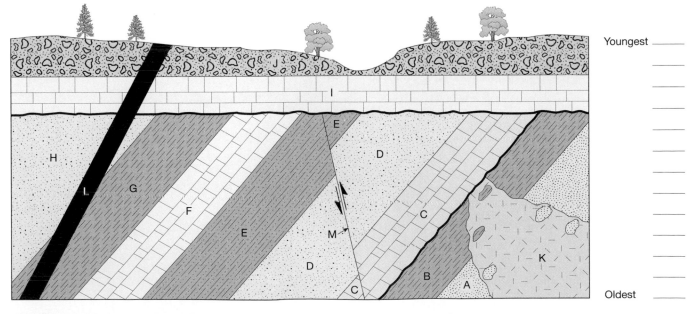

FIGURE 6.8 Geologic cross section of a hypothetical area showing igneous intrusive features (K and L), a fault (M), and sedimentary rocks.

or principles have you applied to determine your answer?

_____ and _____

22. The age of the igneous intrusion L is (older, younger) than layers J, I, H, G, and F.
23. The igneous intrusion K occurred (before, after) the deposition of layer C.
24. List the entire sequence of events, in order from oldest to youngest, by writing the appropriate letter in the space provided on the figure.

RADIOMETRIC DATING

The discovery of radioactivity and its subsequent understanding has provided a reliable means for calculating the *absolute age* in years of many earth materials. Radioactive atoms, such as the isotope uranium-238, emit particles from their nuclei that we detect as radiation. Ultimately, this process of decay produces an atom that is stable and no longer radioactive. For example, eventually the stable atom lead-206 is produced from the radioactive decay of uranium-238.

Determining Radiometric Ages. The radioactive isotope used to determine a **radiometric date** is referred to as the *parent isotope*. The amount of time it takes for one-half of the radioactive nuclei in a sample to change to their stable end product is referred to as

the **half-life** of the isotope. The isotopes resulting from the decay of the parent are termed the *daughter products*. For example, if we begin with one gram of radioactive material, half a gram would decay and become a daughter product after one half-life. After the second half-life, one half of the remaining radioactive isotope, 0.25 g or 1/4 of the original amount (1/2 of 1/2), would still exist. With each successive half-life, the remaining parent isotope would be reduced by half.

Figure 6.9 graphically illustrates how the ratio of a parent isotope to its stable daughter product continually changes with time. Use Figure 6.9 to help answer questions 25–29.

25. What fraction of the original parent isotope still exists after each of the following half-lives has elapsed?

**Fraction of
Parent Isotope Remaining**

One half-life: _____

Two half-lives: _____

Three half-lives: _____

Four half-lives: _____

FIGURE 6.9 Decay of a radioactive isotope. (The Tasa Collection: Geologic Time. Published by Merrill Publishing Co., Columbus, OH. Copyright © 1986, by Tasa Graphic Arts, Inc. All rights reserved. Reproduced with permission.)

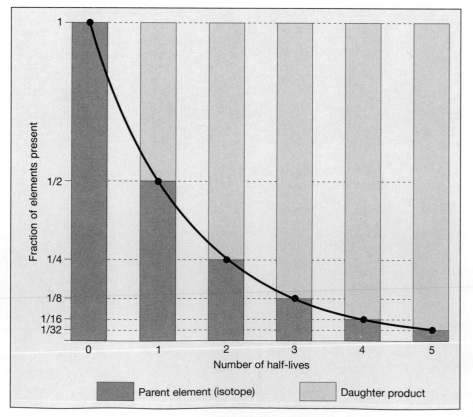

26. Assume you begin with 10.0 g of a radioactive parent isotope. How many grams of parent isotope will be present in the sample after each of the following half-lives?

Remaining Parent Isotope

One half-life: _____ grams

Four half-lives: _____ grams

27. If a radioactive isotope has a half-life of 400 million years, how long will it take for 50% of the material to change to the daughter product?

_____ years

28. A sample is brought to the laboratory and the chemist determines that the fraction of the parent isotope remaining is 1/8 of the total amount that was originally present. If the half-life of the material is 600 million years, how old is the sample?

_____ years old

29. Determine the absolute ages of rock samples that contain a parent isotope with a half-life of 100 million years and have the following fractions of parent isotope and daughter product.

1/2 parent and 1/2 daughter: Age = _____

1/8 parent and 7/8 daughter: Age = _____

1/32 parent and 31/32 daughter: Age = _____

Applying Radiometric Dates. When used in conjunction with relative dates, radiometric dates help earth scientists refine their interpretation of the geologic history of an area. Completing questions 30–35 will aid in understanding how both types of dates are often used together.

Previously in the exercise you determined the geologic history of the area represented in Figure 6.8 using relative dating techniques. Assume that the rock layers H and I in Figure 6.8 each contain radioactive materials with known half-lives.

▬▬▬

30. An analysis of a sample of rock from layer H in Figure 6.8 indicates an equal proportion of parent isotope and daughter produced from the parent. The half-life of the parent is known to be 425 million years.
   a. One (half, third, fourth) of the original parent has decayed to the daughter product. Circle your answer.
   b. How many half-lives of the parent isotope have elapsed since rock H formed?

_____

   c. What is the absolute age of rock layer H? Write your answer below and at rock layer H on Figure 6.8.

Age of rock layer H = _____ years

31. The analysis of a sample of rock from layer I in Figure 6.8 indicates its age to be 400 million years. Write the absolute age of layer I on Figure 6.8.

Refer to the relative and absolute ages you determined for the rocks in Figure 6.8 to answer the following questions.

▬▬▬

32. How many years long is the interval of time represented by the unconformity that separates rock layer H from layer I? Explain how you arrived at your answer.

The unconformity represents an interval of time

that was _____ million years long.

Explanation: _____

_____

33. The age of fault M is (older, younger) than 400 million years. Circle your answer. Explain how you arrived at your answer.

Explanation: _____

_____

34. What is the approximate maximum absolute age of the igneous intrusion L?

The igneous intrusion L formed more recently

than _____ million years ago.

35. Complete the following general statement describing the absolute ages of rock layers G, F, and E.

All of the rock layers are (younger, older) than

_____ million years.

THE GEOLOGIC TIME SCALE

Applying the techniques of geologic dating, the history of the earth has been subdivided into several different units which provide a meaningful time frame within which the events of the geologic past are arranged. Since the span of a human life is but a "blink of an eye" compared to the age of the earth, it is often difficult to comprehend the magnitude of geologic time. By completing questions 36–40, you will be better able to grasp the great age of the earth and appreciate the sequence of events that have brought it to this point in time.

36. Obtain a piece of adding machine paper slightly longer than 5 meters and a meterstick or metric measuring tape from your instructor. Draw a line at one end of the paper and label it "PRESENT." Using the following scale, construct a time line by completing the indicated steps.

Scale

1 meter = 1 billion years

10 centimeters = 100 million years

1 centimeter = 10 million years

1 millimeter = 1 million years

Step 1. Using the geologic time scale, Figure 6.10, as a reference, divide your time line into the **eons** and **eras** of geologic time. Label each division with its name and indicate its absolute age.

Step 2. Using the scale, plot and label the events listed in Table 6.1 on your time line.

TABLE 6.1 Generalized summary of a few geologic events that have taken place on the earth.

Age in Years Ago	Event
10,000	Pleistocene ice age ends
500,000	*Homo sapiens* (humans) evolve
1,600,000	Pleistocene ice age begins
60 million	Mammals dominate the land
66 million	Dinosaurs die out, Rocky Mountains forming
80 million	Flowering plants evolve
200 million	Reptiles evolve, approximate age of oldest rocks in the ocean basins
245 million	Trilobites die out
250 million	Uplift of Appalachian Mountains
300 million	Coal swamps exist in large areas of the northeast and midwest United States and other parts of the world
350 million	Amphibians are established
400 million	Plants move to the land
500 million	Oldest vertebrates
570 million	First animals with hard parts (shells) and therefore the first abundant fossil evidence
3100 million	Bacteria and blue-green algae evolve
4600 million	Formation of the earth

After completing your time line, answer questions 37–40.

37. Did you have difficulty plotting any of the events listed in Table 6.1? If so, which ones?

_____

38. If the scale of the time line was 1 millimeter=500 years, would this have made it easier to plot some of the more recent events? If you had used a scale of 1 millimeter=500 years, how many meters long would the time line have been?

_____

39. How many times longer is the whole of geologic time than the time represented by recorded history, about 5000 years?

Geologic time is _____ times longer than recorded history.

40. For what fraction or percent of geologic time has the human species, *Homo sapiens,* been present on the earth?

Approximately _____ of geologic time.

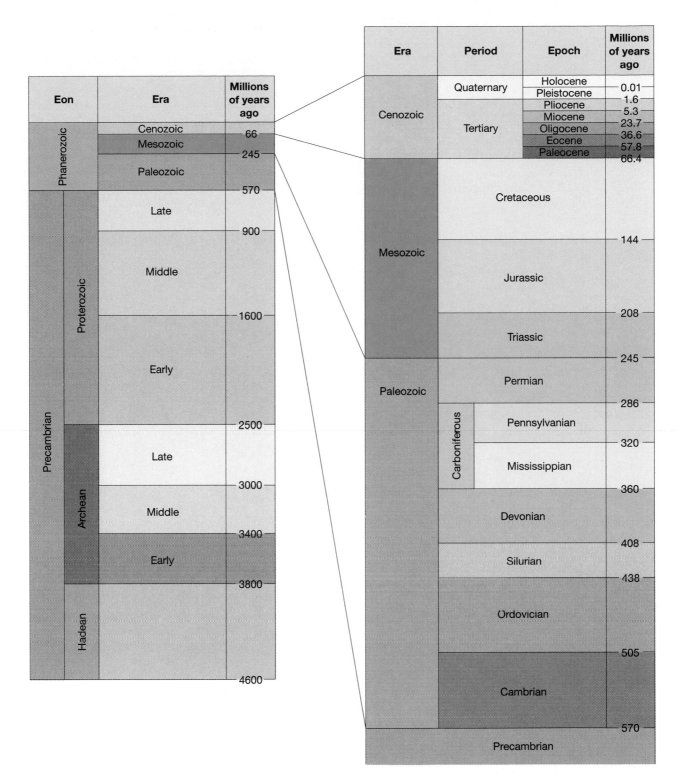

FIGURE 6.10   The geologic time scale. (Data from the Geologic Society of America)

REVIEW

Having completed the exercise, you should know the following:

1. The order and ages of events in earth history are determined by using both relative and radiometric dates.
2. Relative dates are used to establish the sequence of occurrence of geologic events.
3. Geologists use various laws, principles, and doctrines to arrive at relative dates.
4. Fossilization can be accomplished in several ways, such as casts and petrification.
5. Fossils are used by geologists to establish ages and correlate rock units.
6. Radiometric dates give the number of years ago that geologic events occurred using the radioactive decay of certain elements.
7. The half-life of a radioactive isotope is the amount of time it takes for one-half of the remaining parent isotope to change to the stable daughter product.
8. The broad divisions of geologic time are called eons, which are further subdivided into eras.

Determining Geologic Ages

SUMMARY/REPORT PAGE

Name: _____

Date Due: _____

Date: _____

Class: _____

After you have finished Exercise Six, complete the following questions. You may have to refer to the exercise for assistance or to locate specific answers. Be prepared to submit this summary/report to your instructor at the designated time.

1. Determine the sequence of geologic events that have occurred at the hypothetical area illustrated in Figure 6.11. List your answers from oldest to youngest in the space provided by the figure. Letter J is a fault, K is an igneous intrusion, and all other layers are sedimentary rocks.

2. The following questions refer to Figure 6.11.

a. What type of unconformity separates layer H from layer G?

_____

b. Which law, principle, or doctrine of relative dating did you apply to determine that rock layer H is older than layer I?

_____

c. Which law, principle, or doctrine of relative dating did you apply to determine that fault J is older than rock layer F?

_____

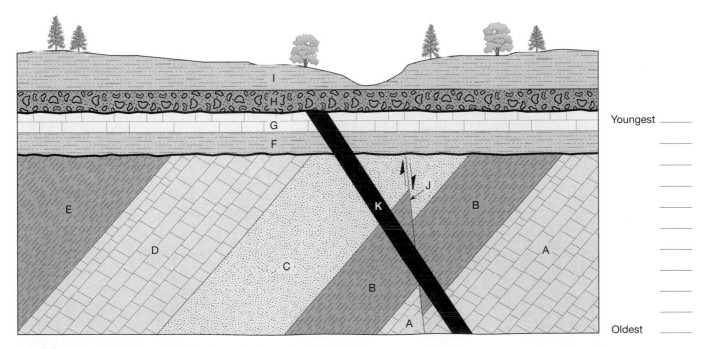

FIGURE 6.11 Geologic cross section of a hypothetical region.

d. Explain why you know that fault J is older than the igneous intrusion K.

_____

e. If rock layer H is 150 million years old and layer G is 160 million years old, what is the approximate age of the igneous intrusion K?

_____ million years

f. The analysis of samples from layers F and E indicates the following proportions of parent isotope to the daughter product produced from it. If the half-life of the parent is known to be 200 million years, what are the ages of the two layers?

	Parent	**Daughter**	**Age**
Layer F:	1/2	1/2	_____
Layer E:	1/4	3/4	_____

g. What absolute time interval is represented by the unconformity at the base of rock layer F?

From _____ to _____ million years

3. List the sequence of geologic events that you determined took place in the area represented by Figure 6.8, question 24, in the exercise.

Oldest _____ Youngest

4. Use the time line you constructed in the exercise to estimate what fraction of time is represented by each of the following geologic eons.

Phanerozoic eon: _____ Precambrian eon: _____

5. How many meters long would the time line you constructed in the exercise have been if you had used a scale of 1 millimeter equals 500 years?

_____ meters

EXERCISE SEVEN

Geologic Maps and Structures

The Earth is a dynamic planet. Its internal forces result in crustal movements that are continuously causing rocks to deform (Figure 7.1). If the magnitude of the force exceeds the strength of the rocks, the rocks will yield and folding or breaking may occur. To help understand these forces and the structures they create, many geologic investigations begin by preparing geologic maps of the study area. In this exercise you will investigate some of the common rock structures that are produced during crustal deformation and learn how earth scientists map and analyze these geologic features to gain insights into the nature of the evolving Earth.

OBJECTIVES

After you have completed this exercise, you should be able to

1. Explain how geologists describe the orientation of folded rocks and faults using the measurements, strike and dip.
2. Draw and interpret a simple geologic block diagram.
3. Describe the various types of folds and how they form.

FIGURE 7.1 Deformed sedimentary strata. (Photo by E. J. Tarbuck)

4. Recognize and diagram anticlines, synclines, domes, and basins in both geologic map and cross-sectional views.
5. Discuss the formation and types of dip-slip and strike-slip faults.
6. Recognize and diagram the various types of faults in both geologic map and cross-sectional views.
7. Interpret a simplified geologic map and use it to construct a geologic cross section.

TEXTBOOK REFERENCE

Chapter 8

MATERIALS

ruler
colored pencils
protractor

TERMS

geologic map	strike	dip-slip fault
stress	dip	strike-slip fault
compressional stress	cross section	normal fault
tensional stress	anticline	reverse fault
plastic deformation	syncline	right-lateral fault
fold	axial plane	left-lateral fault
fault	dome	
joint	basin	

INTRODUCTION

Geologists are continuously attempting to understand the nature of the forces that cause rocks to deform. They often begin their study by preparing a **geologic map** that shows the types, ages, distribution, and orientation of rocks on the surface of the earth. Working from these maps, geologists can interpret the nature of the rocks below the surface and assess any forces that may have caused their deformation.

Stress is the term used to describe the force that acts on a rock unit to change its shape and/or volume. The forces that act to shorten a rock body are known as **compressional stresses,** whereas those that elongate or pull apart a rock unit are called **tensional stresses** (Figure 7.2).

If the stress applied to a rock unit exceeds its strength, the rock may undergo plastic deformation or fracture. **Plastic deformation** takes place at high temperatures and pressures within the earth and results in permanent changes in the rock unit. A rock's size and shape may be altered through folding or flowing. Plastic deformation often produces wavelike undulations called **folds** in formerly flat-lying rocks. Under the lower temperature and pressure conditions found near the surface, most rocks behave like a brittle solid and fracture or break when stressed beyond their limit. If the rocks on either side of the fracture move, the geologic feature is called a **fault.** A **joint** is a fracture or break in a rock along which there has been no displacement.

STRIKE AND DIP

Geologists use measurements called **strike** (trend) and **dip** (inclination) to help define the orientation or atti-

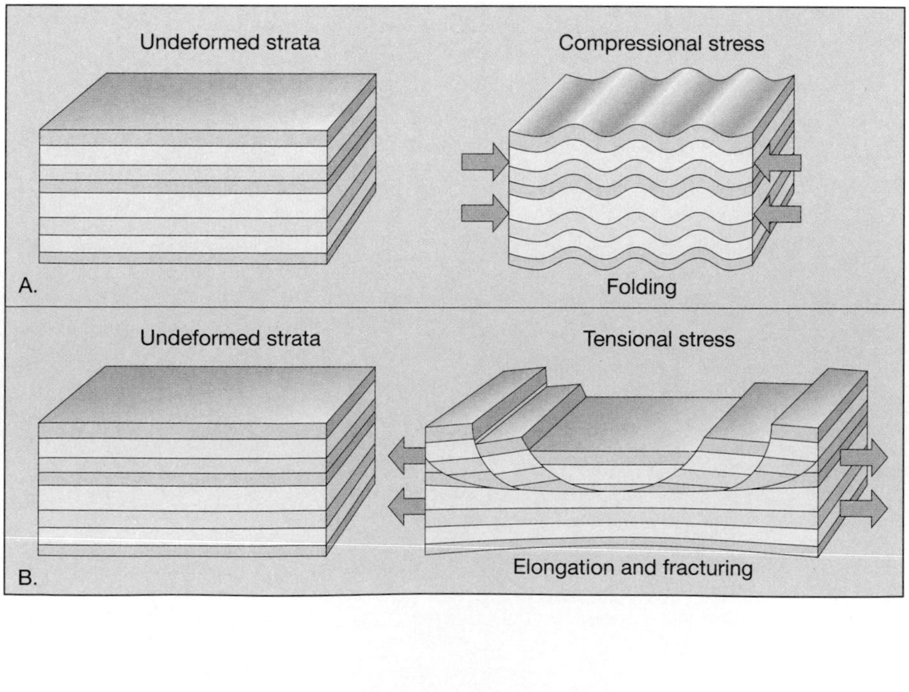

FIGURE 7.2 Simplified diagram showing the deformation of rock layers. **A.** Compressional stresses tend to shorten a rock body, often by folding. **B.** Tensional stresses act to elongate, or pull apart, a rock unit.

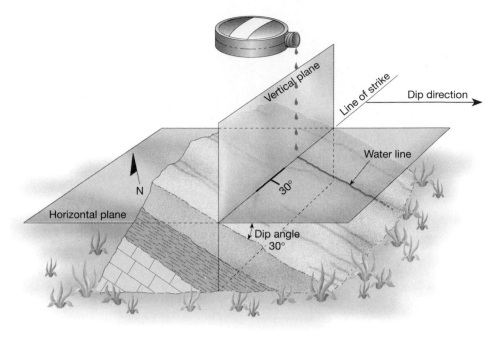

FIGURE 7.3 Determining the strike and dip of a rock layer.

tude of a rock layer or fault (Figure 7.3). By knowing the strike and dip of rocks at the surface, geologists can predict the nature and structure of rock units and faults that are hidden beneath the surface, beyond their view.

Strike is the compass direction of the line produced by the intersection of an inclined rock layer or fault with a horizontal plane at the surface (see Figure 7.3). The strike, or compass bearing, is generally expressed as an angle relative to north. For example, "north 10° east" (N10°E) means the line of strike is ten degrees to the east of north. The strike of the rock units illustrated in Figure 7.3 is approximately north 60° east (N60°E).

Dip is the angle of inclination of the surface of the rock unit or fault from the horizontal plane. Dip includes both an angle of inclination and a direction toward which the rock is inclined. In Figure 7.3 the dip angle of the rock layer is 30°. A good way to visualize dip is to imagine that a water line will always run down the rock surface parallel to the dip. The direction of dip will always be at a 90° angle to the strike. (To illustrate this fact, hold your closed textbook at an angle to the table top. The upper edge of your text represents the strike. Regardless of the way you point the text, the direction of dip of the book is always at 90°, or a right angle, to the strike.)

Typically, the strike and dip of rock units are shown on geologic maps. The standard map symbol for strike and dip is $\overline{}_{20°}$. The long line shows the strike direction and the short line points in the direction of the dip. The number written at the end of the short dip line is the angle of dip. In Figure 7.3, the strike-dip symbol indicates that the rocks are dipping toward the southeast at a 30° angle from the horizontal plane (30°SE).

1. Diagram 7.1 illustrates geologic map views (views from directly overhead) of two hypothetical areas showing a strike-dip symbol for a rock layer in each area. Complete the information requested below each map and draw a single large arrow on each map illustrating the direction of dip of the rock layer.

Block Diagrams. Block diagrams allow geologists to illustrate both the geologic map view and geologic **cross section** (view from the side or beneath the surface) of an area (Figure 7.4).

Use the block diagram illustrated in Figure 7.4 to answer questions 2 and 3.

2. In Diagram 7.2, prepare a sketch illustrating the rock units in Figure 7.4 as they would appear in a geologic map view *with north oriented toward the top of the map*. Draw a strike-dip symbol for each rock layer on your sketch.

3. From the information presented in Diagram 7.2 and Figure 7.4, supply the following answers.

Strike of the rock units:

N _____ ° _____

Direction of dip of the rock units:

to the _____

Angle of dip of the rock units: _____°

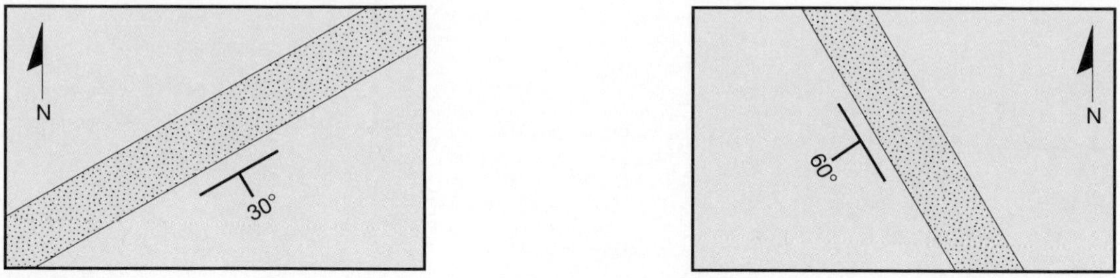

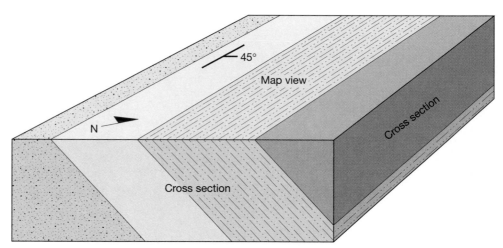

Strike: North _____° _____.

Direction of dip: _____.

Angle of dip: _____°

Strike: North _____° _____.

Direction of dip: _____.

Angle of dip: _____°

DIAGRAM 7.1

FIGURE 7.4 Block diagram illustrating inclined rock layers as they would appear in a geologic map view (top of block) and cross sections (front and side of block).

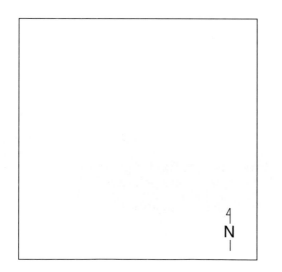

DIAGRAM 7.2 Geologic Map View.

4. Use the information presented to complete the cross sections on the front and side of the block diagram illustrated in Diagram 7.3.

5. Use the information supplied on the geologic map views to complete each of the block diagrams illustrated in Diagram 7.4.

6. On Diagram 7.5, use these guidelines to sketch both a geologic map view and block diagram of an area. Show strike-dip symbols for each rock layer on both illustrations.

Guidelines:
   a. Four sedimentary layers of equal thickness exposed at the surface.
   b. The strike of each layer is N45°W.
   c. The direction of dip of each layer is to the northeast.
   d. The angle of dip of each layer is 60°.

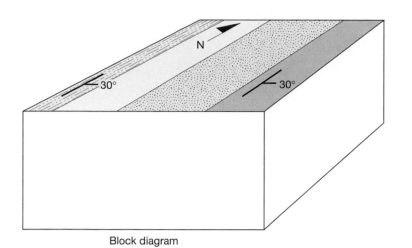

Block diagram

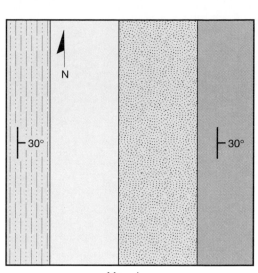

Map view

DIAGRAM 7.3

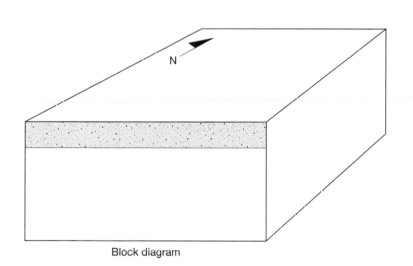

Block diagram

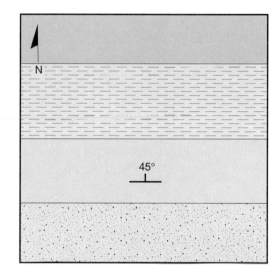

Map view

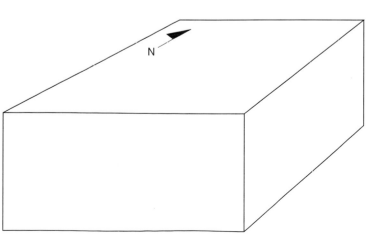

Block diagram

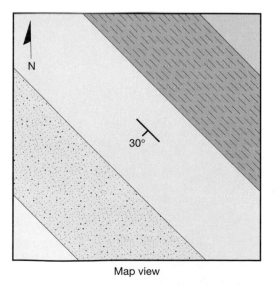

Map view

DIAGRAM 7.4

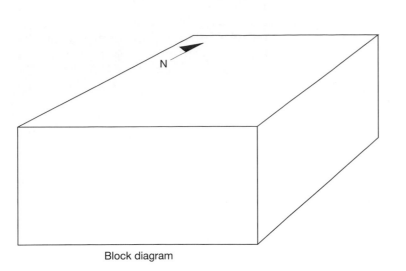

Block diagram　　　　　　　　　　Map view

DIAGRAM 7.5

TYPES OF FOLDS

During the process of mountain building, formerly flat-lying rocks are often compressed into a series of waves called *folds*. **Anticlines** and **synclines** are the two most common types of folds (Figure 7.5). Rock layers that fold upward forming an arch are called anticlines. Often associated with anticlines are downfolds, or troughs, called synclines.

The **axial plane** is an imaginary plane drawn through the long axis of a fold that divides it as equally as possible into two halves called *limbs* (Figure 7.6). In a *symmetrical fold,* the limbs are mirror images of each other and diverge at the same angle (see Figure 7.5). In an *asymmetrical fold* the limbs each have different angles of dip. A fold where one limb is tilted beyond the vertical is referred to as an *overturned fold* (see Figure 7.5).

Folds do not continue forever. Where folds 'die out' and end, the axis is no longer horizontal and the fold is said to be plunging (Figure 7.6B and Figure 7.7).

In an anticline, rock layers dip away from the axial plane. If erosion levels an anticline, then the formerly deepest, and hence oldest, rock layers will be exposed on the surface at the axial plane (see Figures 7.5 and 7.7). In an eroded syncline, the layers dip toward the axial plane where the youngest, most recently deposited rock layer occurs (see Figures 7.5 and 7.7).

Figure 7.8 illustrates an eroded anticline and an eroded syncline. Use the figure to answer questions 7–16.

7. On each block diagram in Figure 7.8, label the type of fold, anticline or syncline, illustrated.

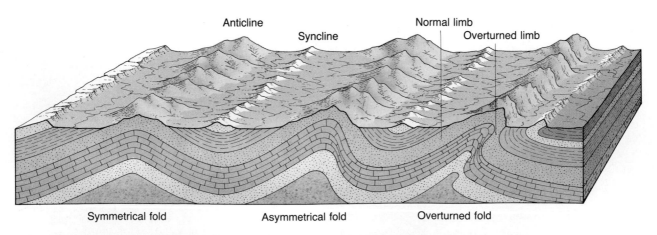

FIGURE 7.5　Block diagram illustrating the principal types of folded geologic structures.

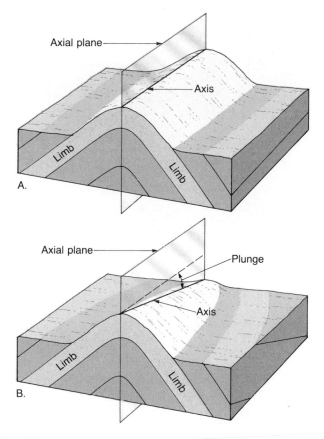

FIGURE 7.6 Features of simple folds. **A.** Horizontal axis. **B.** Plunging axis.

8. A geologic law, called the *law of superposition,* states that, in most situations concerning layered rocks, the oldest rocks are at the bottom. With this in mind, list by number the rock layers in each block diagram in Figure 7.8 from oldest to youngest.
Anticline:
Oldest _____ Youngest
Syncline:
Oldest _____ Youngest

9. The near-vertical plane through each block diagram represents each fold's _____

_____.

10. In Figure 7.8, draw appropriate strike and dip symbols on the surface of each block diagram for a rock layer on both sides of the axial plane.

11. In the anticline, all the rock layers are dipping (toward, away from) the axial plane. Circle your answer.

12. In the syncline, all the rock layers are dipping (toward, away from) the axial plane. Circle your answer.

13. As illustrated, the anticline in Figure 7.8 is (symmetrical, asymmetrical), and the syncline is (symmetrical, asymmetrical). Circle your answers.

14. In Figure 7.8, both folds, as illustrated, are (plunging, nonplunging) folds. Circle your answer.

15. Using the word 'old,' label the area(s) of the oldest rocks exposed on the surface of the map view portion of both block diagrams.

16. By circling the correct response, complete the following statements that describe what happens to the ages of the surface rocks as you walk away from the axial plane of each of the following structures.
   a. On an *eroded anticline* the surface rocks get (older, younger) as you walk away from the axial plane.
   b. On an *eroded syncline* the surface rocks get (older, younger) as you walk away from the axial plane.

17. On Diagram 7.6, complete the block diagram using the information provided on the map view. Rocks illustrated with the same pattern are part of the same rock layer. Assume all dip angles to be 30°. Write the names of the two types of geologic structures illustrated at the appropriate place on the block diagram.

Anticlines and synclines are linear features caused by compressional forces. Two other types of folds, **domes** and **basins,** are often nearly circular features that result from vertical displacement. Upwarping of sedimentary rocks produces a dome, whereas a basin is a downwarped structure. Using your text as a reference, answer question 18.

18. On Diagrams 7.7 and 7.8, complete the geologic block diagrams for the indicated features. Draw a minimum of four rock layers on each diagram and label the oldest and youngest rocks. Then describe the directions of dip and map view locations of the oldest and youngest surface rocks that comprise each feature.

Eroded dome: _____

_____

Eroded basin: _____

_____

TYPES OF FAULTS

Faults are fractures or breaks in rocks along which movement has occurred. As with folds, geologists also use strike and dip to describe the attitude of faults. Faults in which the relative movement of rock units is primarily

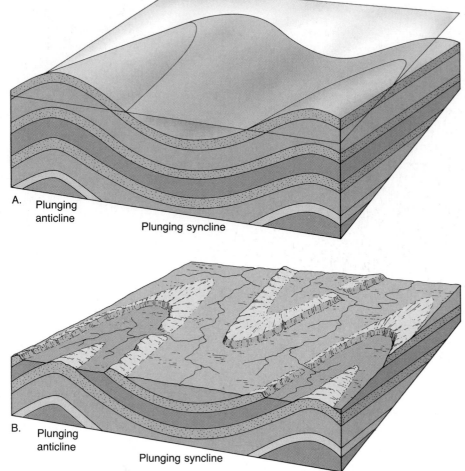

A. Plunging
anticline
Plunging syncline

B. Plunging
anticline
Plunging syncline

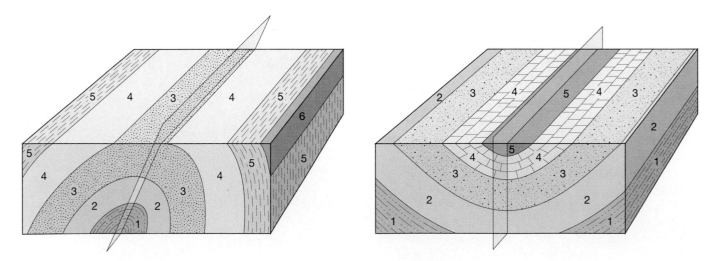

FIGURE 7.8 Idealized eroded folds.

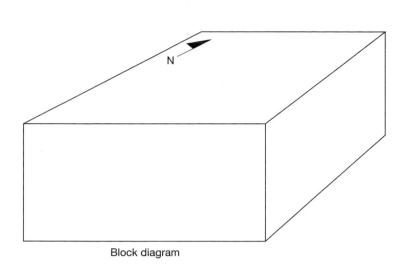

Block diagram

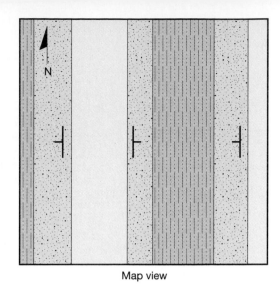

Map view

DIAGRAM 7.6

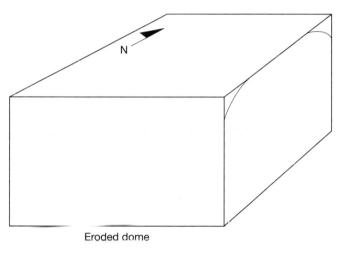

Eroded dome

DIAGRAM 7.7

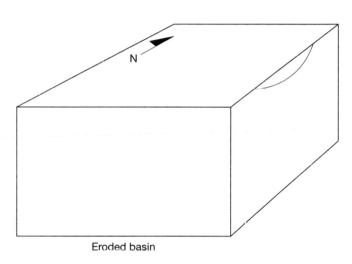

Eroded basin

DIAGRAM 7.8

vertical (along the dip of the fault plane) are called **dip-slip faults.** When the dominant motion of rock units is horizontal (in the direction of the strike), the faults are called **strike-slip faults.** Although most faults exhibit both vertical and horizontal displacement, this investigation will only consider the two fundamental motions.

Dip-slip Faults. Since very few faults have a vertical dip, geologists describe the different types of dip-slip faults by noting the relative motion of the blocks on top and below the fault. If the hanging wall (the top of the fault) moves down relative to the footwall (the bottom of the fault), the fault is classified as a **normal fault** (Figure 7.9A). Normal faults result from tensional forces. Compressional forces will often result in a **reverse fault,** where the hanging wall moves up relative to the footwall (see Figure 7.9B).

Use Figure 7.9 to answer question 19 and 20.

19. Draw the appropriate strike-dip symbol for each fault on the surface of the block diagrams in Figure 7.9.
20. On each block diagram in Figure 7.9, write the word "older" or "younger" to indicate the relative ages of the surface rocks on both sides of the fault. Then complete the following statements by circling the correct response.
   a. On an eroded normal fault, the rocks at the surface on the direction of dip side of the fault are the (youngest, oldest) surface rocks.
   b. On an eroded reverse fault, the rocks at the surface on the direction of dip side of the fault are the (youngest, oldest) surface rocks.

In questions 21 and 22, complete each of the AFTER FAULTING block diagrams by illustrating the type

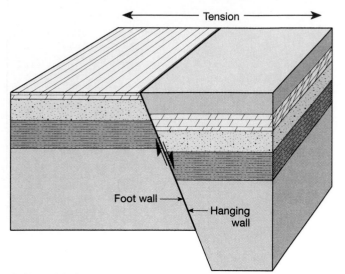

A. Normal fault

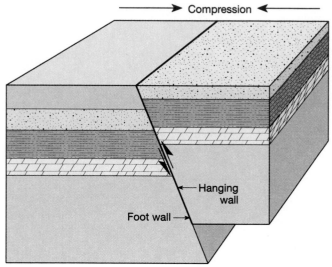

B. Reverse fault

FIGURE 7.9 Block diagrams illustrating simplified eroded versions of the two common types of dip-slip faults. **A.** Normal fault. **B.** Reverse fault. The arrows by each fault show the relative motions of the blocks on both sides of the fault.

of fault indicated. Faults are shown with bold lines and a rock layer is illustrated with a "dot" pattern.

For each block diagram: (1) draw an appropriate strike-dip symbol for the fault; (2) label the hanging wall and footwall sides of the fault; and (3) on the front or side cross section of the AFTER FAULTING block diagram, draw arrows on both sides of the fault to illustrate the relative motion of the blocks.

---

21. On Diagram 7.9, complete each of the AFTER FAULTING block diagrams by illustrating an eroded normal fault.
22. On Diagram 7.10, complete each of the AFTER FAULTING block diagrams by illustrating an eroded reverse fault.

Strike-slip Faults. Strike-slip faults result from horizontal displacement of rock units along the strike or trend of a fault (Figure 7.10). Many strike-slip faults, such as the famous San Andreas fault system in California, are near-vertical faults that exhibit tens or hundreds of kilometers of displacement along their boundaries.

A strike-slip fault is described as being either a **right-lateral fault** or a **left-lateral fault** depending on the relative motion of the blocks on either side of the fault. If you are standing on one side of the fault, looking across the fault to the other side, and the opposite side has been displaced to your right, the fault is called a right-lateral fault (see Figure 7.10). On the other hand,

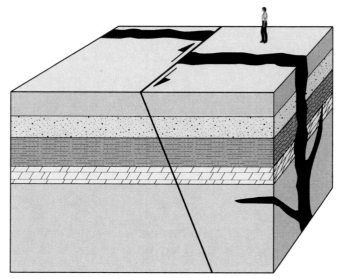

Right-lateral fault

FIGURE 7.10 Block diagram illustrating a simplified eroded version of a right-lateral strike-slip fault. The arrows on the surface show the relative motions of the blocks on both sides of the fault.

if the side opposite the fault has moved to the left, the fault is a left-lateral fault.

---

23. On Diagram 7.11, complete the AFTER FAULTING block diagram by illustrating a left-lateral fault. The fault is shown with a bold line and a rock layer is illustrated with a "dot" pattern. Draw arrows on both sides of the fault on the surface to illustrate the relative motion of each block.

Before faulting   After faulting

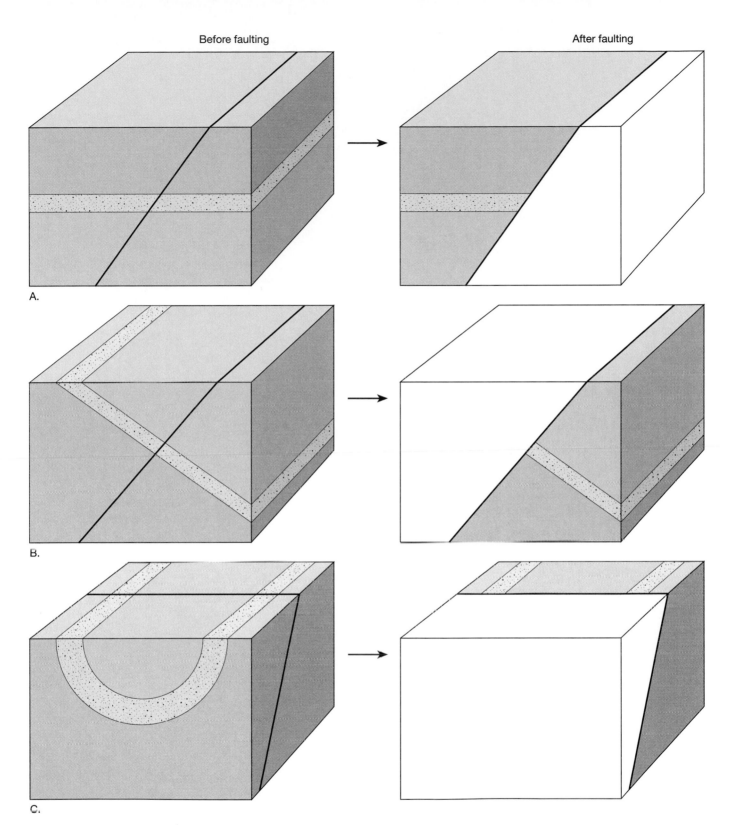

A.

B.

C.

DIAGRAM 7.9

Before faulting

After faulting

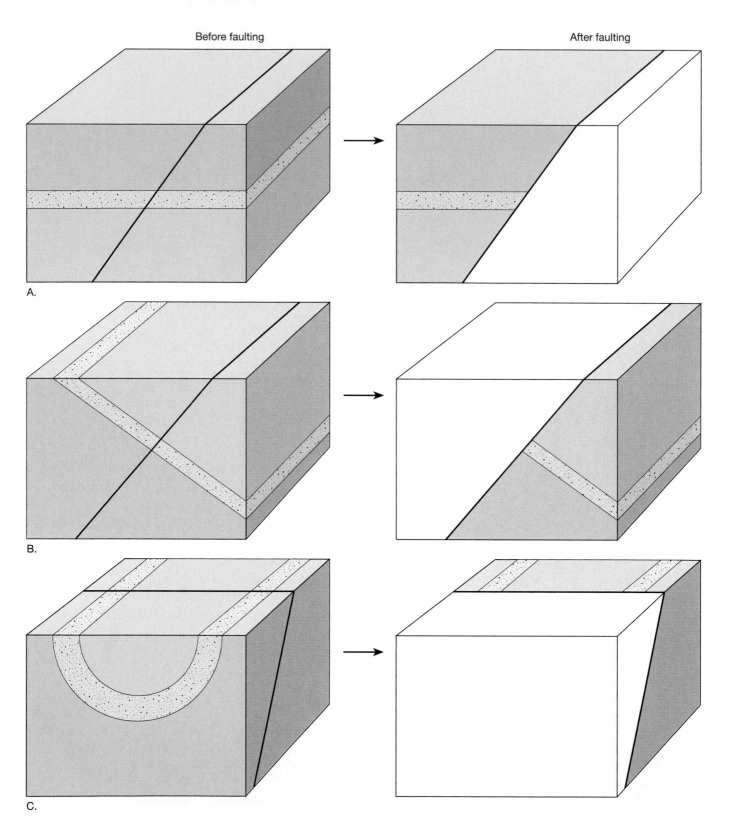

A.

B.

C.

DIAGRAM 7.10

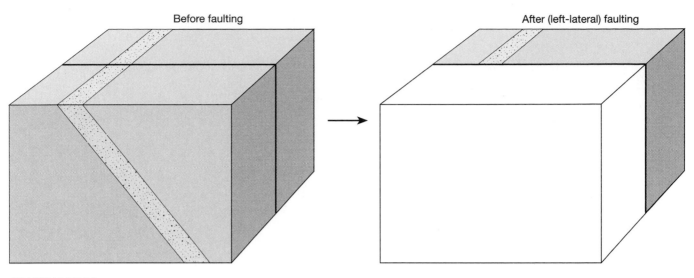

Before faulting   After (left-lateral) faulting

DIAGRAM 7.11

A SIMPLIFIED GEOLOGIC MAP

Two of the essential skills all geologists learn are how to examine rocks in the field and how to prepare geologic maps. Using the information presented on a map, geologists can determine the attitude of the rocks beneath the surface and the forces responsible for rock deformation.

The primary purpose of a geologic map is to illustrate the types, ages, distribution, and strike and dip of rocks exposed at the earth's surface. Like topographic maps, many geologic maps will also show surface elevations using contour lines. Geologic maps will also have an "Explanation" that shows the patterns used to represent rock units, types and ages of rocks, and descriptions of symbols. The youngest rocks are listed at the top of the "Explanation" and the oldest at the bottom.

Figure 7.11 is a simplified geologic map of an eroded hypothetical area, similar to one that might be prepared by a geologist from data collected in the field. Use Figure 7.11 to answer questions 24–28.

24. From the information presented on the map, what type of fold extends north to south through the center of the map? Give two lines of evidence that support your choice.

25. The fold you identified in question 24 is a (plunging, nonplunging) structure. Circle your answer.
26. Label the hanging wall and footwall sides of the fault that is located in the northwest corner of the map area. Then answer questions 26a and 26b.
   a. Based upon the evidence presented, the (northwest, southeast) side of the fault exhibits upward relative motion. Circle your answer.
   b. The fault is a (normal, reverse) fault. Circle your answer and give the reason for your choice.

27. In the space indicated for the cross section in Figure 7.11, sketch a geologic cross section that extends along the line from A to A' on the geologic map.
28. (Compression, Tension) is the force that is most likely responsible for producing the features in the map area. Circle your answer.

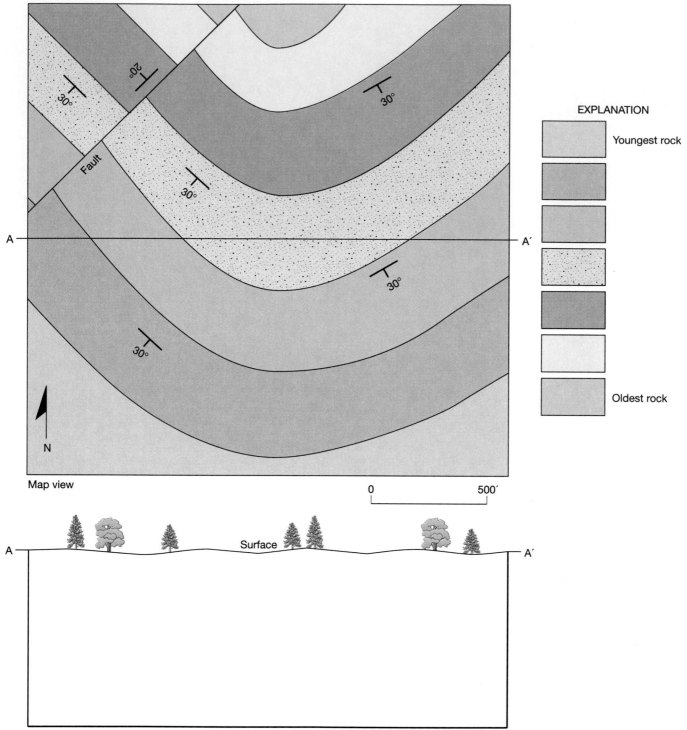

Map view

0   500´

EXPLANATION

Youngest rock

Oldest rock

Cross section

FIGURE 7.11 Simplified geologic map illustrating folded sedimentary rocks and a fault.

EXAMINING A GEOLOGIC MAP

Figure 7.12 is a portion of the Devils Fence, Montana, geologic map with an abbreviated "Explanation" for the map shown in Figure 7.13. Examine the map and explanation, then answer questions 29–37. (*Note:* You may find the geologic time scale, Figure 6.10 in Exercise Six, helpful.)

━━━━━

29. What is the scale of the Devils Fence map?

    Scale: _____ : _____ ,

    which means one inch on the map equals

    approximately _____ mile(s).

30. What are the names and approximate age in years of the youngest and oldest *sedimentary* rock units shown in the map "Explanation"?

    Youngest sedimentary rock unit shown in the "Explanation":

    Name: _____

    Age in years: _____ million

    Oldest sedimentary rock unit shown in the "Explanation":

    Name: _____

    Age in years: _____ million

31. On the geologic map, write the word "oldest" where the oldest sedimentary rock unit is exposed at the surface and the word "youngest" where the youngest sedimentary rocks occur.

32. The igneous intrusive rocks are (younger, older) than the youngest sedimentary rocks. Circle your answer.

33. Examine the strike and dip of the rock units on the Devils Fence geologic map. Draw several large arrows on the map pointing in the direction of dip of several of the rock units. Make sure you examine the entire map. Then answer questions 33a and 33b.
    a. Near the center of the map, in sections 11 and 14, the rocks are dipping to the (northwest, southeast). Circle your answer.
    b. The same rocks that occur in sections 11 and 14 are dipping to the (south, east) in the east-central area of the map, in the western half of section 18.

34. What is the approximate angle of dip of the rock units in section 15?

    Angle of dip = _____°

35. Draw a dashed line representing the axial plane of the large geologic structure illustrated in the eastern half of the map. Label the line "axial plane."

36. The geologic structure present over the majority of the eastern half of the map is a plunging (anticline, syncline). Circle your answer and give two lines of evidence that support your choice.

    _____

    _____

Examine the faults marked with bold black lines that occur in section 14 on the northwest limb of the fold. The relative motion of the faults is shown with a "U" on the side of the fault that has moved up and a "D" on the downward side. Although strike and dip of the faults is not given on the map, assume the direction of dip is to the southwest.

━━━━━

37. Label an example of a normal fault and a reverse fault in section 14 on the geologic map.

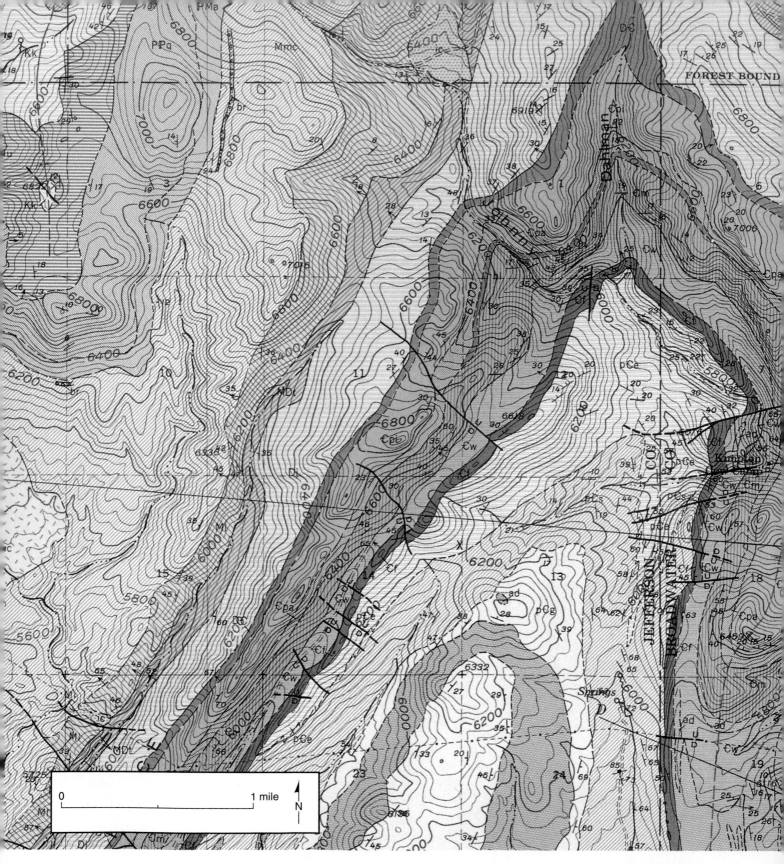

FIGURE 7.12 Geologic Map—Devils Fence, Montana.
(Source: U.S. Geologic Survey, Professional Paper 292)

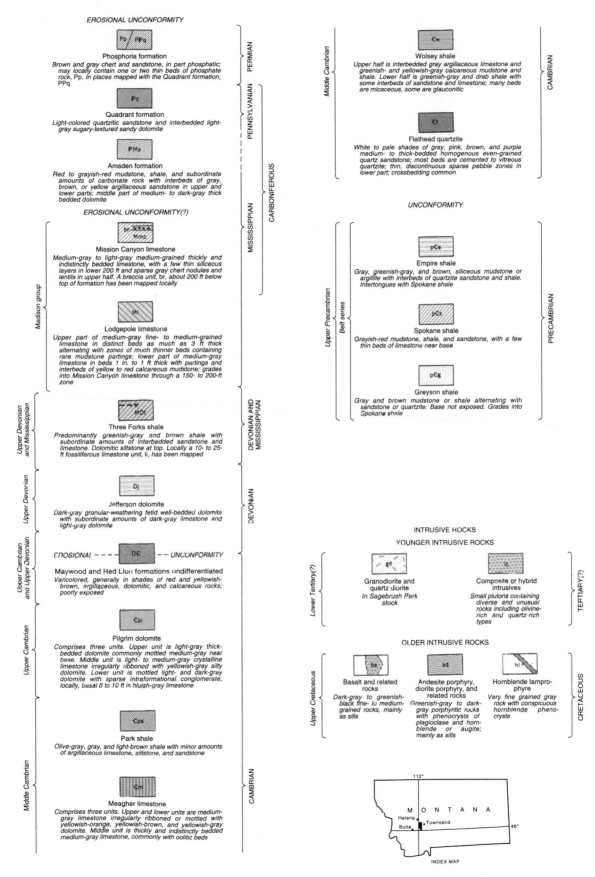

EROSIONAL UNCONFORMITY

Phosphoria formation
Brown and gray chert and sandstone, in part phosphatic; may locally contain one or two thin beds of phosphate rock, Pp. In places mapped with the Quadrant formation, PPq

Quadrant formation
Light-colored quartzitic sandstone and interbedded light-gray sugary-textured sandy dolomite

Amsden formation
Red to grayish-red mudstone, shale, and subordinate amounts of carbonate rock with interbeds of gray, brown, or yellow argillaceous sandstone in upper and lower parts; middle part of medium- to dark-gray thick bedded dolomite

EROSIONAL UNCONFORMITY(?)

Mission Canyon limestone
Medium-gray to light-gray medium-grained thickly and indistinctly bedded limestone, with a few thin siliceous layers in lower 200 ft and sparse gray chert nodules and lentils in upper half. A breccia unit, br, about 200 ft below top of formation has been mapped locally

Lodgepole limestone
Upper part of medium-gray fine- to medium-grained limestone in distinct beds as much as 3 ft thick alternating with zones of much thinner beds containing rare mudstone partings; lower part of medium-gray limestone in beds 1 in. to 1 ft thick with partings and interbeds of yellow to red calcareous mudstone; grades into Mission Canyon limestone through a 150- to 200-ft zone

Three Forks shale
Predominantly greenish-gray and brown shale with subordinate amounts of interbedded sandstone and limestone. Dolomitic siltstone at top. Locally a 10- to 25-ft fossiliferous limestone unit, li, has been mapped

Jefferson dolomite
Dark-gray granular-weathering fetid well-bedded dolomite with subordinate amounts of dark-gray limestone and light-gray dolomite

EROSIONAL — — — — UNCONFORMITY

Maywood and Red Lion formations undifferentiated
Varicolored, generally in shades of red and yellowish-brown, argillaceous, dolomitic, and calcareous rocks; poorly exposed

Pilgrim dolomite
Comprises three units. Upper unit is light-gray thick-bedded dolomite commonly mottled medium-gray near base. Middle unit is light- to medium-gray crystalline limestone irregularly ribboned with yellowish-gray silty dolomite. Lower unit is mottled light- and dark-gray dolomite with sparse intraformational conglomerate; locally, basal 8 to 10 ft is bluish-gray limestone

Park shale
Olive-gray, gray, and light-brown shale with minor amounts of argillaceous limestone, siltstone, and sandstone

Meagher limestone
Comprises three units. Upper and lower units are medium-gray limestone irregularly ribboned or mottled with yellowish-orange, yellowish-brown, and yellowish-gray dolomite. Middle unit is thickly and indistinctly bedded medium-gray limestone, commonly with oolitic beds

PERMIAN / PENNSYLVANIAN / CARBONIFEROUS / MISSISSIPPIAN

Madison group

Upper Devonian and Mississippian / Upper Devonian / Upper Cambrian and Upper Devonian / Upper Cambrian / Middle Cambrian

DEVONIAN AND MISSISSIPPIAN / DEVONIAN / CAMBRIAN

Wolsey shale
Upper half is interbedded gray argillaceous limestone and greenish- and yellowish-gray calcareous mudstone and shale. Lower half is greenish-gray and drab shale with some interbeds of sandstone and limestone; many beds are micaceous, some are glauconitic

Flathead quartzite
White to pale shades of gray, pink, brown, and purple medium- to thick-bedded homogenous even-grained quartz sandstone; most beds are cemented to vitreous quartzite; thin, discontinuous sparse pebble zones in lower part; crossbedding common

Middle Cambrian / CAMBRIAN

UNCONFORMITY

Empire shale
Gray, greenish-gray, and brown, siliceous mudstone or argillite with interbeds of quartzite sandstone and shale. Intertongues with Spokane shale

Spokane shale
Grayish-red mudstone, shale, and sandstone, with a few thin beds of limestone near base

Greyson shale
Gray and brown mudstone or shale alternating with sandstone or quartzite. Base not exposed. Grades into Spokane shale

Upper Precambrian / Belt series / PRECAMBRIAN

INTRUSIVE ROCKS
YOUNGER INTRUSIVE ROCKS

Granodiorite and quartz diorite
In Sagebrush Park stock

Composite or hybrid intrusives
Small plutons containing diverse and unusual rocks including olivine-rich and quartz-rich types

Lower Tertiary(?) / TERTIARY(?)

OLDER INTRUSIVE ROCKS

Basalt and related rocks
Dark-gray to greenish-black fine- to medium-grained rocks, mainly as sills

Andesite porphyry, diorite porphyry, and related rocks
Greenish-gray to dark-gray porphyritic rocks with phenocrysts of plagioclase and hornblende or augite; mainly as sills

Hornblende lamprophyre
Very fine grained gray rock with conspicuous hornblende phenocrysts

Upper Cretaceous / CRETACEOUS

MONTANA
Helena · Townsend
Butte
112° / 46°
INDEX MAP

FIGURE 7.13 "Explanation" for the geologic map of Devils Fence, Montana, Figure 7.12.

REVIEW

Having completed the exercise, you should know the following:

1. Compression and tension are the stresses that cause rocks within the earth to deform.
2. At high temperatures and pressures within the earth, stress often causes rocks to behave like a plastic and fold.
3. Near the earth's surface, rocks behave like a brittle solid and fracture when stressed beyond their limit.
4. Strike is the compass trend of a line that results when an inclined rock layer or fault intersects a horizontal plane.
5. Dip is the inclination of a rock layer or fault from a horizontal plane. Dip has both direction and amount of inclination.
6. Geologists use block diagrams to show both the geologic map view and cross-sectional view of an area.
7. Geological structures with upward folded rocks are called anticlines, while structures with down folded rocks are called synclines.
8. In an eroded anticline, the oldest rocks occur on the surface at the axial plane.
9. In an eroded syncline, the youngest rocks occur on the surface at the axial plane.
10. The two basic types of faults are dip-slip faults and strike-slip faults.
11. Dip-slip faults exhibit vertical movement.
12. Normal faults are dip-slip faults that result from tensional stress. In a normal fault the hanging wall moves down relative to the footwall.
13. Reverse faults are dip-slip faults that result from compressional stress. In a reverse fault the hanging wall moves up relative to the footwall.
14. Strike-slip faults exhibit horizontal movement.
15. The two basic types of strike-slip faults are left-lateral and right-lateral.
16. Most faults are the result of both vertical and horizontal motions.
17. Geologic maps often show the ages, types, distribution, and strike and dip of rocks exposed at the earth's surface. They are used by geologists to determine the nature of the rocks beneath the surface and the forces that are responsible for causing rock deformation.

Geologic Maps and Structures

SUMMARY/REPORT PAGE

Date Due: _____

Name: _____

Date: _____

Class: _____

After you have finished Exercise Seven, complete the following questions. You may have to refer to the exercise for assistance or to locate specific answers. Be prepared to submit this summary/report to your instructor at the designated time.

1. Diagram 7.12 is a simplified geologic map. In the spaces provided, describe the strike and dip of the rock layer illustrated on the map.

    Strike: North _____ ° _____

    Angle of dip: _____ ° _____

    Direction of dip: _____

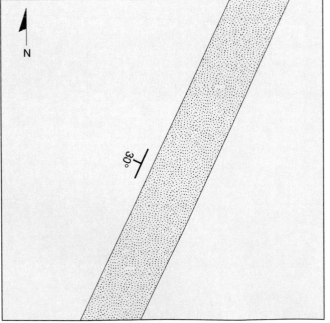

Map view

DIAGRAM 7.12

2. What type of geologic structure is illustrated in the eastern half of the Devils Fence geologic map, Figure 7.12? List two lines of evidence in support of your choice.

    _____

    _____

3. List the two types of geologic structures that are illustrated in Figure 7.11.

    _____

4. (Compression, Tension) was the type of stress responsible for producing the geologic structures in Figure 7.11. Circle your answer.

5. Define the following terms.

    Strike: _____

    _____

    Dip: _____

    _____

    Symmetrical anticline: _____

    _____

    Normal fault: _____

    _____

    Axial plane: _____

    _____

    Left-lateral fault: _____

    _____

6. On Diagram 7.13, complete each of the block diagrams by illustrating the indicated geologic structure. Show a minimum of four rock layers and draw the appropriate strike-dip symbols on the surface of each block diagram.

7. With reference to the axial plane, where are the youngest rocks located in an eroded syncline?

_____

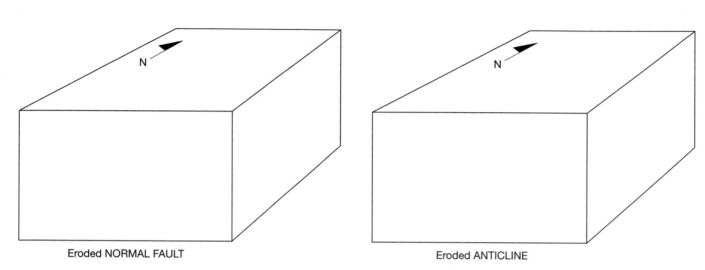

Eroded NORMAL FAULT   Eroded ANTICLINE

DIAGRAM 7.13

Earthquakes and the Earth's Interior

Almost all of the earth lies beneath us, yet its accessibility to direct examination is limited. Therefore, one of the most difficult problems faced by earth scientists is determining the physical properties of the earth's interior. The branch of earth science called **seismology** combines mathematics and physics to explain the nature of earthquakes and how they can be used to gather information about the earth beyond our view. This exercise introduces some of the techniques that are used by seismologists to determine the location of an earthquake and to investigate the structure of the earth's interior.

OBJECTIVES

After you have completed this exercise, you should be able to

1. Examine an earthquake seismogram and recognize the P waves, S waves, and surface waves.
2. Use a seismogram and travel-time graph to determine how far a seismic station is from the epicenter of an earthquake.
3. Determine the actual time that an earthquake occurred using a seismogram and travel-time graph.
4. Locate the epicenter of an earthquake by plotting seismic data from three seismic stations.
5. Explain how earthquakes are used to determine the structure of the earth's interior.
6. List the name, depth, composition, and state of matter of each of the earth's interior zones.
7. Describe the temperature gradient of the upper earth.
8. Explain why earth scientists think that the asthenosphere consists of partly melted, plastic material at a depth of about 100 kilometers.
9. Explain how earthquakes and the earth's temperature gradient have been used to explain the fact that

large, rigid slabs of the lithosphere are descending into the mantle at various locations on the earth.

TEXTBOOK REFERENCE

Chapter 5

MATERIALS

calculator   colored pencils
drawing compass   ruler

Materials Supplied by Your Instructor

atlas or wall map

TERMS

seismology	P wave	asthenosphere
lithosphere	S wave	mantle
focus	surface wave	outer core
seismic wave	amplitude	inner core
seismograph	period	geothermal gradient
seismogram	epicenter	

INTRODUCTION

Most of the solid earth is beyond our direct observation. However, earth scientists have devised ingenious methods to provide answers concerning the internal constitution of the earth. The information that is used to solve many of these problems has come from the analysis of earthquake records.

EARTHQUAKES

Earthquakes are vibrations of the earth that occur when the rigid materials of the **lithosphere** are strained beyond their limit, yield, and "spring back" to their original shape, rapidly releasing stored energy. This energy radiates in all directions from the source of the earthquake, called the **focus,** in the form of **seismic waves.** **Seismograph** instruments located throughout the world amplify and record the ground motions produced by passing seismic waves on **seismograms.** The seismograms are then used to determine the time of occurrence and location of an earthquake, as well as to define the internal structure of the earth.

Examining Seismograms.
The three basic types of seismic waves generated by an earthquake at its focus are **P waves, S waves,** and **surface,** or **L, waves.** P and S waves travel through the earth while L waves are transmitted along the outer layer. Of the three wave types, P waves have the greatest velocity and, therefore, reach the seismograph station first. Surface waves arrive at the seismograph station last because they must travel a longer distance along the earth's outer layer. P waves also have smaller **amplitudes** (range from the mean, or average, to the extreme) and shorter **periods** (time interval between the arrival of successive wave crests) than S and surface waves.

Figure 8.1 illustrates a typical earthquake recording on a seismogram. Each vertical line of dots marks a one minute time interval. After you have reviewed Chapter 5 of the text, answer questions 1–6 by referring to Figure 8.1.

---

1. It took approximately (5, 7, 11) minutes to record the entire earthquake from the first recording of the P waves to the end of the surface waves. Circle your answer.
2. (Five, Seven) minutes elapsed between the arrival of the first P wave and the arrival of the first S wave on the seismogram. Circle your answer.
3. (Five, Seven) minutes elapsed between the arrival of the first P wave and the first surface wave.

4. The maximum amplitude of the surface waves is approximately (5, 9) times greater than the maximum amplitude of the P waves.
5. The approximate period of the surface waves is (10, 30, 60) seconds.
6. The period of the surface waves is (greater, less) than the period for P waves.

Locating an Earthquake.
The focus of an earthquake is the actual place within the earth where the earthquake originates. Earthquake foci have been recorded to depths as great as 600 km. When locating an earthquake on a map, seismologists plot the **epicenter,** the point on the earth's surface directly above the focus.

The difference in the velocities of P and S waves provides a method for locating the epicenter of an earthquake. Both P and S waves leave the earthquake focus at the same instant. Since the P wave has a greater velocity, the further away the recording instrument is from the focus, the greater will be the difference in the arrival times of the first P wave compared to the first S wave.

To determine the distance between a recording station and an earthquake epicenter, find the place on the travel-time graph, Figure 8.2, where the vertical separation between the P and S curves is equal to the number of minutes' difference in the arrival times between the first P and first S waves on the seismogram. From this position, a vertical line is drawn that extends to the bottom of the graph and the distance to the epicenter is read from the lower axis.

To accurately locate an earthquake epicenter, records from three different seismograph stations are needed. First, the distance that each station is from the epicenter is determined using Figure 8.2. Then, for each station, a circle centered on the station with a radius equal to the station's distance from the epicenter is drawn. The geographic point where all three circles, one for each station, intersect is the earthquake epicenter.

Answering questions 7–16 will help you understand the process used to determine an earthquake epicenter.

FIGURE 8.1 Typical earthquake seismogram.

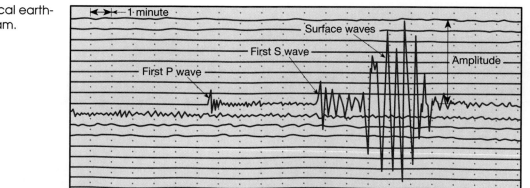

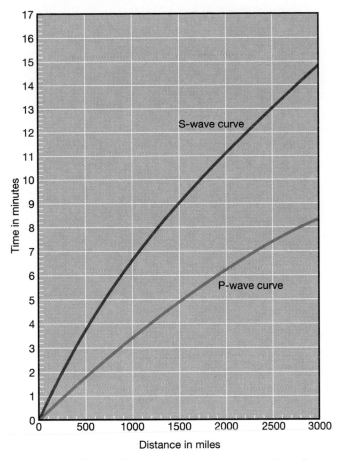

FIGURE 8.2 Travel-time graph used to determine distance to an earthquake epicenter.

7. An examination of Figure 8.2 shows that the difference in the arrival times of the first P and the first S waves on a seismogram (increases, decreases) the farther a station is from the epicenter. Circle your answer.

8. Use Figure 8.2 to determine the difference in arrival times (in minutes) between the first P wave and first S wave for stations that are the following distances from an epicenter.

   1000 miles: _____ minutes difference

   2000 miles: _____ minutes difference

   3000 miles: _____ minutes difference

On the seismogram in Figure 8.1, you determined the difference in the arrival times between the first P and the first S waves to be five minutes.

9. Refer to the travel-time graph. What is the distance from the epicenter to the station that recorded the earthquake in Figure 8.1?

_____ miles

10. From the travel-time (minutes) axis of the travel-time graph, the first P waves from the seismogram in Figure 8.1 arrived at the recording station approximately (3, 7, 14) minutes after the earthquake occurred. Circle your answer.

11. If the first P wave was recorded at 10:39 PM local time at the station in Figure 8.1, what was the local time when the earthquake actually occurred?

_____ PM local time

Figure 8.3 illustrates seismograms for the same earthquake recorded at New York, Seattle, and Mexico City. Use this information to answer questions 12–16.

12. Use the travel-time graph, Figure 8.2, to determine the distance that each station in Figure 8.3 is from the epicenter. Write your answers in the epicenter data table, Table 8.1.

13. After referring to an atlas or wall map, accurately place a small dot showing the location of each of the three stations on the map provided in Figure 8.4. Also label each of the three cities.

14. On Figure 8.4, use a drawing compass to draw a circle around each of the three stations with a radius, in miles, equal to its distance from the epicenter. (*Note:* Use the distance scale provided on the map to set the distance on the drawing compass for each station.)

15. What is the approximate latitude and longitude of the epicenter of the earthquake that was recorded by the three stations?

_____ latitude and _____ longitude

16. Note on the seismograms that the first P wave was recorded in New York at 9:01 Greenwich Mean Time (GMT). At what time (GMT) did the earthquake actually occur?

_____ GMT

World Distribution of Earthquakes. Earth scientists have determined that the global distribution of earthquakes is not random but follows a few relatively narrow belts that wind around the earth. Figure 8.5 illustrates the world distribution of earthquakes for a 9-year period. Using Figure 8.5 (page 132) and your text as references, answer questions 17 and 18.

17. List the locations of the three major belts of concentrated earthquake activity on the earth.

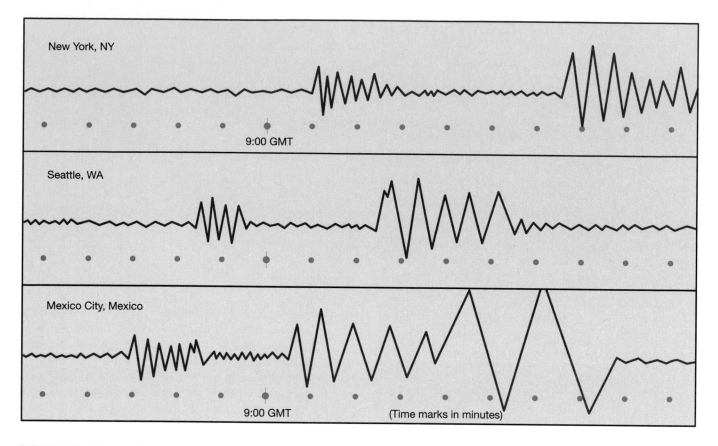

New York, NY

9:00 GMT

Seattle, WA

9:00 GMT

Mexico City, Mexico

9:00 GMT   (Time marks in minutes)

FIGURE 8.3   Three seismograms of the same earthquake recorded at three different locations.

TABLE 8.1   Epicenter data table.

	New York	Seattle	Mexico City
Elapsed time between first P and first S waves Distance from epicenter in miles			

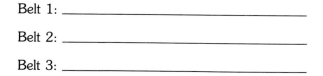

Belt 1: _____

Belt 2: _____

Belt 3: _____

18. According to the text, with what earth phenomenon is the location of earthquake epicenters closely correlated?

_____

THE EARTH BEYOND OUR VIEW

The Earth's Interior Structure.   The study of earthquakes has contributed greatly to earth scientists' un-

derstanding of the internal structure of the earth. Variations in the travel times of P and S waves as they journey through the earth provide scientists with an indication of changes in rock properties. Also, since S waves cannot travel through fluids, the fact that they are not present in seismic waves that penetrate deep into the earth suggests a fluid zone near the earth's center.

In addition to the lithosphere, the other major zones of the earth's interior include the **asthenosphere, mantle, outer core,** and **inner core.** After you have reviewed these zones and the general structure of the earth's interior in Chapter 5 of your text, use Figure 8.6 to answer questions 19–24.

⬛▬▬▬

19. The layer labeled A on Figure 8.6 is the solid, rigid, upper zone of the earth that extends from the surface to a depth of about (100, 500, 1000) kilometers. Circle your answer.

a. Zone A is called the (core, mantle, lithosphere).

b. What are the approximate velocities of P and S waves in zone A?

P wave velocity: _____ km/sec
S wave velocity: _____ km/sec

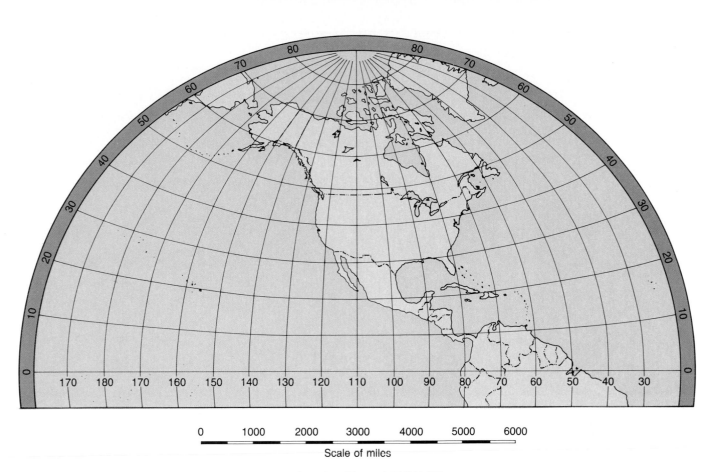

FIGURE 8.4 Map for locating an earthquake epicenter. (From AGI/NAGT, *Laboratory Manual in Physical Geology*, 3rd ed., New York: Macmillan Publishing Company, 1993.)

c. The velocity of both P and S waves (increases, decreases) with increased depth in zone A. Circle your answer.

d. List the two parts of the earth's *crust* that are included in zone A and briefly describe the composition of each.

1) _____ , _____

2) _____ , _____

20. Zone B is the part of the earth's upper mantle that extends from the base of zone A to a depth of up to (100, 700, 2000) kilometers in some regions of the earth. Circle your answer.

a. Zone B is called the (crust, asthenosphere, core).

b. The velocity of P and S waves (increases, decreases) immediately below zone A in the upper part of zone B.

c. According to the text, the change in velocity of the S waves in zone B indicates that it consists of (partially molten, entirely liquid) material.

21. Zone C (which includes the lower part of zone A and zone B) extends to a depth of

_____ kilometers.

a. Zone C is called the earth's _____ .

b. What fact concerning S waves indicates that zone C is not liquid?

_____

c. What is the probable composition of zone C?

_____

22. Zone D extends from 2885 km to about (5100, 6100) kilometers.

a. Zone D is the earth's _____ _____.

b. What happens to S waves when they reach zone D and what does this indicate about the zone?

_____

_____

c. The velocity of P waves (increases, decreases) as they enter zone D. Circle your answer.

23. Zone E is the earth's _____ _____ .

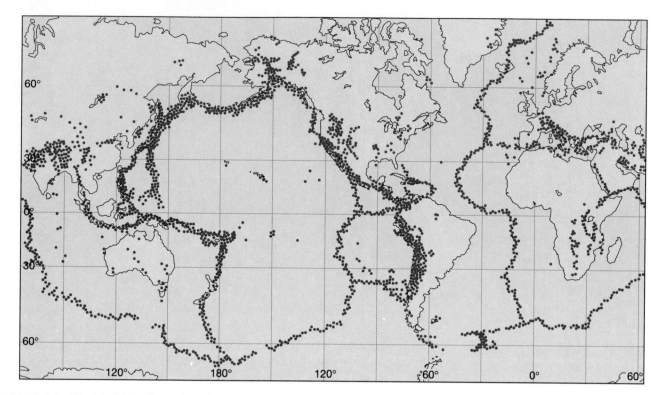

FIGURE 8.5 World distribution of earthquakes for a nine-year period. (Data from NOAA)

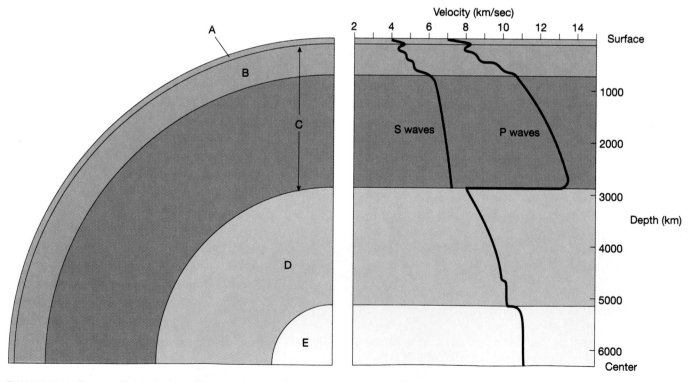

FIGURE 8.6 The earth's interior with variations in P and S wave velocities. (Data from Bruce A. Bolt)

a. Zone E extends from a depth of _____ km to the _____ of the earth.

b. What change in velocity do P waves exhibit at the top of zone E and what does this suggest about the zone?

_____

_____

c. What is the probable composition of the earth's core?

_____

24. Label Figure 8.6 by writing the name of each interior zone at the appropriate letter.

The Earth's Internal Temperature.
Measurements of temperatures in wells and mines have shown that earth temperatures increase with depth. The rate of temperature increase is called the **geothermal gradient.** Although the geothermal gradient varies from place to place, it is possible to calculate an average. Table 8.2 shows an idealized average temperature gradient for the upper earth compiled from many different sources. Use the information in Table 8.2 to answer questions 25–29.

▬▬▬▬

25. Plot the temperature values from Table 8.2 on the graph in Figure 8.7. Then draw a single line

TABLE 8.2 Idealized internal temperatures of the earth compiled from several sources.

Depth (kilometers)	Temperature (°C)
0	20°
25	600°
50	1000°
75	1250°
100	1400°
150	1700°
200	1800°

that fits the pattern of points from the surface to 200 km. Label the line, "temperature gradient."

26. Refer to the graph in Figure 8.7. The rate of increase of the earth's internal temperature (is constant, changes) with increasing depth. Circle your answer.

27. The rate of temperature increase from the surface to 100 km is (greater, less) than the rate of increase below 100 km.

28. The temperature at the base of the lithosphere, which is about 100 kilometers below the surface, is approximately (600, 1400, 1800) degrees Celcius.

29. Use the data and graph to calculate the average temperature gradient (temperature change per unit of depth) for the upper 100 km of the earth in °C/100 km and °C/km.

°C/100 km: _____ , °C/km: _____

FIGURE 8.7 Graph for plotting temperature curves.

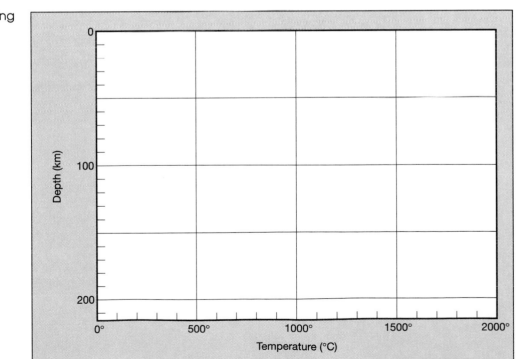

TABLE 8.3 Melting temperatures of granite (with water) and basalt at various depths within the earth.

Granite (with water)		Basalt	
Depth (km)	Melting temp. (°C)	Depth (km)	Melting temp. (°C)
0	950°	0	1100°
5	700°	25	1160°
10	660°	50	1250°
20	625°	100	1400°
40	600°	150	1600°

Melting Temperatures of Rocks. Geologists have always been concerned with the conditions required for pockets of molten rock (magma) to form near the surface, as well as at what depth within the earth a general melting of rock may occur. The melting temperature of a rock changes as pressure increases deeper within the earth. The approximate melting points of the igneous rocks, granite and basalt, under various pressures (depths) have been determined in the laboratory and are shown in Table 8.3. Granite and basalt have been selected because they are the common materials of the upper earth. Use the data in Table 8.3 to answer questions 30–35.

30. Plot the melting temperature data from Table 8.3 on the earth's internal temperature graph you have prepared in Figure 8.7. Draw a different colored line for each set of points and label them "melting curve for wet granite" and "melting curve for basalt."

Use the graphs you have drawn in Figure 8.7 to help answer questions 31–33.

31. Use Figure 8.7 and assume your earth temperature gradient is accurate. At approximately what depth within the earth would wet granite reach its melting temperature and form granitic magma?

_____ km within the earth

Evidence suggests that the oceanic crust and the remaining lithosphere down to a depth of about 100 km are similar in composition to basalt.

32. The melting curve for basalt in Figure 8.7 indicates that the lithosphere above approximately 100 km

(has, has not) reached the melting temperature for basalt and therefore should be (solid, molten). Circle your answers.

33. Figure 8.7 indicates that basalt reaches its melting temperature within the earth at a depth of approximately _____ kilometers. (Solid, Partly melted) basaltic material would be expected to occur below this depth. Circle your answer.

34. What is the name of the zone within the earth that begins at a depth of about 100 km and may extend to approximately 700 kilometers?

_____

35. Why do scientists believe that the zone in question 34 is capable of "flowing," carrying the rigid lithosphere along with it?

_____

_____

EARTHQUAKES AND EARTH TEMPERATURES—A PRACTICAL APPLICATION

The study of earthquakes and the earth's internal temperature has contributed greatly to the understanding of plate tectonics. One part of the theory is that large, rigid slabs of the lithosphere are descending into the mantle where they break up and melt, generating deep focus earthquakes during the process. Using earthquakes and earth temperatures, earth scientists have confirmed that this major earth process is currently taking place near the island of Tonga in the South Pacific and elsewhere.

Figure 8.8 illustrates the distribution of earthquake foci during a one-year period in the vicinity of Tonga Island. Use the figure to answer questions 36–40.

36. At approximately what depth do the deepest earthquakes occur in the area represented on Figure 8.8?

_____ kilometers

37. The earthquake foci in the area are distributed (in a random manner, nearly along a line). Circle your answer.

38. Draw a line on Figure 8.8 that outlines the area of earthquakes within the earth.

39. Using previous information from this exercise, draw a line on Figure 8.8 at the proper depth that

FIGURE 8.8 Distribution of earthquake foci in 1965 in the vicinity of Tonga Island. (Data from B. Isacks, J. Oliver, and L. R. Sykes)

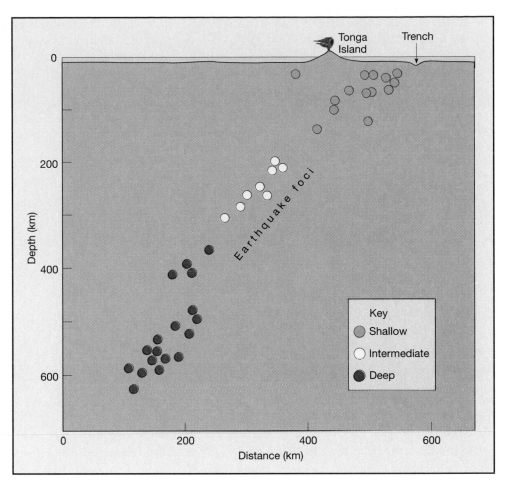

indicates the top of the *asthenosphere*—the zone of partly melted or plastic earth material. Label the line you draw "top of asthenosphere."

40. Remember that earthquakes only occur in solid, rigid material and refer to Figure 8.8. Why have earth scientists been drawn to the conclusion of a descending slab of solid lithosphere being consumed into the mantle near Tonga?

_____

_____

_____

_____

REVIEW

Having completed the exercise, you should know the following:

1. Earthquakes are vibrations of the earth produced by the rapid release of energy from rocks that have been strained beyond their limit.

2. P waves, S waves, and surface waves are seismic waves that are generated at the focus of an earthquake.

3. Of the three types of seismic waves, P waves have the greatest velocity, smallest amplitude, and are the first motions recorded on a seismogram.

4. Seismologists at a seismic station can determine how far they are from an earthquake epicenter by noting the difference in arrival times between the first P waves and first S waves on a seismogram and using a travel-time graph.

5. The epicenter of an earthquake is located on the surface of the earth at the point where the distance circles drawn around three seismic stations intersect.

6. The distribution of earthquakes on the earth follows definite belts that correspond to plate boundaries.

7. The lithosphere of the earth is solid and consists of the crust and rigid upper mantle.

8. The earth's crust consists of oceanic crust and continental crust.

9. The asthenosphere is a plastic zone in the mantle located directly below the lithosphere.

10. The outer core of the earth is fluid because S waves are not transmitted.
11. The inner core of the earth is solid and consists of large amounts of iron and nickel.
12. The internal temperature of the upper 100 km of the earth, as calculated in this exercise, increases about 14°C per kilometer of depth.
13. That the igneous rock basalt would be partly melted at the temperature and pressure that occur at a depth of 100 km supports the presence of a plastic zone, the asthenosphere, beginning at this depth.
14. That earthquakes occur at depths of 600 km or more at some locations (depths where plastic material is normally expected) suggests that rigid slabs of the lithosphere are descending into the mantle.

Earthquakes and the Earth's Interior

SUMMARY/REPORT PAGE

Date Due: _____

Name: _____

Date: _____

Class: _____

After you have finished Exercise Eight, complete the following questions. You may have to refer to the exercise for assistance or to locate specific answers. Be prepared to submit this summary/report to your instructor at the designated time.

1. Use the minute marks provided below to sketch a typical seismogram where the first P wave arrives 3 minutes ahead of the first S wave. Show and label P waves, S waves, and surface (L) waves.

. . . .

(minute marks)

2. How far from the earthquake epicenter is the seismic station that recorded the seismogram in question 1 of this Summary/Report page?

_____ miles

3. Use a diagram to explain how the epicenter of an earthquake is located.

Explanation: _____

_____

_____

_____

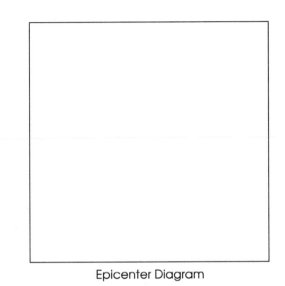

Epicenter Diagram

4. Where are the locations of the three major earthquake belts on the earth?

_____

_____

5. The change in velocity of S waves below the base of the lithosphere indicates that the asthenosphere is made of (solid, partly melted, gaseous) material. Circle your answer.

6. What happens to the S waves from an earthquake when they reach the earth's outer core?

_____

7. List the depths and probable compositions of the following interior zones of the earth.

Mantle: depth (km) from _____ to

_____

Composition: _____

Outer core: depth (km) from _____ to

_____

Composition: _____

8. On the internal temperature graph you constructed in Figure 8.7, at what depth did you determine granitic magma should form?

_____ kilometers

9. Why do earth scientists think that rigid slabs of the lithosphere are descending into the mantle near the island of Tonga?

_____

_____

10. Define the following terms:

Earthquake focus: _____

_____

Earthquake epicenter: _____

_____

Seismogram: _____

_____

Mantle: _____

_____

Geothermal gradient: _____

_____

Lithosphere: _____

_____

Introduction to Oceanography

The global ocean covers nearly three-quarters of the earth's surface and is an important focus of earth science studies. This exercise investigates some of the physical characteristics of the oceans. To establish a foundation for reference, the extent, depths, and distribution of the world oceans are the first topics examined. Salinity and temperature, two of the most important variables of seawater, are studied to ascertain how they influence the density of water and, hence, the deep ocean circulation.

OBJECTIVES

After you have completed this exercise, you should be able to

1. Locate and name major water bodies on the earth.
2. Discuss the distribution of land and water in each hemisphere.
3. Locate and describe the general features of ocean basins.
4. Explain the relation between salinity and the density of seawater.
5. Describe how seawater salinity varies with latitude and depth in the oceans.
6. Explain the relation between temperature and the density of seawater.
7. Describe how seawater temperature varies with latitude and depth in the oceans.

TEXTBOOK REFERENCE

Chapter 10

MATERIALS

colored pencils   ruler

Materials Supplied by Your Instructor

measuring cylinder (100 ml, clear, pyrex or plastic)	world wall map, globe or atlas
	ice
test tubes	dye
salt	rubber band
salt solutions	beaker

TERMS

oceanography	seamount	density
continental shelf	deep-ocean trench	density current
continental slope	mid-ocean ridge	salinity
abyssal plain		

INTRODUCTION

Oceanography, the application of all sciences to the study of oceans, is an important part of the earth sciences because of the sea's vital role in our physical environment. Chemical oceanographers investigate the composition of seawater and how it varies from place to place and through time. Marine biologists study the relations between marine organisms and their environments. Physical oceanographers attempt to understand waves, currents, and tides, and how they are influenced by and related to ocean water temperatures, densities, depths, and shapes of coastlines. Geologic oceanographers seek answers to such questions as the ages and evolution of the ocean basins by detailing on a global scale their topography, geologic structure, and sediment.

Oceanographic research is being conducted continuously. New discoveries provide a greater understanding about the earth—"the water planet."

EXTENT OF THE OCEANS

Locations. If you have previously completed Exercise Twenty-one, "Location and Distance on the Earth," review the locations of the water bodies plotted on the world map, Figure 21.5. Then examine the list of water bodies in question 1 and write the names on the map of those you have *not* previously identified.

1. If you have not completed Exercise Twenty-one, refer to a globe, wall map of the world, or world map in an atlas and identify each of the oceans and major water bodies listed below. Locate and label each on the world map, Figure 21.5, provided in Exercise Twenty-one. To conserve space, mark only the number or letter of the feature at the appropriate location on the map.

Oceans	Other Major Water Bodies	
A. Pacific	1. Caribbean Sea	9. Black Sea
B. Atlantic	2. North Sea	10. Baltic Sea
C. Indian	3. Coral Sea	11. Arabian Sea
D. Arctic	4. Sea of Japan	12. Weddell Sea
	5. Sea of Okhotsk	13. Bering Sea
	6. Gulf of Mexico	14. Red Sea
	7. Persian Gulf	15. Bay of Bengal
	8. Mediterranean Sea	16. Caspian Sea

Area. The area of the earth is about 510 million square kilometers (197 million square miles). Of this, approximately 360 million square kilometers (140 million square miles) are covered by oceans and marginal seas.

2. What percent of the earth's surface is covered by oceans and marginal seas?

$$\frac{\text{Area of oceans and marginal seas}}{\text{Area of the earth}} \times 100 =$$

_____ % oceans

3. What percent of the earth's surface is land?

_____ % land

Distribution of Land and Water by Hemisphere.

Answer questions 4–6 by examining Figure 9.1 and either a globe, wall map of the world, or world map in an atlas.

4. Which hemisphere, Northern or Southern, could be called the "water" hemisphere and which the "land" hemisphere?

"Water" hemisphere: _____

"Land" hemisphere: _____

5. Follow a line around the earth at the latitudes listed below and estimate what percent of the earth's surface is ocean at each latitude.

	Northern Hemisphere	Southern Hemisphere
40°:	_____ % ocean	_____ % ocean
60°:	_____ % ocean	_____ % ocean

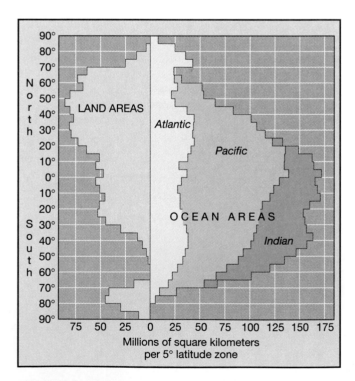

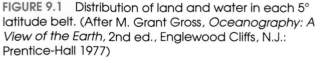

FIGURE 9.1 Distribution of land and water in each 5° latitude belt. (After M. Grant Gross, *Oceanography: A View of the Earth*, 2nd ed., Englewood Cliffs, N.J.: Prentice-Hall 1977)

a. The oceans become (wider, more narrow) as you go from the equator to the pole in the Northern Hemisphere. Circle your answer.

b. In the Southern Hemisphere the width of the oceans (increases, decreases) from the equator to the pole.

6. Which ocean covers the greatest area?

_____

Ocean Basin Topography.   Various features are located along the continental margins and on the ocean basin floor. **Continental shelves** and **continental slopes** are submerged portions of continental margins. **Abyssal plains, seamounts, deep-ocean trenches,** and **mid-ocean ridges** occur on ocean basin floors. Using your text as a reference, briefly describe each of these features in questions 7–12. Label one or more examples of each feature on the ocean floor map of the North Atlantic Ocean basin, Figure 9.2.

7. Continental shelf: _____

_____

a. What is the approximate average ocean depth along the continental shelves bordering North America?

_____

b. Write a brief statement comparing the width of the continental shelf along the east coast, west coast and gulf coast of North America.

_____

_____

8. Continental slope: _____

_____

9. Abyssal plain: _____

_____

a. The general topography of abyssal plains is (flat, irregular). Circle your answer.

b. How do abyssal plains form and what is their composition?

_____

_____

10. Seamount: _____

_____

11. Deep-ocean trench: _____

_____

a. Approximately how deep is the Puerto Rico trench?

_____ meters

b. Use a map or globe to locate three deep-ocean trenches in the Western Pacific Ocean. Give the name, location, and depth of each.

Trench 1: _____

_____

Trench 2: _____

_____

Trench 3: _____

_____

12. Mid-ocean ridge: _____

_____

a. Examine the mid-ocean ridge system on the two-page ocean floor map in Chapter 10 of the text. Follow the ridge eastward from the Atlantic Ocean into the Indian Ocean and then into the Pacific. What happens to the ridge along the southwest coast of North America?

_____

_____

b. Approximately how high above the adjacent ocean floor does the Mid-Atlantic Ridge rise?

_____ meters

CHARACTERISTICS OF OCEAN WATER

Ocean circulation has two primary components: surface ocean currents and deep-ocean circulation. Both are examined with greater detail in Exercise Eleven "Waves, Currents, and Tides." While surface currents like the famous Gulf Stream are driven primarily by the prevailing world winds, the deep-ocean circulation is largely the result of differences in ocean water **density** (mass

FIGURE 9.2 North Atlantic basin. (North Atlantic reproduced from *World Ocean Floor* by Bruce C. Heezen and Marie Tharp, 1977. Copyright by Marie Tharp, 1977. Reproduced by permission of Marie Tharp, 1 Washington Avenue, South Nyack, NY 10960.)

0   500 miles

0   800 km

Approximate scale

per unit volume of a substance). A **density current** is the movement (flow) of one body of water over, under, or through another caused by density differences and gravity. Variations in **salinity** and temperature are the two most important factors in creating the density differences that result in the deep-ocean circulation.

Salinity. Salinity is the amount of dissolved solid material in water, expressed as parts per thousand parts of water. The symbol for parts per thousand is 0/00. Although there are many dissolved salts in seawater, sodium chloride (common table salt) is the most abundant.

Variations in the salinity of seawater are primarily a consequence of changes in the water content of the solution. In regions where evaporation is high, the proportionate amount of dissolved material in seawater is increased by removing the water and leaving behind the salts. On the other hand, in areas of high precipitation and high runoff, the additional water dilutes seawater and lowers the salinity. Since the factors that determine the concentration of salts in seawater are not constant from the equator to the poles, the salinity of seawater also varies with latitude and depth.

Salinity-Density Experiment. To gain a better understanding of how salinity affects the density of water, examine the equipment in the lab (see Figure 9.3) and conduct the following experiment by completing each of the indicated steps.

Step 1. Fill the measuring cylinder provided with cool tap water up to the rubber band.

Step 2. Fill a test tube about 1/2 full of solution A (saltwater) and pour it slowly into the cylinder. Observe and describe what happens.

Observations: _____

_____

Step 3. Repeat steps 1 and 2 two additional times and measure the time required for the front edge of the saltwater to travel from the rubber band to the bottom of the cylinder. Record the times for each test in the data table, Table 9.1. *Make certain* that you drain the cylinder after each trial and refill it with fresh water.

Step 4. Determine the travel time two times for solution B as you did with solution A and enter your measurements in Table 9.1.

Step 5. Fill a test tube about 1/2 full of solution B and add to it some additional salt. Then shake the test tube vigorously. Determine the travel time of this solution and enter your results in Table 9.1.

FIGURE 9.3 Lab setup for salinity-density experiment.

TABLE 9.1 Salinity-density experiment data table

Solution	Timed Trial #1	Timed Trial #2	Average of Both Trials
A			
B			
Solution B plus salt		XXXX	XXXX

Step 6. Clean all your glassware.

13. Questions 13a and 13b refer to the salinity-density experiment.
    a. Since the solution that traveled fastest has the greatest density, solution (A, B) is most dense. Circle your answer.
    b. Write a brief summary of the results of your salinity-density experiment.

_____

_____

Table 9.2 lists the approximate surface water salinity at various latitudes in the Atlantic and Pacific Oceans. Using the data, construct a salinity curve for each ocean on the graph, Figure 9.4. Use a different colored pencil for each ocean. Then, using your text as a reference and the graph, answer questions 14–18.

14. At which latitudes are the highest surface salinities found?

_____

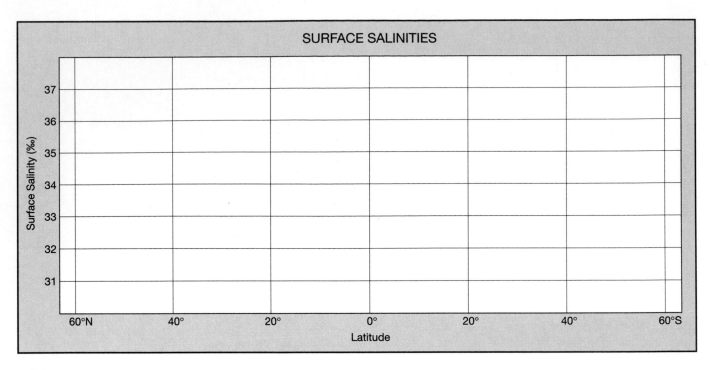

FIGURE 9.4 Graph for plotting surface salinities.

TABLE 9.2 Ocean surface water salinity in parts per thousand (0/00) at various latitudes in the Atlantic and Pacific Oceans.

Latitude	Atlantic Ocean	Pacific Ocean
60°N	33.0 0/00	31.0 0/00
50°	33.7	32.5
40°	34.8	33.2
30°	36.7	34.2
20°	36.8	34.2
10°	36.0	34.4
0°	35.0	34.3
10°	35.9	35.2
20°	36.7	35.6
30°	36.2	35.7
40°	35.3	35.0
50°	34.3	34.4
60°S	33.9	34.0

15. What are two factors that control the concentration of salts in seawater?

_____ and _____

16. Refer to the factors listed in question 15. What is the cause of the difference in surface water salinity between equatorial and subtropical regions in the Atlantic Ocean?

_____
_____

17. Of the two oceans, the (Atlantic, Pacific) Ocean has higher average surface salinities. Circle your answer.

18. Suggest a reason for the difference in average surface salinities between the oceans.

_____
_____

Figure 9.5 shows how ocean water salinity varies with depth at different latitudes. Use the figure and the appropriate section of the text to answer questions 19–22.

19. In general, salinity (increases, decreases) with depth in the equatorial and tropical regions and (increases, decreases) with depth at high latitudes. Circle your answers.

20. Why are the surface salinities higher than the deep-water salinities in the lower latitudes?

_____

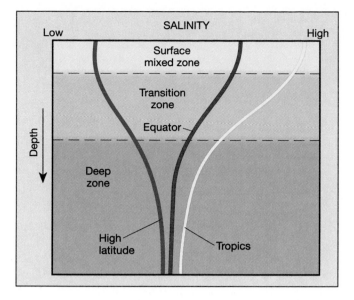

FIGURE 9.5 Ocean water salinity changes with depth at high latitudes, equatorial regions, and the tropics.

21. What is the *halocline*? Label the halocline on Figure 9.5.

_____

22. Below the halocline the salinity of ocean water (increases rapidly, remains fairly constant, decreases rapidly). Circle your answer.

Ocean Water Temperatures. Seawater temperature is the most extensively determined variable of the oceans because it is easily measured and has an important influence on marine life. Like salinity, ocean water temperatures vary from equator to pole and change with depth.

Temperature, like salinity, also affects the density of seawater. However, the density of seawater is more sensitive to temperature fluctuations than salinity. Cool surface water, which has a greater density than warm water, forms in the polar regions, sinks, and flows toward the tropics.

Temperature-Density Experiment. To illustrate the effects of temperature on the density of water, examine the equipment in the lab (see Figure 9.6) and then conduct the following experiment by completing each of the indicated steps.

Step 1. Fill a measuring cylinder with cold tap water up to the rubber band.

Step 2. Put 2–3 drops of dye in a test tube and fill it 1/2 full with hot tap water.

Step 3. Pour the contents of the test tube slowly into the cylinder and then record your observations.

Observations: _____

_____

Step 4. Empty the cylinder and refill it with hot water.

Step 5. Add a test tube full of cold water and 2–3 drops of dye to some ice in a beaker. Stir the solution for a few seconds. Fill the test tube three-fourths full with some liquid (no ice) from your beaker. Pour this cold liquid slowly into the cylinder. Then record your observations.

Observations: _____

_____

Step 6. Clean the glassware and return it along with the other materials to your instructor.

23. Questions 23a and 23b refer to the temperature-density experiment.
   a. Given equal salinities, (cold, warm) seawater would have the greatest density. Circle your answer.
   b. Write a brief summary of your temperature-density experiment.

_____

_____

_____

FIGURE 9.6 Lab setup for temperature-density experiment.

TABLE 9.3 Idealized ocean surface water temperatures and densities at various latitudes.

Latitude	Surface Temperature (C°)	Surface Density (g/cm³)
60°N	5	1.0258
40°	13	1.0259
20°	24	1.0237
0°	27	1.0238
20°	24	1.0241
40°	15	1.0261
60°S	2	1.0272

Table 9.3 shows the average surface temperature and density of seawater at various latitudes. Using the data in Table 9.3, plot a line on the graph in Figure 9.7 for temperature and a separate, different color, line for density. Then, using your text as a reference and the graph, answer questions 24–26.

24. (Warm, Cool) surface temperatures and (high, low) surface densities occur in the equatorial regions. While at high latitudes, (warm, cool) surface temperatures and (high, low) surface densities are found. Circle your answers.

25. What is the reason for the fact that higher average surface densities are found in the Southern Hemisphere?

_____

In question 14 you concluded that surface salinities were greatest at about latitudes 30°N and 30°S.

26. Refer to the density curve in Figure 9.7. What evidence supports the fact that the temperature of seawater is more of a controlling factor of density than salinity?

_____

_____

Figure 9.8 shows how ocean water temperature varies with depth at different latitudes. Use the figure and the appropriate section of the text to answer questions 27–29.

27. Temperature decreases most rapidly with depth at (high, low) latitudes. Circle your answer and give the reason that the decrease with depth is most rapid at these latitudes.

_____

28. What is the *thermocline*? Label the thermocline on Figure 9.8.

_____

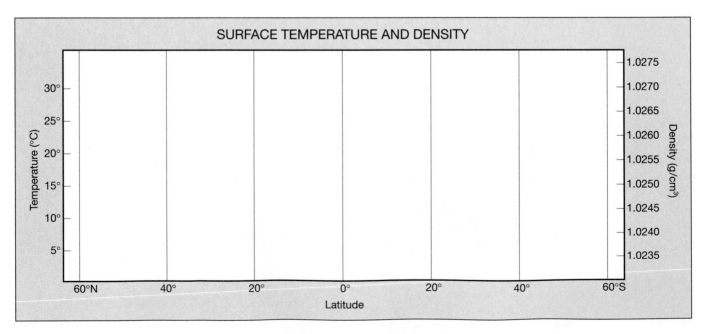

FIGURE 9.7 Graph for plotting surface temperatures and densities.

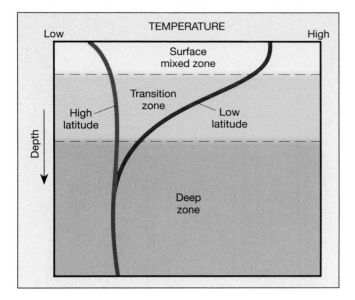

FIGURE 9.8 Ocean water temperature changes with depth at high and low latitudes.

29. Below the thermocline the temperature of ocean water (increases rapidly, remains fairly constant, decreases rapidly). Circle your answer.

REVIEW

Having completed the exercise, you should know the following:

1. Where each of the major water bodies on the earth is located.
2. The Southern Hemisphere could be called the "water" hemisphere.
3. Oceans become more narrow from the equator to the pole in the Northern Hemisphere.
4. Oceans become wider from the equator to the pole in the Southern Hemisphere.
5. The Pacific Ocean is the largest ocean on the earth.
6. Where many of the continental shelves, continental slopes, abyssal plains, mid-ocean ridges, and deep-ocean trenches are located on the earth.
7. Deep-ocean circulation is largely the result of density differences of ocean water.
8. Salinity and temperature are the two most important factors that determine the density of seawater.
9. Evaporation and precipitation influence the surface salinity of seawater.
10. Temperature and salinity both vary with latitude in the ocean.
11. Temperature and salinity both vary with depth in the ocean.

Introduction to Oceanography

SUMMARY/REPORT PAGE

Date Due: _____

Name: _____
Date: _____
Class: _____

After you have finished Exercise Nine, complete the following questions. You may have to refer to the exercise for assistance or to locate specific answers. Be prepared to submit this summary/report to your instructor at the designated time.

1. Give the approximate latitude and longitude of the centers of each of the following water bodies.

   Mediterranean Sea: _____

   Sea of Japan: _____

   Indian Ocean: _____

2. Write a brief statement comparing the distribution of water and land in the Northern Hemisphere to the distribution in the Southern Hemisphere.

   _____

   _____

3. On the ocean basin profile in Diagram 9.1, label the continental shelves, continental slopes, abyssal plains, seamounts, mid-ocean ridge, deep-ocean trench.

4. List the names and depths of two Pacific Ocean trenches.

Name	Depth
_____	_____
_____	_____

5. The following are some short statements. Circle the most appropriate response.
   a. The higher the salinity of seawater, the (lower, higher) the density.
   b. The lower the temperature of seawater, the (lower, higher) the density.
   c. Surface salinity is greatest in (polar, subtropical, equatorial) regions.
   d. (Temperature, Salinity) has the greatest influence on the density of seawater.
   e. (Warm, Cold) seawater with (high, low) salinity would have the greatest density.
   f. Vertical movements of ocean water are most likely to begin in (equatorial, subtropical, polar) regions, because the surface water there is (most, least) dense.

6. Why is the surface salinity of an ocean higher in the subtropics than in the equatorial regions?

   _____

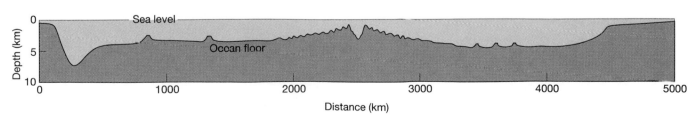

DIAGRAM 9.1

7. Refer to the salinity-density experiment you conducted. Solution (A, B) had the greatest density. Circle your answer.

8. Describe the change in salinity *and* temperature with depth that occurs at low latitudes.

Salinity: _____

Temperature: _____

9. Are the following statements true or false? Circle your response.

T F a. The Atlantic Ocean covers the greatest area of all the world oceans.

T F b. Continental shelves are part of the deep-ocean floor.

T F c. Deep-ocean trenches are located in the middle of oceans.

T F d. High evaporation rates in the subtropics cause the surface ocean water to have a lower than average salinity.

The Dynamic Ocean Floor

One of the most significant scientific revelations of the century is the fact that the ocean basins are young, ephemeral features. Based upon this discovery, a revolutionary theory called *plate tectonics* has been developed that helps to explain and interrelate earthquakes, mountain building, and other geologic events and processes. This exercise examines some of the lines of evidence that have been used to verify this comprehensive model of the way earth scientists view the restless earth.

OBJECTIVES

After you have completed this exercise, you should be able to

1. List and explain the lines of evidence that support the theory of plate tectonics.
2. Locate and describe the mid-ocean ridge system and deep-ocean trenches.
3. Describe the relation between earthquakes and plate boundaries.
4. Describe the magnetic polarity reversals that have taken place on the earth in the last four million years.
5. Determine the rate of sea-floor spreading that occurs along a mid-ocean ridge by using paleomagnetic evidence.
6. Use the rate of sea-floor spreading for an ocean basin to determine its age.
7. Describe the three types of plate boundaries and the motion of the plates that occurs along each boundary.

TEXTBOOK REFERENCE

Chapters 6 and 10

MATERIALS

calculator
ruler
colored pencils

Materials Supplied by Your Instructor

atlas, globe, or world wall map

TERMS

plate tectonics	mid-ocean ridge	Pangaea
lithosphere	deep-ocean trench	paleomagnetism
plates	continental drift	sea-floor spreading

INTRODUCTION

Since the early days of sailing vessels, the investigation of the ocean floor has been both a challenge and a source of new understanding about the mechanisms of the dynamic earth. Using the information gathered over the years concerning the topography, age, and method of evolution of the ocean basins, earth scientists have developed an exciting theory called **plate tectonics.** Plate tectonics is the foundation used by modern geology to help explain the origin of mountains and continents, the occurrence of earthquakes, the evolution and distribution of plants and animals, as well as many other geologic processes.

Plate tectonics postulates that the **lithosphere** of the earth is broken into several large, rigid slabs called **plates.** These plates and the continents on them are moving. Where the plates are separating along the **mid-ocean ridges,** new ocean floor crust is forming. Along the plate margins, earthquakes are generated as plates slide past each other, collide to form mountains, or override each other causing **deep-ocean trenches.**

The idea that our present-day continents at one time were parts of a single supercontinent was stated in 1912

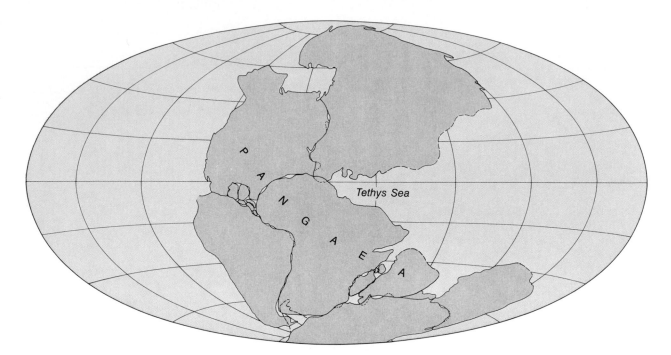

FIGURE 10.1   Pangaea, as it is thought to have appeared 200 million years ago. (After R. S. Dietz and J. C. Holden, *Journal of Geophysical Research* 75: 4943. Copyright by American Geophysical Union)

by Alfred Wegener. His hypothesis, called **continental drift,** proposed that long ago a supercontinent called **Pangaea** broke apart into the present continents (Figure 10.1). Over a period of millions of years, each of the continents drifted across the earth's surface to its current position. Among the lines of evidence used to support this hypothesis were (1) the geometric fit of the continents, (2) the fit of geologic structures (mountains, etc.) and rock ages across oceans, (3) the global distribution of fossils *(paleontology)*, and (4) ancient climates *(paleoclimatology)*.

In the 1960s great advances resulting from new technologies in oceanographic research, such as the ocean research vessel *Glomar Challenger,* brought new evidence in support of many of Wegener's ideas. Research concerning rock magnetism, the cause and distribution of earthquakes, and the age of ocean sediments led to the development of a much broader theory than continental drift. The expanded theory is known as *plate tectonics.*

THE EVIDENCE FOR PLATE TECTONICS

Features of the Ocean Floor.   Prior to the plate tectonics theory, the origin of many ocean floor features was uncertain. However, today earth scientists are beginning to comprehend these features, especially when viewed as parts of a dynamic, evolving lithosphere. Two of these features, mid-ocean ridges and deep-ocean

trenches, are of major importance in understanding plate tectonics.

1. After referring to Chapter 10 of your text, write a brief statement that describes each of the following ocean floor features.

    Mid-ocean ridge: _____

    _____

    _____

    Mid-ocean ridge rift valley: _____

    _____

    _____

    Deep-ocean trench: _____

    _____

    _____

2. Use the two-page ocean floor map in Chapter 10 of your text, an atlas, and/or world wall map as a reference, and accurately draw the global mid-ocean ridge system on the world map, Figure 10.2.

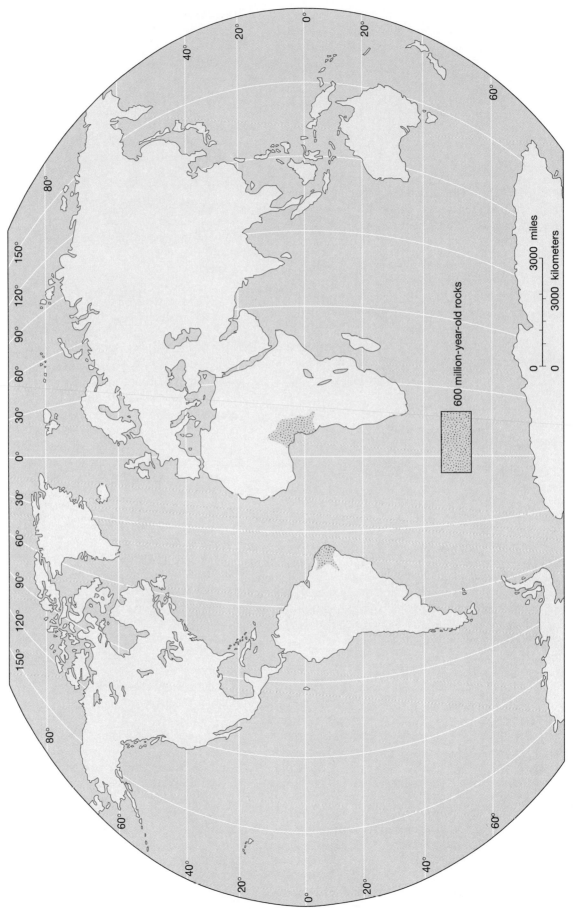

FIGURE 10.2 World map.

3. Use an atlas and/or world wall map to locate and label each of the following deep-ocean trenches on Figure 10.2. Draw a line to represent the trench. To conserve space, write only the letter of each trench on the map.

A. Puerto Rico   F. Japan
B. Cayman   G. Mariana
C. Peru-Chile   H. Tonga
D. Aleutian   I. Kermadec
E. Kuril   J. Java

Fit of the Continents.

Examine the east coast of South America and the west coast of Africa on the world map, Figure 10.2. Then answer questions 4–7.

4. The shape of the east coast of South America conforms to the shape of the (east, west) side of the mid-ocean ridge in the central South Atlantic Ocean. Circle your answer.
5. The shape of the west coast of Africa conforms to the shape of the (east, west) side of the mid-ocean ridge in the South Atlantic Ocean. Circle your answer.
6. If South America and Africa were moved to the mid-ocean ridge, their shapes (would, would not) fit together along the ridge. Circle your answer.

Notice on Figure 10.2 that the ages of the rocks in eastern South America and western Africa are both about 600 million years old. The rocks that comprise the ocean floor separating the two continents are much younger.

7. If the South American and African continents are brought together at the mid-ocean ridge, do the areas of ancient rocks match? Does the match support the idea that these continents were once joined?

_____

_____

(*Note:* Actually, the fit of the continents is more exact on a globe because flat maps have distortions. Also, using the seaward edge of the continental shelf rather than the coastline would be more accurate.)

Earthquakes.

The distribution and depths of earthquakes provide evidence for the mechanics of plate tectonics.

Locate and examine the world map of plates *and* the earthquake distribution map in Chapter 6 of the text. Use the maps and Figure 10.2 to answer questions 8–10.

8. Most deep-focus earthquakes are associated with (mid-ocean ridges, deep-ocean trenches). Circle your answer.
9. Most mid-ocean, shallow-focus earthquakes are associated with (mid-ocean ridges, deep-ocean trenches).

Compare the earthquake distribution map to the map showing the plates. Notice that plate boundaries are outlined by earthquake zones.

10. On the world map, Figure 10.2, label, by name, the major plates.

Paleomagnetism.

Some minerals in igneous rocks develop a slight magnetism in alignment with the earth's magnetic field at the time of their formation. Also, scientists have discovered that the polarity of the earth's magnetic field has periodically reversed and the north magnetic pole becomes the south magnetic pole, while the south magnetic pole becomes the north magnetic pole. Putting these facts together provides additional support for plate tectonics.

The ancient magnetism, called **paleomagnetism**, present in rocks on the ocean floor can be used to determine the rate at which the plates are separating and, consequently, the time when they began to separate. Where plates separate along the mid-ocean ridge, magma from the mantle rises to the surface and creates new ocean floor. As the magma cools, the minerals assume a magnetism equal to the prevailing magnetic field. As the plates continue to separate and the earth's magnetic field reverses polarity, new material forming at the ridge is magnetized in an opposite (or reversed) direction.

Scientists have reconstructed the earth's magnetic polarity reversals over the past several million years. A generalized record of these polarity reversals is shown in graphic form in Figure 10.3. The periods of normal polarity, when a compass needle would have pointed North as it does today, are shown in color and labeled a–f for reference. Use Figure 10.3 to answer questions 11–16.

11. The magnetic field of the earth has had (3, 5, 7) intervals of reversed polarity during the past 4 million years. Circle your answer.

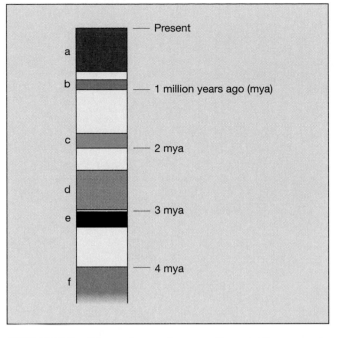

FIGURE 10.3 Chronology of magnetic polarity reversals on the earth during the last four million years. Periods of normal polarity, when a compass would have pointed North as it does today, are shown in color. Periods of reverse polarity are shown in white. (Data from Allan Cox and G. Dalrymple)

12. Approximately how long ago did the current normal polarity begin?

_____ years ago

13. One and a half million years ago the indicator on a compass needle would have pointed to the (North, South). Circle your answer.

14. The period of normal polarity, c, began (1, 2, 3) million years ago.

15. During the past 4 million years, each interval of reverse polarity has lasted (more, less) than 1 million years. Circle your answer.

16. Based upon the pattern of magnetic changes exhibited in Figure 10.3, does it appear as though the earth is due for another magnetic polarity reversal in the near future?

Paleomagnetism and the Ocean Floor. The records of the magnetic polarity reversals that have been determined from the oceanic crust across sections of the mid-ocean ridges in the Pacific, South Atlantic, and North Atlantic Oceans are illustrated in Figure 10.4. As new ocean crust forms along the mid-ocean ridge, it spreads out equally on both sides of the ridge. Therefore, a record of the reversals is repeated (mirrored). Notice that the general pattern of polarity reversals pre-

sented in Figure 10.3 can be matched with the polarity of the rocks on either side of the ridge for each ocean basin in Figure 10.4.

Use Figures 10.3 and 10.4 to answer questions 17–22.

▬▬▬

17. On Figure 10.4, identify and mark the periods of normal polarity with the letters a–f (as in Figure 10.3) along each ocean floor. Begin at the ridge crest and label along both sides of each ridge. (*Note:* The left side of the South Atlantic has already been done and can act as a guide. Also, only the periods of normal polarity through c are illustrated in the Pacific basin.)

18. Using the South Atlantic as an example, label the beginning of the normal polarity period c, "2 million years ago," on the left sides of the Pacific and North Atlantic diagrams.

19. Using the distance scale in Figure 10.4, the (Pacific, South Atlantic, North Atlantic) has spread the greatest distance during the last 2 million years. Circle your answer.

20. Refer to the distance scale. Notice that the left side of the South Atlantic basin has spread approximately 39 kilometers from the center of the ridge crest in 2 million years.
    a. How many kilometers has the left side of the Pacific basin spread in 2 million years?

    _____ kilometers

    b. How many kilometers has the left side of the North Atlantic basin spread in 2 million years?

    _____ kilometers

The distances in question 20 are for only one side of the ridge. Assuming that the ridge spreads equally on both sides, the actual distance each ocean basin has opened would be twice this amount.

▬▬▬

21. How many kilometers has each ocean basin opened in the past 2 million years?

Pacific Ocean basin: _____ kilometers

South Atlantic Ocean basin: _____ kilometers

North Atlantic Ocean basin: _____ kilometers

If both the *distance* that each ocean basic has opened and the *time* it took to open that distance are known, the rate of **sea-floor spreading** can be cal-

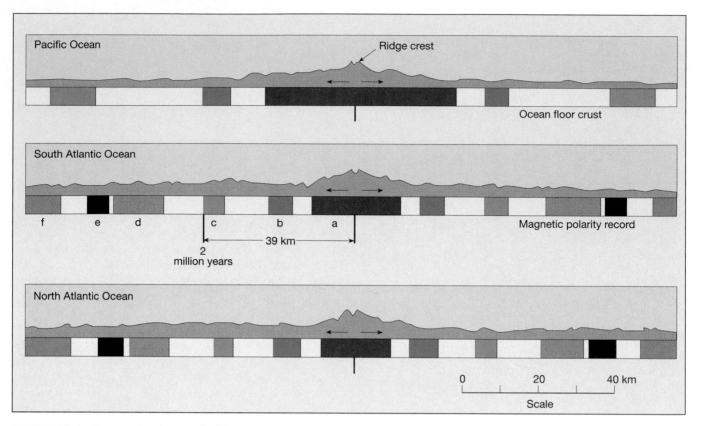

FIGURE 10.4 Generalized record of the magnetic polarity reversals near the mid-ocean ridge in the Pacific, South Atlantic, and North Atlantic Oceans. Periods of normal polarity are shown in color and correspond to those illustrated in Figure 10.3.

culated. To determine the rate of spreading in centimeters per year for each ocean basin, first convert the distance the basin has opened (see question 21) from kilometers to centimeters and then divide this distance by the time, 2 million years.

22. Determine the *rate of sea-floor spreading* for the Pacific and North Atlantic Ocean basins. As an example, the South Atlantic has already been done.
    a. South Atlantic: distance =
       78 km × 100,000 cm/km = 7,800,000 cm

       $$\text{Rate of spreading} = \frac{7,800,000 \text{ cm}}{2,000,000 \text{ yr}} = 3.9 \text{ cm/yr}$$

    b. Pacific: distance = _____ km × 100,000 cm/km

       = _____ cm

       Rate of spreading = _____

       = _____ cm/yr

c. North Atlantic: distance = _____ km

   × 100,000 cm/km = _____ cm

   Rate of spreading = $\dfrac{\rule{3cm}{0.4pt} \text{ cm}}{\text{yr}}$

   = _____ cm/yr

Ocean Basin Ages. To estimate how many millions of years ago the North Atlantic and South Atlantic Ocean basins began forming, complete questions 23–25.

23. On a large wall map or globe, accurately measure the distance between the seaward edges of the continental shelves from eastern North America at North Carolina to northwestern Africa at Mauritania (20°N latitude). Determine the distance in kilometers, then convert to centimeters. (*Note:* If you have completed Exercise Twenty-one, "Location and Distance on the Earth," you may find it eas-

ier to determine the great circle distance between the two.)

Distance = _____ kilometers

= _____ centimeters

24. Divide the distance in centimeters separating the continents that you measured in question 23 by the rate of sea-floor spreading for the North Atlantic basin, calculated in question 22c. Your answer is the approximate age in years of the North Atlantic Ocean basin.

Distance\Rate of sea-floor spreading =

_____ =

Age of the North Atlantic Basin: _____ years

Use the same procedure you used in question 24 to determine the age of the South Atlantic basin. Measure the distance between the continents along an east-west line from the eastern edge of Brazil directly east to Africa. (*Note:* You may want to determine the total number of degrees of longitude separating the two and then use the Table 21.1, "Longitude as distance," in Exercise Twenty-one, to calculate the distance.) Use the rate of sea-floor spreading for the South Atlantic in your calculation.

25. How many years ago did South America and Africa begin to separate?

Age of The South Atlantic Basin: _____ years

Compare your answers for the dates when the continents began to separate with those given in Chapter 6 of the text.

26. How accurate is your assessment of the ages of the ocean basins?

North Atlantic: _____

South Atlantic: _____

27. Write the ages of the North and South Atlantic Ocean basins on the map in Figure 10.2. Also in Figure 10.2, use the world map of plates in Chapter 6 of the text as a reference and draw arrows on all the plates showing their directions of movement.

PLATE BOUNDARIES

Types of Plate Boundaries.
Earth scientists recognize three distinct types of plate boundaries, with each distinguished by the movement of the plates associated with it. The three diagrams in Figure 10.5 illustrate each type of plate boundary. Use the figure and Chapter 6 of the text to answer questions 28–30.

28. Figure 10.5A represents a (convergent, divergent, transform) plate boundary. Circle your answer.
    a. The plates along the boundary are (spreading, colliding). Circle your answer.
    b. This type of plate boundary occurs at (deep-ocean trenches, mid-ocean ridges).
    c. This type of plate boundary results in (construction, destruction) of lithospheric material.

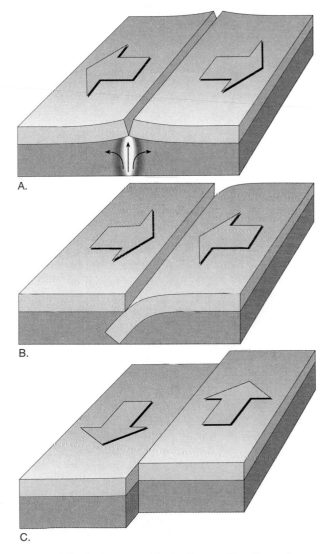

A.

B.

C.

FIGURE 10.5 Schematic illustrations of the three types of plate boundaries.

29. Figure 10.5B represents a (convergent, divergent, transform) plate boundary. Circle your answer.
  a. The plates along this type of boundary are (spreading, colliding).
  b. Briefly describe each of the following types of plate convergence. Give a specific example of a physical feature on the earth that has formed as a result of each type.

  Oceanic-continental convergence: _____

  _____

  Feature: _____

  Oceanic-oceanic convergence: _____

  _____

  Feature: _____

  Continental-continental convergence: _____

  _____

  Feature: _____

30. Figure 10.5C represents a (convergent, divergent, transform) plate boundary. Circle your answer.
  a. Lithospheric material (is being created, is being destroyed, remains unchanged) along this type of boundary. Circle your answer.

  b. What type of faults are likely to parallel the direction of plate movement along this type of boundary? Write a brief description of the fault.

  _____ faults; _____

  _____

Igneous Rocks and Plate Boundaries.

The theory of plate tectonics provides a framework for explaining the occurrence of many igneous rocks and the features they compose. The formation of magma is most frequently associated with weaknesses along the boundaries between lithospheric plates.

Figure 10.6 illustrates the geologic events that are happening along some plate boundaries. Examine the figure closely. Then, use Figure 10.6 to complete questions 31–36.

31. On Figure 10.6, indicate the direction of movement of the plates on both sides of the mid-ocean ridge by drawing arrows on the ocean floor.

32. The ocean floor is composed of the igneous rock (granite, basalt). Circle your answer.

33. The eruption of basaltic magma from the part of the mantle called the *asthenosphere* is associated with what ocean features?

  _____ and _____

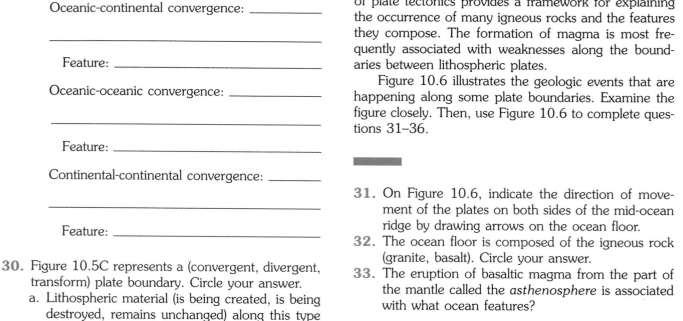

FIGURE 10.6 Ocean floor features and the occurrence of magma along plate boundaries.

34. Volcanic islands in the ocean, such as the Hawaiian Islands, form over (trenches, hot spots) in the asthenosphere and are made of the igneous rock (granite, basalt). Circle your answers.

35. (Trenches, Mid-ocean ridges) form on the ocean floor where oceanic plates are descending beneath continents.

36. As oceanic plates descend beneath the continents, the plates (remain solid, partially melt) and produce pockets of magma in the continents. These magma chambers are the sources of lava that form (ridges, volcanoes) on the continents. Circle your answers.

REVIEW

Having completed the exercise, you should know the following:

1. The plate tectonics theory proposes that the earth's lithosphere consists of large, rigid plates that are moving and interacting in various ways.

2. Mid-ocean ridges and deep-ocean trenches are two of the ocean floor features directly related to plate tectonics.

3. Most earthquakes are associated with plate boundaries.

4. The fit of continents and paleomagnetism of the ocean floor are two of the lines of evidence used in support of plate tectonics.

5. The polarity of the earth's magnetic field has reversed several times during the last four million years.

6. The minerals in molten, igneous material assume the magnetism of the earth as the rock solidifies at the mid-ocean ridge.

7. Using the evidence of magnetic polarity reversals present in the igneous rock of the ocean floor, scientists can calculate the rate of sea-floor spreading for the ocean basins.

8. The time when an ocean basin first began forming can be determined using the rate of sea-floor spreading.

9. The three types of plate boundaries are divergent, convergent, and transform.

10. Basaltic magma is erupted from the earth's asthenosphere along mid-ocean ridges and volcanic islands.

11. The partial melting and oceanic plates is the source of magma for volcanoes on the continents.

The Dynamic Ocean Floor

SUMMARY/REPORT PAGE

Date Due: _____

Name: _____

Date: _____

Class: _____

_____

_____

_____

After you have finished Exercise Ten, complete the following questions. You may have to refer to the exercise for assistance or to locate specific answers. Be prepared to submit this summary/report to your instructor at the designated time.

1. In the following space, sketch a general profile of an ocean floor between two continents illustrating a mid-ocean ridge and a deep-ocean trench. Label each of the features and draw arrows showing plate motion.

Ocean Floor Profile

2. Deep-focus earthquakes are associated with what ocean floor features?

_____

3. Shallow-focus earthquakes are associated with what ocean floor features?

_____

4. Earthquake foci can be used to mark the boundaries of what earth features?

_____

5. Describe how paleomagnetism is used to calculate the rate of sea-floor spreading.

6. From question 22 in the exercise what were your calculated rates of sea-floor spreading for the following ocean basins?

Pacific Ocean basin: _____ cm/yr

North Atlantic Ocean basin: _____ cm/yr

7. From questions 24 and 25 in the exercise, what were your calculated ages for the North and South Atlantic Ocean basins?

North Atlantic: _____ million years old

South Atlantic: _____ million years old

8. How well did your calculated ages for the North and South Atlantic Ocean basins compare to the ages given in the text?

9. In the following space, sketch a profile of a typical divergent plate boundary. Show the motion of the plates with arrows. Explain what will happen along the boundary between the plates.

Profile of a Divergent Plate Boundary

Explanation: _____

_____

10. Describe the origin of volcanoes on the ocean floor and along continental margins bordering ocean trenches.

Ocean floor volcanoes: _____

_____

Volcanoes along continental margins bordering ocean trenches: _____

_____

Waves, Currents, and Tides

Waves, currents, and tides are among the most observable phenomena of the sea; however, only recently has their study focused on scientific understanding rather than on practical knowledge. Investigating the causes, mechanics, and results of these ocean water movements will provide a greater understanding of some important systems that operate over seventy percent of the earth's surface—the world oceans (Figure 11.1).

OBJECTIVES

After you have completed this exercise, you should be able to

1. Explain how waves and currents are generated in the ocean.
2. Name the parts of a wave and describe the motion of water particles in a deep-water and shallow-water wave.
3. Use a formula to calculate wavelength, wave velocity, and wave period.
4. Explain why waves are refracted and what causes them to break and form surf.
5. Locate each of the major surface ocean currents.
6. List the names and characteristics of the principal deep-water masses.
7. Identify the features of erosion and deposition that occur along shorelines and explain how each is formed.
8. Explain the cause of tides and identify the different types of tides.

TEXTBOOK REFERENCE

Chapter 11

FIGURE 11.1   Cliff undercut by wave erosion along the Oregon coast. (Photo by E. J. Tarbuck)

MATERIALS

colored pencils
calculator

163

atlas or world wall map

TERMS

wave crest	surface current	estuary
wave trough	Coriolis effect	beach
wave height	density current	spit
wavelength	longshore current	tombolo
wave period	tidal current	baymouth bar
surf zone	emergent coast	diurnal tide
tsunami	wave-cut cliff	semidiurnal tide
refraction	platform	mixed tide
headland	submergent coast	

INTRODUCTION

The world's ocean waters are in constant motion via waves, currents, and tides. The immediate cause of each of these movements varies. However, the ultimate source of energy is the sun. In fact, the earth's oceans and atmosphere are often described as being huge "engines" powered by the sun's energy. Ocean water movements and global winds are the parts of the "engine" that transport excess heat from equatorial areas to the heat-deficient polar regions.

WAVES

Most waves are set in motion when friction with wind begins rotating water particles in circular orbits. If you were to watch a ball floating on the surface, you would notice that while the wave form moves forward, the ball, and hence the water, does not. On the surface, as wa-

ter particles reach the highest point in their circular orbits, a **wave crest** is formed, while particles at their lowest orbital points form **wave troughs** (Figure 11.2). **Wave height** is the vertical distance between the crest and trough of a wave.

Beneath the surface in deep water the circular orbits of water particles become progressively smaller with depth. At a depth equal to about one-half the **wavelength** (the horizontal distance separating two successive wave crests) the circular motion of water particles becomes negligible.

Answer questions 1–3 after you have reviewed Chapter 11 of your text.

1. From Figure 11.2 in this exercise, select the letter that identifies each of the following.

	Letter		**Letter**
wave crest	_____	wavelength	_____
wave trough	_____	depth of negligible water particle	
wave height	_____	motion	_____

2. Below what depth would a submarine have to submerge so that it would not be swayed by surface waves with a wavelength of 24 meters?

   Below _____ meters

Wave Mechanics. In deep water, where the depth is greater than one-half the wavelength, the velocity (V) of a wave depends upon the **wave period** (T) (the time

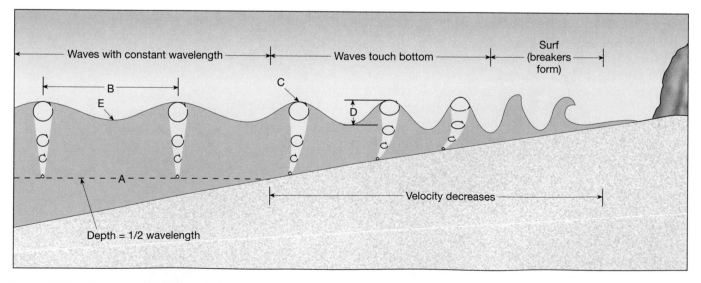

FIGURE 11.2 Deep and shallow water waves.

interval between successive wave crests, measured from a stationary point) and the wavelength (L). The mathematical equation that expresses the relation between these variables is velocity = wavelength divided by wave period ($V = L/T$).

As a wave approaches the shore and the depth of water becomes less than one-half the deep-water wavelength, the ocean bottom begins to interfere with the orbital motion of water particles, and the wave begins to "feel bottom." Interference between the bottom and water particle motion causes changes to occur in the wave. At a depth of water equal to about one-twentieth of the deep-water wavelength ($1/20L$ or $0.05L$), the top of the wave begins to fall forward and the wave breaks. In the **surf zone,** where waves are breaking and releasing energy, a significant amount of water is transported toward the shoreline.

3. According to the text, what are the three wind factors that determine the height, length, and period of waves?

   Factor 1: _____

   Factor 2: _____

   Factor 3: _____

   Refer to Figure 11.2 to answer questions 4–7.

4. The shape of the orbits of surface water particles in deep-water waves is (circular, elliptical). Near the shore in shallow water, the shapes become (circular, elliptical). Circle your answers.

5. In shallow water, water particles in the wave crest are (ahead of, behind) those at the bottom of the wave.

6. As waves approach the shore in shallow water, their heights (increase, decrease) and the wavelength becomes (longer, shorter.)

7. In the surf zone, water particles in the crest of a wave are (falling forward, standing still).

8. What would be the velocity of deep-water waves with a wavelength of 40 meters and a wave period of 6.3 seconds?

$$\text{Velocity} = \frac{\text{wavelength}(L)}{\text{wave period}(T)} = \frac{40 \text{ m}}{6.3 \text{ sec}} =$$

_____ m/sec

9. What would be the wavelength of deep-water waves that have a period of 8 seconds and a velocity of 2 meters/sec? (HINT: $V \times T = L$)

   Wavelength(L) = _____ meters

   a. What would be the *wave base* (depth below which water particle motion in the wave ceases) for the waves in question 9?

      Wave base = _____ meters

   b. The waves in question 9 will begin to break at a water depth of about (1, 3, 5) meters. Circle your answer.

10. What factor(s) determine the distance between where waves begin to break and the shoreline?

   _____

11. Imagine that you are standing on the shore considering walking out into the surf zone where the waves are beginning to break, but you cannot swim. You estimate the wavelength of the incoming deep-water waves to be 80 meters. Would it be safe to walk out to where the waves are breaking? Explain how you arrived at your answer.

   _____

   _____

12. Along some shorelines, why does the water simply rise and fall as the waves reach the shore rather than forming a surf zone?

   _____

13. What effect will *breakwaters* (walls of concrete or rock built offshore and parallel to the beach) have on waves?

   _____

Tsunamis (ocean waves produced by a submarine earthquake, sometimes mistakenly called "tidal waves") can have a wavelength of 125 miles and a wave period of 20 minutes.

14. If a tsunami had a wavelength of 125 miles and a period of 20 minutes, what would be its velocity?

   Velocity = _____ miles per hour

Wave Refraction. Waves that approach the shore at an angle are **refracted** (bent) because that part of the wave that feels bottom first is slowed down, while the remaining part of the wave continues to move forward. Refraction causes most waves to reach the shore approximately parallel to the shoreline.

Figure 11.3 illustrates a **headland** along a coastline with water depths shown by contour lines. Assume that waves, with a wavelength of 80 feet, are approaching the shoreline from the lower margin of the figure.

Use Figure 11.3 to answer questions 15–21.

15. In Figure 11.3, the approaching waves will begin to feel bottom and slow down at a water depth of about (10, 20, 30, 40) feet. Circle your answer.
16. At a water depth of approximately (4, 8, 12, 16) feet, the waves will begin to break.
17. On Figure 11.3, indicate where the waves will begin to break with a dashed line. Write the words "surf zone" along the line.

18. On Figure 11.3, beginning with the wave shown, sketch a succession of lines to illustrate the wave refraction that will take place as the waves approach shore.
19. Use arrows to indicate on Figure 11.3 where most of the wave energy will be concentrated as the waves are refracted and impact the shore.
20. Erosion by waves will be most severe (on the headland, in the bay). Circle your answer.
21. What effect will the concentrated energy from wave impact eventually have on the shape of the coastline?

_____

_____

CURRENTS

Moving masses of water on the surface or within the ocean are called *currents*. The primary generating force for surface currents is wind, whereas deep-ocean circulation is a response to density differences among water masses.

FIGURE 11.3 Coastline with depth of water contours and an approaching wave.

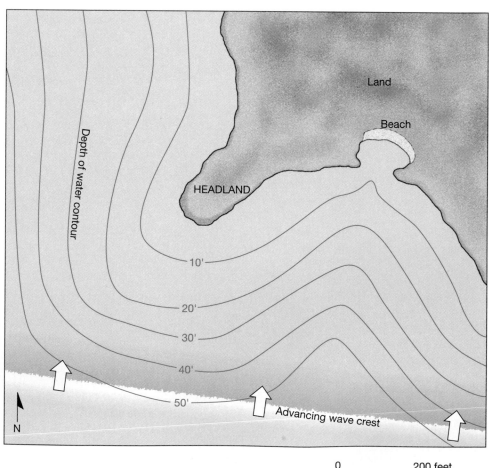

Surface Currents. **Surface currents** develop when friction between the moving atmosphere and the water causes the surface layer of the ocean to move as a single, large mass. Once set in motion, surface currents are influenced by the **Coriolis effect,** which deflects the path of the moving water to the right in the Northern Hemisphere and to the left in the Southern Hemisphere. *Warm currents* carry equatorial water toward the poles while *cold currents* move water from higher latitudes toward the equator.

22. Many surface ocean currents flow with great persistence. On the world map in Exercise Twenty-one, "Location and Distance on the Earth," Figure 21.5, draw arrows representing each of the following principal surface ocean currents. Use your text, an atlas or, if available, a large wall map that depicts surface currents as a reference. *Show warm currents with red arrows and cold currents with blue arrows.* To conserve space on the map, indicate the name of each current by writing the number that has been assigned to it.

Principal Surface Ocean Currents

1. Equatorial	7. Kuro Siwo
2. Gulf Stream	8. West Wind Drift
3. California	9. Labrador
4. Canaries	10. North Atlantic Drift
5. Brazilian	11. North Pacific Drift
6. Benguela	12. Peruvian

Using Figure 21.5, Exercise Twenty-one, or a world map of surface ocean currents, answer questions 23–26.

23. Which surface ocean current travels completely around the globe, west to east, without interruption?

_____

24. Which surface ocean current flows along the eastern coast of the United States? The current is a (warm, cold) current. Circle your answer.

_____

25. What is the name of the surface ocean current located along the western coast of the United States? The current is a (warm, cold) current.

_____

26. The general circulation of the surface currents in the North Atlantic Ocean is (clockwise, counterclockwise). In the South Atlantic, circulation is (clockwise, counterclockwise). Circle your answers.

Density Currents. **Density currents** result when water of greater density flows under or through water of a lower density. At any given depth, the density of water is influenced by its temperature and salinity—factors which you may have investigated in Exercise Nine, "Introduction to Oceanography."

Figure 11.4 is a cross section of the Atlantic Ocean illustrating the deep circulation. Use the figure and corresponding text information to answer questions 27 and 28.

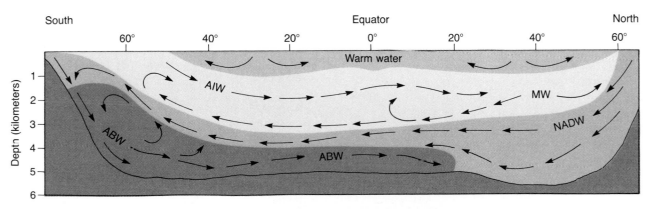

FIGURE 11.4 Cross section of the deep circulation of the Atlantic Ocean. (After Gerhard Neumann and Willard J. Pierson, Jr., *Principles of Physical Oceanography,* 1966. Reprinted by permission of Gerhard Neumann)

27. What are the temperature and/or salinity characteristics and general movements of each of the following water masses?

ABW (Antarctic Bottom Water): _____

_____

_____

NADW (North Atlantic Deep Water): _____

_____

_____

AIW (Antarctic Intermediate Water): _____

_____

_____

MW (Mediterranean Water): _____

_____

_____

28. What is the mechanism responsible for causing the very high density of Antarctic Bottom Water?

_____

_____

Currents Generated by Waves and Tides. Movements of water that result from waves and tides constitute a third class of currents. Whenever waves or tides push water against a shore, currents form that transport the water along the coast and return it seaward. Two of these currents are longshore currents and tidal currents.

Longshore currents form when waves strike the coast at an angle and the water moves in a "zig-zag" pattern parallel to the shore in the surf zone. These currents transport tremendous amounts of sediment which, when deposited, form many types of coastal features.

Tidal currents, which reverse their direction of flow with each tide, submerge and then expose low-lying coastal zones.

In Figure 11.3 you completed a diagram illustrating wave refraction. Answer questions 29–32 by referring to Figure 11.3.

29. On Figure 11.3, indicate with arrows the directions of the longshore currents.
30. What effect will the small bay have on the longshore current and its transportation of sediment?

_____

31. On Figure 11.3, write the word "deposition" where you would most likely find sediment being deposited by the longshore current.
32. Explain the cause of the sandy beach deposit at the head of the small bay.

_____

_____

SHORELINE FEATURES

The nature of shorelines varies considerably from place to place. One way that geologists classify coasts is based upon changes that have occurred with respect to sea level. This very general classification divides coasts into two types, emergent and submergent.

Emergent coasts have been raised above the sea as a result of rising land or falling sea level and are characterized by **wave-cut cliffs** or **platforms.**

Submergent coasts, resulting from a rising sea level or subsiding land, are often irregular due to the fact that many river mouths are flooded and become **estuaries.**

Nevertheless, whether along the rugged New England coast or the steep coastlines of California, the effects of wave erosion and sediment deposition by currents produce many similar features. Some of the more common depositional features include **beaches, spits, tombolos,** and **baymouth bars.**

Features of Emergent and Submergent Coasts. Figure 11.5 illustrates several erosional and depositional features of emergent and submergent coastlines. Using your text as a reference, complete questions 33–35.

33. Identify each of the following coastal features on Figure 11.5 by writing their name above the appropriate vertical line.

Features Caused by Erosion	Features Produced by Deposition	
wave-cut cliff	beach	baymouth bar
sea stack	spit	barrier island
	tombolo	

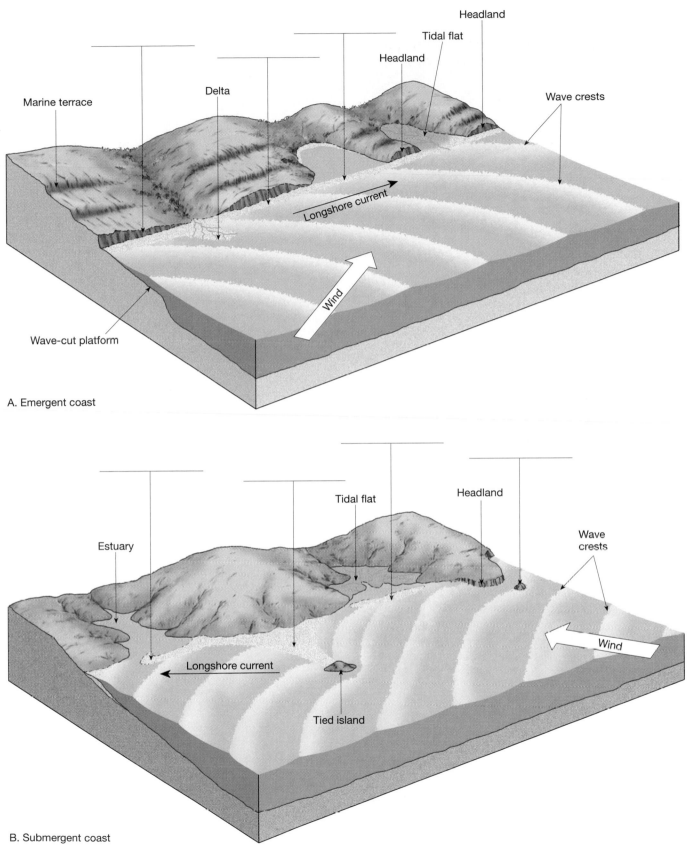

Marine terrace

Delta

Headland

Tidal flat

Headland

Wave crests

Longshore current

Wind

Wave-cut platform

A. Emergent coast

Estuary

Tidal flat

Headland

Wave crests

Longshore current

Wind

Tied island

B. Submergent coast

FIGURE 11.5 Hypothetical illustrations showing general features of (**A.**) emergent and (**B.**) submergent coastlines.

FIGURE 11.6 Portion of the Point Reyes, California, topographic map. (Map source: United States Department of the Interior, Geological Survey)

SCALE 1:62500
CONTOUR INTERVAL 80 FEET
DOTTED LINES REPRESENT 40-FOOT CONTOURS
DATUM IS MEAN SEA LEVEL

QUADRANGLE LOCATION

34. What is the purpose for constructing each of the following artificial features along a coast?

a *groin*: _____

a pair of *jetties*: _____

35. Draw and label a pair of jetties at the most appropriate location along the emergent coast illustrated in Figure 11.5A.

Identifying Coastal Features on a Topographic Map.

Figure 11.6 is a portion of the Point Reyes, CA, topographic map. Compare the map with the high-altitude image of the same area in Figure 11.7. Then use your text, Figure 11.6, Figure 11.7, and Figure 11.5 to answer questions 36–39.

36. The features along the shoreline of Drakes Bay suggest that the coast is (emergent, submergent). Circle your answer.

37. Drakes Estero and other bays shown on the map are (estuaries, headlands).

38. Point Reyes, a typical headland, is undergoing severe wave erosion. What type of feature is Chimney Rock and the other rocks located off the shore of Point Reyes? How have they formed?

_____

_____

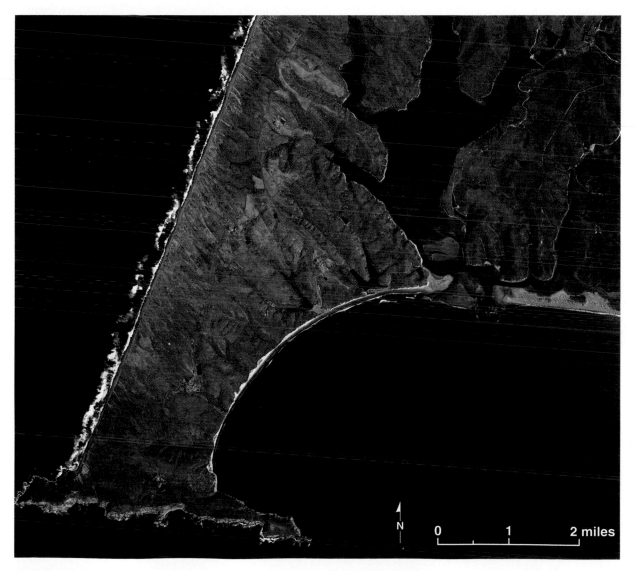

FIGURE 11.7   High-altitude image of Point Reyes area north of San Francisco, California. (Courtesy of USDA-ASCS)

39. Several depositional features in Drakes Bay are related to the movement of sediment by longshore currents. The feature labeled A on the map is one of these features, called a (spit, tombolo).

TIDES

Tides are the cyclical rise and fall of sea level caused by the gravitational attraction of the moon and, to a lesser extent, by the sun. Gravitational pull creates a bulge in the ocean on the side of the earth nearest the moon and on the opposite side of the earth from the moon. Tides develop as the rotating earth moves through these bulges causing periods of high and low water. Using tidal information from many sources, tides are classified into three types: **diurnal tides, semidiurnal tides,** and **mixed tides.**

40. Briefly describe the high and low water characteristics of each of the three types of tides. Use your text as a reference.

Diurnal tides: _____

_____

Semidiurnal tides: _____

_____

Mixed tides: _____

_____

Identifying Types of Tides. Tidal curves for the month of September at several locations are illustrated in Figure 11.8. Use the figure to answer questions 41–43.

41. Classify each of the tidal curves shown in Figure 11.8 as to the most appropriate type.

Diurnal tides occur at: _____

_____

Semidiurnal tides: _____

_____

Mixed tides: _____

_____

42. Of the locations you classified as having a *mixed* tide, (Port Adelaide, Seattle, Los Angeles) had the greatest inequality between successive low water heights on September 5. Circle your answer.

43. Write a general statement comparing the type of tide that occurs along the Pacific coast of the United States to the type found along the Atlantic coast.

_____

_____

Tidal Variations. In Figure 11.8, notice that during the month of September, at any given location, the heights of the tides were not constant. Two important factors that influence this variation are (1) the alignment of the sun, Earth, and moon, and (2) the distance between the earth and moon. Although these two controls are significant, they alone cannot be used to predict the height or time of actual tides at a particular place. Other factors, such as the shape of the coastline and the configuration of ocean basins, are also important. Consequently, tides at various locations respond differently to the tide-producing forces.

Use Figure 11.8 and your text to answer questions 44–49.

44. As shown in Figure 11.8, the *tidal range* (difference in height between high tide and the following low tide) at any one location (changes, remains the same) throughout September. Circle your answer.

The lunar phases for the month are shown at the top of Figure 11.8 (new moon on the 8th and full moon on the 23rd of the month).

45. What general relation seems to exist between the phases of the moon and the tidal ranges at New York?

_____

_____

FIGURE 11.8 Tidal curves for the month of September at various locations. (Source: U.S. Navy Hydrographic Office, *Oceanography*, U.S. Government Printing Office, 1966)

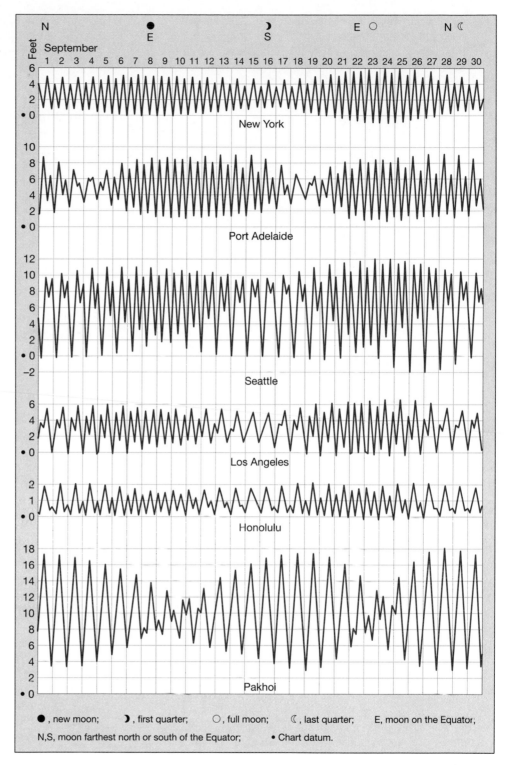

●, new moon; ☽, first quarter; ○, full moon; ☾, last quarter; E, moon on the Equator; N,S, moon farthest north or south of the Equator; ● Chart datum.

46. After referring to Chapter 11 of your text, explain the cause of *spring tides* and *neap tides*. Sketch a diagram showing the relative positions of the Earth, moon, and sun, viewed from above, that would cause each situation.

Spring tide

Spring tides: _____

_____

_____

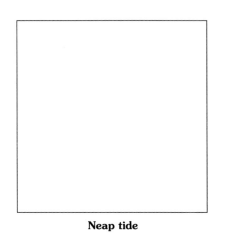

Neap tide

Neap tides: _____

_____

_____

47. On Figure 11.8, label the times of spring tide and times of neap tide for New York.

48. Do the tides at Pakhoi, China, have the same relations to the lunar phases as those that occur at New York? What are some other factors that may be influencing the tides at Pakhoi?

_____

_____

At some locations, tidal power is being considered as a means of generating electricity.

49. According to your text, what are two criteria that a bay must meet before its tidal energy can be economically harnessed?

Criterion 1: _____

Criterion 2: _____

REVIEW

Having completed the exercise, you should know the following:

1. Most surface waves are produced by friction from the wind causing water particles to move in circular orbits.
2. The circular motion of water particles in a deep-water wave becomes progressively smaller with depth and ceases at a depth equal to about one-half the wavelength.
3. Waves are often refracted as they approach a shoreline because, at a water depth equal to one-half the wavelength, the bottom interferes with the motion of water particles and parts of the wave are slowed down.
4. Waves begin to break and form a surf at a depth of water equal to about one-twentieth of the wave's deep-water wavelength.
5. The velocity of a deep-water wave is equal to the wavelength divided by the wave period.
6. Surface currents are the result of the wind moving large masses of water.
7. The general circulation of the major surface currents in the North Atlantic Ocean is clockwise.
8. Density differences in ocean water are responsible for the deep-ocean circulation.
9. Emergent and submergent coasts are recognized by their different shapes and features.
10. Longshore currents are responsible for forming depositional features such as tombolos and baymouth bars.
11. The gravitational attractions of the sun, Earth, and moon are the primary causes of tides, but other factors also cause local tidal variations.
12. Each of the three types of tides has characteristic periods of high and low water.
13. Spring tides have the largest tidal ranges whereas neap tides have the smallest tidal ranges.

EXERCISE ELEVEN
Waves, Currents, and Tides

SUMMARY/REPORT PAGE

Date Due: _____

Name: _____

Date: _____

Class: _____

After you have finished Exercise Eleven, complete the following questions. You may have to refer to the exercise for assistance or to locate specific answers. Be prepared to submit this summary/report to your instructor at the designated time.

1. Describe the changes in water particle motion, wavelength, and height that waves undergo as they move from deep water into shallow water.

_____

_____

2. What will happen to the shapes of waves as they approach a headland that is surrounded by shallow water?

_____

3. Describe the formation and appearance of each of the following features.

    Spit: _____

    _____

    Stack: _____

    _____

    Tombolo: _____

    _____

    Estuary: _____

    _____

4. Refer to Figure 11.6. What types of coastal features are Point Reyes, Drakes Estero, and Chimney Rock?

    Point Reyes: _____

    Drakes Estero: _____

    Chimney Rock: _____

5. The circulation of the surface currents in the South Atlantic Ocean is (clockwise, counterclockwise). Circle your answer.

6. What are the names of the surface currents that are located along the east and west coasts of the United States? Is each a warm or a cold current?

    East coast: _____

    West coast: _____

7. List the characteristics and describe the movement of the following deep-ocean water masses in the Atlantic Ocean.

    NADW: _____

    _____

    ABW: _____

    _____

8. Spring tides are most likely to occur during which lunar phase(s)?

    _____

9. Which of the tidal curves illustrated in Figure 11.8 exhibits the greatest tidal range?

_____

10. Refer to Figure 11.8. Of the three types of tides, name the type that occurs at each of the following locations.

   Pakhoi: _____

   Honolulu: _____

   New York: _____

EXERCISE TWELVE

Earth–Sun Relations

To life on this planet, the relations between the earth and the sun are perhaps the most important of all astronomical phenomena. The variations in solar energy striking the earth as it rotates and revolves around the sun cause the seasons and therefore are an appropriate starting point for studying weather and climate.

In this exercise you will investigate the reasons why the amount of solar radiation intercepted by the earth varies for different latitudes and changes throughout the year at a particular place. The next exercise, Exercise Thirteen, examines how the atmosphere is warmed by this radiation.

OBJECTIVES

After you have completed this exercise, you should be able to

1. Describe the effect that sun angle has on the amount of solar radiation a place receives.
2. Explain why the intensity and duration of solar radiation varies with latitude.
3. Explain why the intensity and duration of solar radiation varies at any one place throughout the year.
4. Describe the significance of these special parallels of latitude: Tropic of Cancer, Tropic of Capricorn, Arctic Circle, Antarctic Circle, and equator.
5. Diagram the relation between the earth and the sun on the dates of the solstices and equinoxes.
6. Determine the latitude where the overhead sun is located on any day of the year.
7. Calculate the noon sun angle for any place on the earth on any day.
8. Calculate the latitude of a place using the noon sun angle.

TEXTBOOK REFERENCE

Chapter 12

MATERIALS

metric ruler
colored pencils
protractor
calculator

Materials Supplied by Your Instructor
globe
large rubber band or string

TERMS

weather
weather element
weather control
solar intensity
solar duration
langley

calorie
solar constant
equator
Tropic of Cancer
Tropic of Capricorn
Arctic Circle

Antarctic Circle
solstice
equinox
analemma
noon sun angle

INTRODUCTION

Weather is the state of the atmosphere at a particular place for a short period of time. The condition of the atmosphere at any place and time is described by measuring the four basic **elements** of weather: temperature, moisture, air pressure, and wind. Of all the **controls** that are responsible for causing global variations in the weather elements, the amount of solar radiation received at any location is the most important.

SOLAR RADIATION AND THE SEASONS

The amount of solar energy (radiation) striking the outer edge of the atmosphere is not uniform over the face of the earth at any one time, nor is it constant throughout the year at any particular place. The amount of solar radiation received at any particular place on the earth, on any given day, is determined by the sun's **intensity** and **duration.** Intensity is the angle at which the rays of sunlight strike a surface, whereas duration refers to the length of daylight.

The standard unit of solar radiation is the **langley,** equal to one **calorie** per square centimeter. The **solar constant,** or average intensity of solar radiation falling on a surface perpendicular to the solar beam at the outer edge of the atmosphere, is about 2 langleys per minute. As the radiation passes through the atmosphere, it undergoes absorption, reflection, and scattering. Therefore, at any one location, less radiation reaches the surface of the earth than was originally intercepted at the upper atmosphere.

Solar Radiation and Latitude.

The amount of radiation striking a square meter at the outer edge of the atmosphere, and eventually the earth's surface, varies with latitude because of a changing sun angle. To illustrate this fact, answer questions 1–11 using the appropriate figure.

1. On Figure 12.1, extend the 1 cm wide beam of sunlight from the sun vertically to point A on the surface. Extend the second 1 cm wide beam, beginning at the sun, to the surface at point B.

Notice in Figure 12.1 that the sun is directly overhead (vertical) at point A and the beam of sunlight strikes the surface at a 90° angle above the horizon.

Using Figure 12.1, answer questions 2–5.

2. Using a protractor, measure the angle between the

surface and the beam of sunlight coming from the sun to point B.

_____ ° = angle of the sun above the surface (horizon) at point B.

3. What are the lengths of the line segments on the surface covered by the sun beam at point A and point B?

Point A: _____ mm   point B: _____ mm

4. Of the two beams, beam (A, B) is more spread out at the surface and covers a larger area. Circle your answer.

5. More langleys per minute would be received by a square centimeter on the surface at point (A, B). Circle your answer.

Use Figure 12.2 to answer questions 6–11 concerning the total amount of solar radiation intercepted by each 30° segment of latitude of the earth.

6. With a metric ruler, measure the total width of incoming rays from point x to point y in Figure 12.2.

The total width is _____ centimeters

(_____ millimeters). Fill in your answers.

7. Assume the total width of the incoming rays from point x to point y equals 100% of the solar radiation that is intercepted by the earth. Each centimeter would equal _____ % and each millimeter would equal _____ %. Fill in your answers.

8. What percent of the total incoming radiation is concentrated in the zone between:

FIGURE 12.1   Vertical and oblique sun beams.

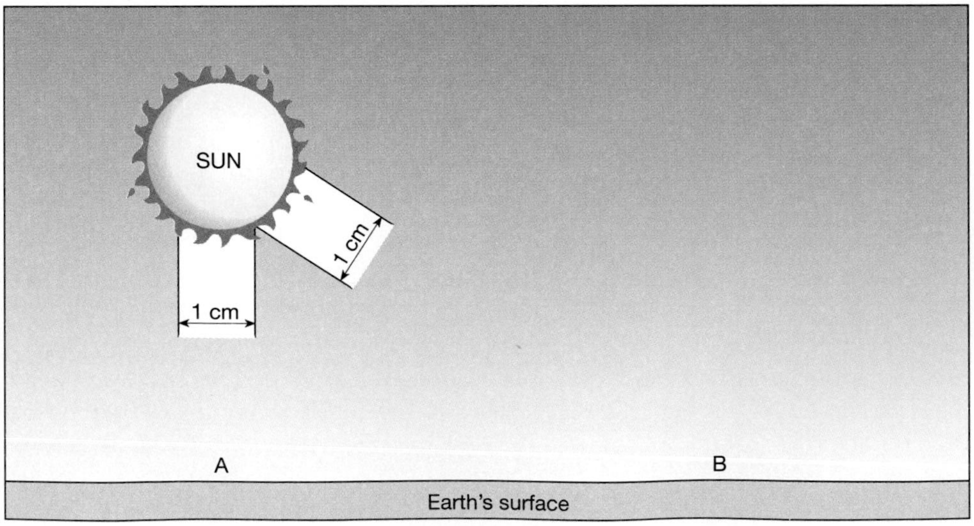

FIGURE 12.2 Distribution of solar radiation per 30° segment of latitude on the earth.

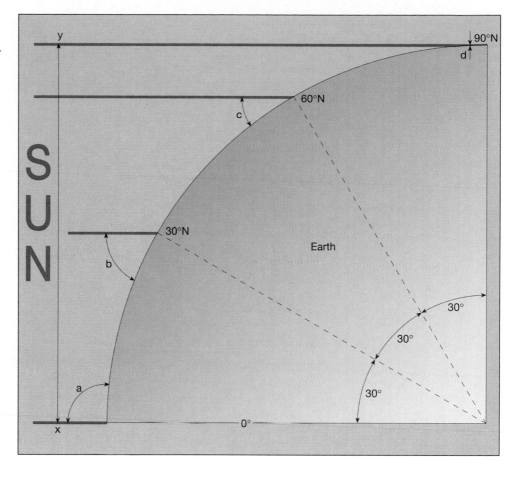

0°–30° = _____ mm = _____ %

30°–60° = _____ mm = _____ %

60°–90° = _____ mm = _____ %

9. Use a protractor to measure the angle between the surface and sun ray at each of the following locations. Angle b is already done as an example.

   Angle a: _____ °   angle c: _____ °

   Angle b: ___60°___   angle d: _____ °

10. What is the general relation between the amount of radiation received in each 30° segment and the angle of the sun's rays?

   _____

   _____

11. Explain in your own words what fact about the earth creates the unequal distribution of solar energy, even though each zone represents an equal 30° segment of latitude.

_____

_____

_____

Yearly Variation in Solar Energy. The amount of solar radiation received at a particular place would remain constant throughout the year if it were not for these facts:

   a. The earth rotates on its axis and revolves around the sun.

   b. The axis of the earth is inclined 23½° from the perpendicular to the plane of its orbit.

   c. Throughout the year, the axis of the earth points to the same place in the sky, which causes the overhead (vertical or 90°) noon sun to cross over the **equator** twice as it migrates from the **Tropic of Cancer** (23½°N latitude) to the **Tropic of Capricorn** (23½°S latitude) and back to the Tropic of Cancer.

As a consequence, the position of the vertical or overhead noon sun shifts between the hemispheres, causing variations in the intensity of solar radiation and changes in the length of daylight and darkness. *The seasons are the result of this changing intensity and duration of solar energy and subsequent heating of the atmosphere.*

To help understand how the intensity and duration of solar radiation varies throughout the year, answer questions 12–31 after you have thoroughly reviewed Chapter 12 of the text and have examined the location of the Tropic of Cancer, Tropic of Capricorn, **Arctic Circle,** and **Antarctic Circle** on a globe or world map.

12. List some of the countries each of the following passes through.

Tropic of Cancer: _____

Tropic of Capricorn: _____

Arctic Circle: _____

13. Write the date represented by each position of the earth at the appropriate place in Figure 12.3. Then label the following on each position.

North Pole and South Pole
Axis of the earth
Equator, Tropic of Cancer, Tropic of Capricorn
Arctic Circle and Antarctic Circle
Circle of illumination (day–night line)

Questions 14–19 refer to the June **solstice** position of the earth in Figure 12.3.

14. What term is used to describe the June 21–22 date in each hemisphere?

Northern Hemisphere: _____ solstice

Southern Hemisphere: _____ solstice

15. On June 21–22 the sun's rays are perpendicular to the earth's surface at noon at the (Tropic of Cancer, equator, Tropic of Capricorn). Circle your answer.

16. What latitude is receiving the most intense solar energy on June 21–22?

Latitude: _____

17. Toward what direction, north or south, would you look to see the sun at noon on June 21–22 if you lived at the following latitudes?

40°N latitude: _____

10°N latitude: _____

18. Position a rubber band or string on a globe corresponding to the circle of illumination on June 21–22. Then determine the length of daylight at

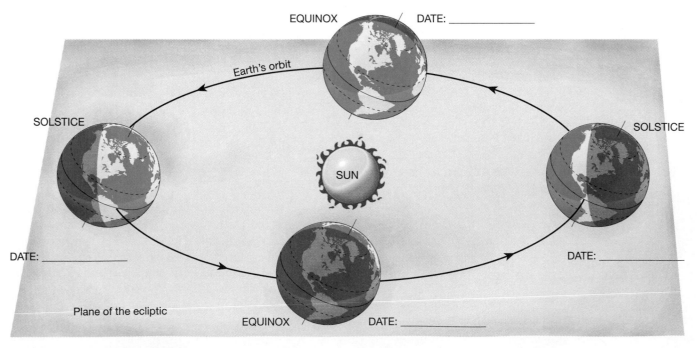

FIGURE 12.3 Earth–sun relations.

the following latitudes by examining the proportionate number of degrees of longitude a place located at each latitude spends in daylight as the earth rotates. (*Note:* The earth rotates a total of 360° of longitude per day. Therefore, each 15° of longitude is equivalent to one hour.)

70°N latitude: _____ hrs

40°S latitude: _____ hrs

40°N latitude: _____ hrs

90°S latitude: _____ hrs

0° latitude: _____ hrs

19. On June 21–22, latitudes north of the Arctic Circle are receiving (6, 12, 24) hours of daylight, while latitudes south of the Antarctic Circle are experiencing (6, 12, 24) hours of darkness. Circle your answers.

Questions 20–24 refer to the December solstice position of the earth in Figure 12.3.

20. What name is used to describe the December 21–22 date in each hemisphere?

Northern Hemisphere: _____ solstice

Southern Hemisphere: _____ solstice

21. On December 21–22 the sun's rays are perpendicular to the earth's surface at noon on the (Tropic of Cancer, equator, Tropic of Capricorn). Circle your answer.

22. On December 21–22 the (Northern, Southern) Hemisphere is receiving the most intense solar energy.

23. If you lived at the equator, on December 21–22 you would look (north, south) to see the sun at noon.

24. Refer to Table 12.1, "Length of daylight." How many hours of daylight would there be at each of the following latitudes on December 21–22?

90°N latitude: _____ hrs

40°S latitude: _____ hrs

40°N latitude: _____ hrs

90°S latitude: _____ hrs

0° latitude: _____ hrs

TABLE 12.1 Length of daylight

Latitude (degrees)	Summer Solstice	Winter Solstice	Equinoxes
0	12 h	12 h	12 h
10	12 h 35 min	11 h 25 min	12
20	13 12	10 48	12
30	13 56	10 04	12
40	14 52	9 08	12
50	16 18	7 42	12
60	18 27	5 33	12
70	2 mo	0 00	12
80	4 mo	0 00	12
90	6 mo	0 00	12

Questions 25–31 refer to the March and September **equinox** positions of the earth in Figure 12.3.

25. For those living in the Northern Hemisphere, what terms are used to describe the following dates?

March 21: _____ equinox

September 22: _____ equinox

26. For those living in the Southern Hemisphere, what terms are used to describe the following dates?

March 21: _____ equinox

September 22: _____ equinox

27. On March 21 and September 22 the sun's rays are perpendicular to the earth's surface at noon at the (Tropic of Cancer, equator, Tropic of Capricorn). Circle your answer.

28. What latitude is receiving the most intense solar energy on March 21 and September 22?

Latitude: _____

29. If you lived at 20°S latitude, you would look (north, south) to see the sun at noon on March 21 and September 22. Circle your answer.

30. What is the relation between the North and South poles and the circle of illumination on March 21 and September 22?

_____

31. Write a brief statement describing the length of daylight everywhere on the earth on March 21 and September 22.

_____

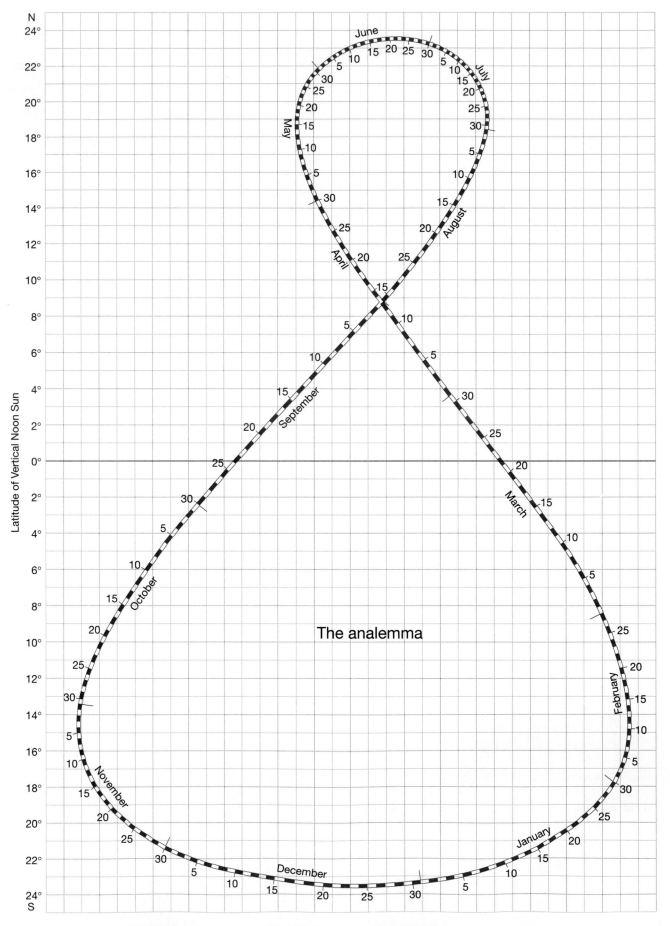

FIGURE 12.4 The analemma, a graph illustrating the latitude of the overhead (vertical) noon sun throughout the year.

As you have seen, the latitude where the noon sun is directly overhead (vertical, or 90° above the horizon) is easily determined for the solstices and equinoxes.

Figure 12.4 is a graph, called an **analemma,** that can be used to determine the latitude where the overhead noon sun is located for any date. To determine the latitude of the overhead noon sun from the analemma, find the desired date on the graph and read the coinciding latitude along the left axis. Don't forget to indicate North or South when writing latitude.

32. Using a colored pencil, draw lines on Figure 12.4 that correspond to the equator, Tropic of Cancer, and Tropic of Capricorn. Label each of these special parallels of latitude on the figure.

33. Using the analemma, Figure 12.4, determine where the sun is overhead at noon on the following dates.

    December 10: _____

    March 21: _____

    May 5: _____

    June 22: _____

    August 10: _____

    October 15: _____

34. The position of the overhead noon sun is always located on or between which two parallels of latitude?

    _____ °N (named the Tropic of _____)

    and _____°S (named the Tropic of _____)

35. The overhead noon sun is located at the equator

    on September_____ and March_____ . To-

    gether, these two days are called the _____.
    Fill in your answers.

36. Write a brief paragraph summarizing the yearly movement of the overhead noon sun and how the intensity and duration of solar radiation varies over the earth's surface throughout the year.

_____

_____

Calculating Noon Sun Angle. Knowing where the noon sun is overhead on any given date (the analemma), you can determine the angle above the horizon of the noon sun at any other latitude on that same day. The relation between latitude and **noon sun angle** is

> For each degree of latitude the place is away from the latitude where the noon sun is overhead, the angle of the noon sun becomes one degree *lower* from being vertical or 90° above the horizon.

For example, a place which is 30° of latitude away from the latitude where the overhead noon sun is located would have a noon sun angle of 60° (90° − 30° = 60°). A place 60° of latitude away would have a 30° noon sun angle (see Figure 12.2).

37. Complete Table 12.2 by calculating the noon sun angle for each of the indicated latitudes on the dates given. Some of the calculations have already been done.

38. From Table 12.2, the highest average noon sun angle occurs at (40°N, 0°, 20°S). Circle your answer.

39. Calculate the noon sun angle for your latitude on today's date.

    Date: _____

    Latitude of overhead noon sun: _____

    Your latitude: _____

    Your noon sun angle: _____

TABLE 12.2 Noon sun angle calculations

Latitude of overhead = noon sun	Mar 21 (__)	Apr 11 (__)	Jun 21 (__)	Dec 22 (__)
	Noon Sun Angle			
90°N				
40°N	50°	___	___	$26\frac{1}{2}°$
0°	___	___	$66\frac{1}{2}°$	___
20°S	___	62°	___	___

40. Calculate the maximum and minimum noon sun angles for your latitude.

Maximum Noon Sun Angle	**Minimum Noon Sun Angle**
Date: _____	Date: _____
Angle: _____°	Angle: _____°

41. Calculate the average noon sun angle (maximum plus minimum, divided by 2) and the range of the noon sun angle (maximum minus minimum) for your location.

Average noon sun angle = _____°

Range of the noon sun angle = _____°

42. Describe some situations in which knowing the noon sun angle might be useful.

_____

_____

Using Noon Sun Angle. One very practical use of noon sun angle is in navigation. You, like a navigator, can determine your latitude if the date and angle of the noon sun at your location are known. As you answer questions 43 and 44, keep in mind the relation between latitude and noon sun angle.

43. What is your latitude if, on March 21, you observe the noon sun to the north at 18° above the horizon?

Latitude: _____

44. What is your latitude if, on October 16, you observe the noon sun to the south at 39° above the horizon?

Latitude: _____

Solar Radiation at the Outer Edge of the Atmosphere. Table 12.3 shows the average daily radiation received at the outer edge of the atmosphere at select latitudes for different months.

To help visualize the pattern, plot the data from Table 12.3 on the graph in Figure 12.5. Using a different color for each latitude, draw lines through the monthly values to obtain yearly curves. Then answer questions 45–48.

TABLE 12.3 Solar radiation at the outer edge of the atmosphere (langleys/day) at various latitudes during select months

Latitude	March	June	September	December
90°N	50	1050	50	0
40°N	700	950	720	325
0°	890	780	880	840

FIGURE 12.5 Graph of solar radiation received at the outer edge of the atmosphere.

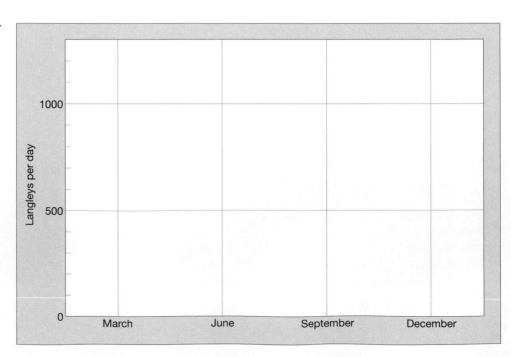

45. Why do two periods of maximum solar radiation occur at the equator?

_____

46. In June, why does the outer edge of the atmosphere at the equator receive less solar radiation than both the North Pole and 40°N latitude?

_____

_____

47. Why does the outer edge of the atmosphere at the North Pole receive no solar radiation in December?

_____

48. What would be the approximate monthly values for solar radiation at the outer edge of the atmosphere at 40°S latitude? Explain how you arrived at the values.

March: _____

June: _____

September: _____

December: _____

Explanation: _____

_____

REVIEW

Having completed the exercise, you should know the following:

1. The curvature of the earth causes the intensity of solar radiation to become less at locations farther from the overhead noon sun.

2. Solar radiation received by the earth is most intense in the zone from 0° to 30° latitude.

3. As the earth rotates and revolves around the sun, the inclination and orientation of the earth's axis causes the intensity and duration of solar radiation received at any place to vary throughout the year.

4. The overhead noon sun is located at $23\frac{1}{2}$°N latitude, the Tropic of Cancer, on June 21–22. This date is the time of the summer solstice in the Northern Hemisphere and winter solstice in the Southern Hemisphere.

5. The overhead noon sun is located at $23\frac{1}{2}$°S latitude, the Tropic of Capricorn, on December 21–22. This date is the time of the winter solstice in the Northern Hemisphere and summer solstice in the Southern Hemisphere.

6. Locations poleward of the Arctic Circle experience twenty-four hours of daylight on the date of the Northern Hemisphere summer solstice, June 21–22.

7. The equator has twelve hours of daylight each day of the year. However, the duration of daylight varies throughout the year everywhere else on the earth because of the orientation of the earth's axis relative to the sun.

8. Knowing the date and latitude of a place, you can calculate the noon sun angle for the location.

9. Knowing the date and noon sun angle, you can calculate the latitude of a place.

10. Two peaks of maximum solar radiation occur at the equator on the days of the equinoxes because the overhead noon sun is located there on each of the days.

11. The North Pole receives more solar radiation than the equator at the outer edge of the atmosphere in June because it is experiencing 24 hours of daylight.

Earth–Sun Relations

SUMMARY/REPORT PAGE

Date Due: _____

Name: _____

Date: _____

Class: _____

After you have finished Exercise Twelve, complete the following questions. You may have to refer to the exercise for assistance or to locate specific answers. Be prepared to submit this summary/report to your instructor at the designated time.

1. From Figure 12.2, what was the calculated percent of solar radiation that is intercepted by each of the following 30° segments of latitude?

    0°–30° _____ %

    30°–60° _____ %

    60°–90° _____ %

2. How many hours of daylight occur at the following locations on the specified dates?

	March 22	December 22
40° N	_____ hrs	_____ hrs
0°	_____ hrs	_____ hrs
90° S	_____ hrs	_____ hrs

3. What is the noon sun angle at these latitudes on April 11?

    40°N _____ °   0° _____ °

4. What is the relation between the angle of the noon sun and the quantity of solar radiation received per square centimeter at the outer edge of the atmosphere?

    _____

    _____

5. Complete Diagram 12.1 showing the earth's relation to the sun on June 22. On the earth, accurately draw and label the following:

    Axis   Tropic of Cancer   Antarctic Circle
    Equator   Tropic of Capricorn Arctic Circle
    Circle of illumination
    Location of the overhead noon sun

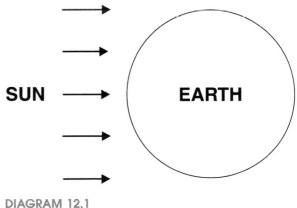

DIAGRAM 12.1

6. What causes the intensity and duration of solar radiation received at any place to vary throughout the year?

_____

_____

7. What are the maximum and minimum noon sun angles at your latitude?

Maximum noon sun angle = _____° on

_____ (date)

Minimum noon sun angle = _____° on

_____ (date)

8. What are the maximum and minimum durations of daylight at your latitude?

Maximum duration of daylight = _____ hrs

Minimum duration of daylight = _____ hrs

9. Write a brief statement describing how the intensity and duration of solar radiation change at your location throughout the year.

_____

_____

_____

10. The day is March 22. You view the noon sun to the south at 35° above the horizon. What is your latitude?

Latitude: _____

EXERCISE THIRTEEN

Atmospheric Heating

The quantity of radiation from the sun that strikes the outer edge of the earth's atmosphere at any one place is not constant but varies with the seasons. This exercise examines, step-by-step, what happens to **solar radiation** as it passes through the atmosphere, is absorbed at the earth's surface, and is reradiated by land and water back to the atmosphere. Investigating the journey of solar radiation and how it is influenced and modified by air, land, and water will provide a better understanding of one of the most basic weather elements, atmospheric temperature.

OBJECTIVES

After you have completed this exercise, you should be able to

1. Explain how the earth's atmosphere is heated.
2. Describe the effect that the atmosphere has on absorbing, scattering, and reflecting incoming solar radiation.
3. List the gases in the atmosphere that are responsible for absorbing long-wave radiation.
4. Explain how the heating of a surface is related to its albedo.
5. Discuss the differences in the heating and cooling of land and water.
6. Summarize the global pattern of surface temperatures for January and July.
7. Describe how the temperature of the atmosphere changes with increasing altitude.
8. List the cause of a surface temperature inversion and the effect it has on atmospheric pollution.
9. Determine the effect that wind speed has on the windchill equivalent temperature.

TEXTBOOK REFERENCE

Chapters 12 and 16

MATERIALS

calculator
colored pencils

Materials Supplied by Your Instructor

light source
black and silver
  containers

two thermometers
wood splint
beaker of sand
beaker of water

TERMS

solar radiation
greenhouse effect
terrestrial
  radiation

albedo
isotherm
environmental
  lapse rate

temperature inversion
windchill equivalent
  temperature

INTRODUCTION

Temperature is an important element of weather and climate because it greatly influences air pressure, wind, and the amount of moisture in the air. The unequal heating that takes place over the surface of the earth is what sets the atmosphere in motion, and the movement of air is what brings changes in our weather.

    The single greatest cause for temperature variations over the surface of the earth is differences in the reception of solar radiation. Secondary factors such as the differential heating of land and water, ocean currents, and altitude can modify local temperatures.

    The amount of solar energy (radiation) striking the earth is not constant throughout the year at any particular place, nor is it uniform over the face of the earth at any one time. However, the total amount of radiation that the earth intercepts from the sun equals the total radiation that it loses back to space. It is this balance between incoming and outgoing radiation that keeps the earth from becoming continuously hotter or colder.

SOLAR RADIATION AT THE OUTER EDGE OF THE ATMOSPHERE

The two factors that control the amount of solar radiation that a square meter receives at the outer edge of the atmosphere, and eventually the earth's surface, are the sun's *intensity* and its *duration*. These variables were examined in detail in Exercise 12, "Earth–Sun Relations." Answer questions 1–3 after you have reviewed Exercise 12 and/or Chapter 12 of your text.

<hr>

1. Briefly define solar intensity and duration.

   Intensity of solar radiation: _____

   _____

   Duration of solar radiation: _____

   _____

2. Complete Table 13.1 by calculating the angle that the noon sun would strike the outer edge of the atmosphere at each of the indicated latitudes on the specified date. How many hours of daylight would each place experience on these dates? (*Hint:* You may find Tables 12.1 and 12.2 helpful.)

3. Explain the reason why the intensity and duration of solar radiation received at the outer edge of the atmosphere is not constant at any particular latitude throughout the year.

   _____

   _____

ATMOSPHERIC HEATING

Atmospheric heating is a function of (1) the ability of atmospheric gases to absorb radiation, (2) the amount of solar radiation that reaches the earth's surface, and (3) the nature of the surface material. Of the three, selective absorption of radiation by the atmosphere provides an insight into the mechanism of atmospheric heating. The quantity of radiation that reaches the earth's surface and the ability of the surface to absorb and reradiate the radiation determine the extent of atmospheric heating.

The atmosphere is rather selective and efficiently absorbs long-wave radiation that we detect as heat while allowing the transmission of most of the short wavelengths—a process called the **greenhouse effect.** The short-wave radiation that reaches the earth's surface and is absorbed ultimately returns to the atmosphere in the form of long-wave, **terrestrial radiation.** As the radiation travels up from the earth's surface through the atmosphere, it is absorbed by atmospheric gases, heating the atmosphere from below. Since terrestrial radiation supplies most of the long-wave radiation to the atmosphere, it is the primary source of heat. The fact that temperature typically decreases with an increase in altitude in the lower atmosphere is clear evidence supporting this mechanism of atmospheric heating.

Solar Radiation Received at the Earth's Surface.
As solar radiation travels through the earth's atmosphere, it may be reflected, scattered, or absorbed. The effect of the atmosphere on incoming solar radiation and the amount of radiation that ultimately reaches the surface is primarily dependent upon the angle at which the solar beam passes through the atmosphere and strikes the earth's surface.

Figure 13.1 illustrates the atmospheric effects on incoming solar radiation for an average noon sun angle. Answer questions 4–7 by examining the figure and supplying the correct response.

<hr>

4. In Figure 13.1, _____ percent of the incoming solar radiation is reflected and scattered back to space.

TABLE 13.1 Noon sun angle and length of day

	March 21		June 21	
	Noon Sun Angle	Length of Day	Noon Sun Angle	Length of Day
40°N:	_____ °	_____ hrs	_____ °	_____ hrs
0°:	_____ °	_____ hrs	_____ °	_____ hrs
40°S:	_____ °	_____ hrs	_____ °	_____ hrs

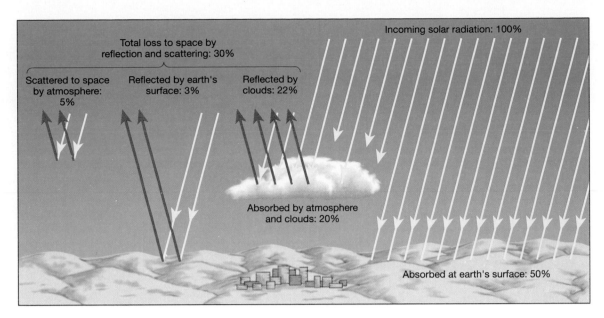

FIGURE 13.1 Solar radiation budget of the atmosphere and earth.

5. _____ percent of the incoming solar radiation is absorbed by gases in the atmosphere and clouds.

6. _____ percent of the incoming solar radiation is absorbed at the earth's surface.

7. (Two and a half, Four) times as much incoming radiation is absorbed by the earth's surface than by the atmosphere and clouds. Circle your answer.

Using Chapter 12 of your text as a reference, answer questions 8–11 by circling the correct response.

8. The short-wave solar radiation that passes through the atmosphere and is absorbed at the earth's surface is primarily in the form of (ultraviolet, visible, infrared) wavelengths.

9. When the earth's surface releases the solar radiation it has absorbed, the long-wave terrestrial radiation is primarily (ultraviolet, visible, infrared) wavelengths.

10. (Ultraviolet, Visible, Infrared) wavelengths of radiation are absorbed most efficiently by the lower atmosphere.

11. (Nitrogen, Carbon dioxide) and (water vapor, argon) are the two principal gases that absorb most of the terrestrial radiation in the lower atmosphere.

Assume Figure 13.1 represents the atmospheric effects on incoming solar radiation for an average noon sun angle of about 50°. Answer questions 12–15 concerning other noon sun angles by circling the appropriate responses.

12. If the noon sun angle is 90°, solar radiation would have to penetrate a (greater, lesser) thickness of atmosphere than with an average noon sun angle.

13. The result of a 90° noon sun angle would be that (more, less) incoming radiation would be reflected, scattered, and absorbed by the atmosphere and (more, less) radiation would be absorbed and reradiated by the earth's surface to heat the atmosphere.

14. If the noon sun angle is 20°, solar radiation would have to penetrate a (greater, lesser) thickness of atmosphere than with an average noon sun angle.

15. The result of a 20° noon sun angle would be that (more, less) incoming radiation would be reflected, scattered, and absorbed by the atmosphere and (more, less) radiation would be absorbed and reradiated by the earth's surface to heat the atmosphere.

16. How is the angle (intensity) at which the solar beam strikes the earth's surface related to the quantity of solar radiation received by each square meter?

_____

_____

17. How is the length of daylight related to the quantity of solar radiation received by each square meter at the earth's surface?

_____

_____

18. Write a brief statement summarizing the mechanism responsible for heating the atmosphere.

_____

_____

_____

The Nature of the Earth's Surface. The various materials that comprise the earth's surface play an important role in determining atmospheric heating. Two significant factors are the **albedo** of the surface and the different abilities of land and water to absorb and reradiate radiation.

Albedo is the reflectivity of a substance, usually expressed as the percentage of radiation that is reflected from the surface. Since surfaces with high albedos are not efficient absorbers of radiation, they cannot return much long-wave radiation to the atmosphere for heating. Most light-colored surfaces have high albedos. Light-colored surfaces (and the air above them) are typically cooler than dark surfaces.

Albedo Experiment. To better understand the effect of color on albedo, observe the equipment in the laboratory (Figure 13.2) and then conduct the following experiment by completing each of the indicated steps.

Step 1. Place the black and silver containers (with lids and thermometers) about six inches away from the light source. Make certain that both containers are of equal distance from the light and are not touching one another.
Step 2. Record the starting temperature of both containers on the albedo experiment data table, Table 13.2.
Step 3. Turn on the light and record the temperature of both containers on the data table at about 30-second intervals for 5 minutes.
Step 4. Plot the temperatures from the data table on the albedo experiment graph, Figure 13.3. Use a different color line to connect the points for each container.

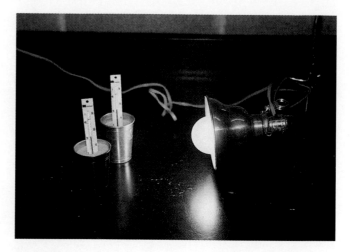

FIGURE 13.2 Albedo experiment lab equipment.

19. Write a statement that summarizes the results of your albedo experiment.

_____

_____

_____

20. What are some earth surfaces that have high albedos and some that have low albedos?

High albedos: _____

Low albedos: _____

21. Given equal amounts of radiation reaching the surface, the air over a snow-covered surface will be (warmer, colder) than air above a dark-colored, barren field. Circle your answer. Then explain your choice fully in terms of what you have learned about albedo.

_____

_____

TABLE 13.2 Albedo experiment data table

	Starting Temperature	30 sec	1 min	1.5 min	2 min	2.5 min	3 min	3.5 min	4 min	4.5 min	5 min
Black Container											
Silver Container											

FIGURE 13.3 Albedo experiment graph.

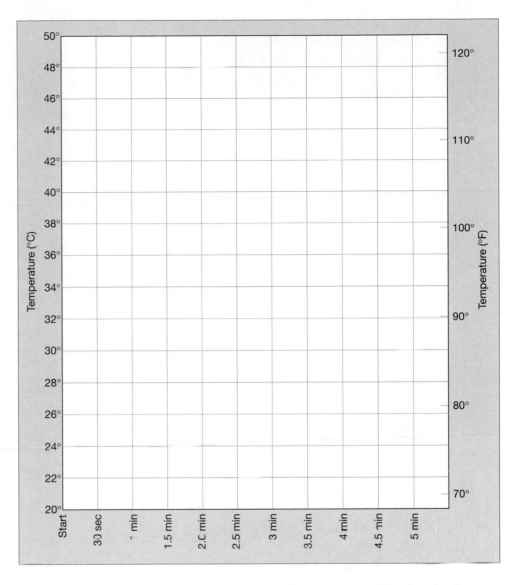

22. Why is it wise to wear light-colored clothes on a sunny, hot day?

_____

23. If you lived in an area with long cold winters, a (light-, dark-) colored roof would be the best choice. Circle your answer.

Land and water influence the air temperatures above them in different manners because they do not absorb and reradiate energy equally.

Land and Water Heating Experiment. Investigate the differential heating of land and water by observing the equipment in the laboratory (Figure 13.4) and conducting the following experiment by completing each of the indicated steps.

Step 1. Hang a light from a stand so it is about 5 inches above the top of the two beakers—one containing dry sand, the other water.

Step 2. Using a wood splint, suspend a thermometer in each beaker so that the bulbs are just below the surfaces of the sand and water.

Step 3. Record the starting temperatures for both the dry sand and water on the land and water heating data table, Table 13.3.

Step 4. Turn on the light and record the temperature on the data table at about 30-second intervals for 5 minutes.

Step 5. Turn off the light for several minutes. Dampen the sand with water and record the starting temperature of the damp sand on the data table. Turn on the light and record the temperature of the damp sand on the data table at about 30-second intervals for 5 minutes.

Step 6. Plot the temperatures for the water, dry sand, and damp sand from the data table on the

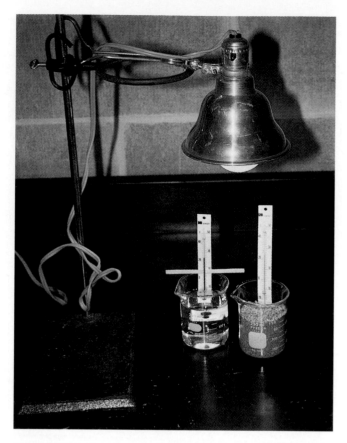

FIGURE 13.4 Land and water heating experiment lab equipment.

b. How do the abilities to change temperature differ for dry sand and damp sand when they are exposed to equal quantities of radiation?

_____

_____

25. Review Chapter 12 of your text and then list the four reasons for the differential heating of land and water.

Reason 1: _____

Reason 2: _____

Reason 3: _____

Reason 4: _____

Figure 13.6 presents the annual temperature curves for two cities, A and B, that are located in North America at approximately 37°N latitude. On any date both cities receive the same intensity and duration of solar radiation. One city is in the center of the continent, while the other is on the west coast. Use Figure 13.6 to answer questions 26–33.

land and water heating graph, Figure 13.5. Use a different color line to connect the points for each material.

24. Questions 24a and 24b refer to the land and water heating experiment.
   a. How do the abilities to change temperature differ for dry sand and water when they are exposed to equal quantities of radiation?

   _____

   _____

26. In Figure 13.6, city (A, B) has the highest monthly temperature. Circle your answer.
27. City (A, B) has the lowest monthly temperature.
28. The greatest _annual temperature range_ (difference between highest and lowest monthly temperatures) occurs at city (A, B).
29. City (A, B) reaches its maximum monthly temperature at an earlier date.
30. City (A, B) maintains a more uniform temperature throughout the year.
31. Of the two cities, city A is most likely located (along a coast, in the center of a continent).
32. The most likely location for city B is (coastal, midcontinent).

TABLE 13.3 Land and water heating data table

	Starting Temperature	30 sec	1 min	1.5 min	2 min	2.5 min	3 min	3.5 min	4 min	4.5 min	5 min
Water											
Dry sand											
Damp sand											

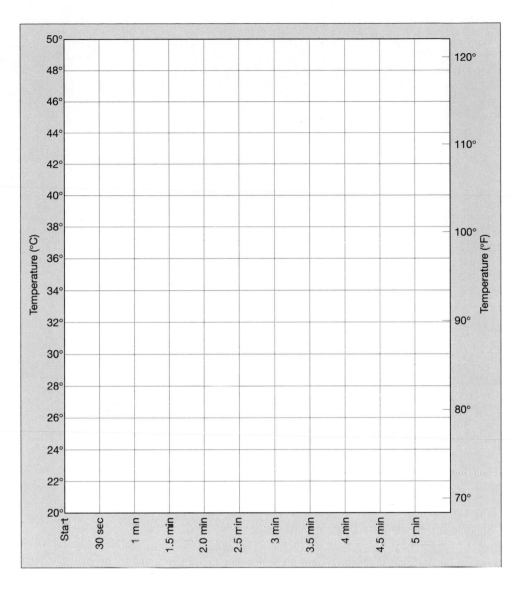

33. Describe the effect that the location, along the coast or in the center of a continent, has on the temperatures of a city.

_____

_____

ATMOSPHERIC TEMPERATURES

Air temperatures are not constant. They normally change through time at any one location, with latitude because of the changing sun angle and length of day light, and with increasing altitude in the lower atmosphere because the atmosphere is primarily heated from the bottom up.

Daily Temperatures. In general, the daily temperatures that occur at any particular place are the result of long-wave radiation being released at the earth's sur-

face. However, secondary factors, such as cloud cover and cold air moving into the area, can also cause variations.

Questions 34–41 refer to the daily temperature graph, Figure 13.7.

━━━

34. In Figure 13.7, the coolest temperature of the day occurs at _____ . Fill in your answer.

35. The warmest temperature occurs at _____ .

36. What is the _daily temperature range_ (difference between maximum and minimum temperatures for the day)?

Daily temperature range: _____°F (_____ °C)

37. What is the _daily temperature mean_ (average of the maximum and minimum temperatures)?

FIGURE 13.6 Mean monthly temperatures for two North American cities located at approximately 37°N latitude.

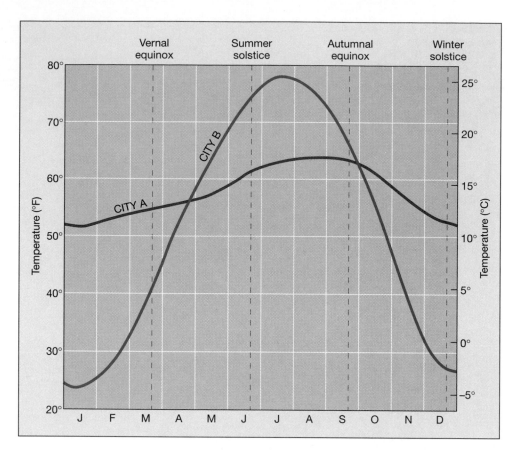

FIGURE 13.7 Typical daily temperature graph for a mid–latitude city during the summer.

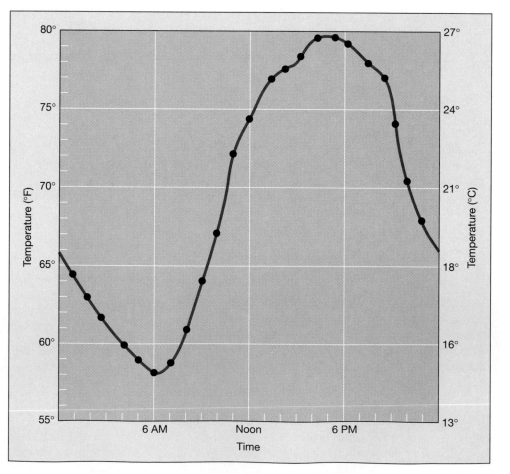

Daily temperature mean: _____°F (_____°C)

38. Refer to the mechanism for heating the atmosphere. Why does the warmest daily temperature occur in mid-to-late afternoon rather than at the time of the highest sun angle?

_____

_____

39. Why does the coolest temperature of the day occur about sunrise?

_____

_____

40. How would cloud cover influence daily maximum and minimum temperatures?

_____

_____

41. On Figure 13.7 sketch and label a colored line that represents a daily temperature graph for a cloudy day.

Global Pattern of Temperature.
The primary reason for global variations in surface temperatures is the unequal distribution of radiation over the earth. Among the most important secondary factors are differential heating of land and water, ocean currents, and differences in altitude.

Questions 42–54 refer to Figure 13.8, "World Distribution of Mean Surface Temperatures (°C) for January and July." The lines on the maps, called **isotherms,** connect places of equal surface temperature.

42. The general trend of the isotherms on the maps is (north-south, east-west). Circle your answer.

43. In general, how do surface temperatures on the earth vary from the equator toward the poles? Why does this variation occur?

_____

_____

_____

44. The highest and lowest temperatures occur over which countries or oceans?

Highest global temperature: _____

_____

Lowest global temperature: _____

_____

45. The locations of the highest and lowest temperatures are over (land, water).

46. Calculate the *annual temperature range* at each of the following locations:

Coastal Norway at 60°N: _____°C (_____°F)

Siberia at 60°N, 120°E: _____°C (_____°F)

On the equator over the center of the Atlantic

Ocean: _____°C (_____°F)

47. Explain the large annual range of temperature in Siberia.

_____

_____

48. Why is the annual temperature range smaller along the coast of Norway than at the same latitude in Siberia?

_____

_____

49. Why is temperature relatively uniform throughout the year in the tropics?

_____

_____

50. Using the two maps in Figure 13.8, calculate the approximate average annual temperature range for your location. How does your temperature range compare with those in the tropics and Siberia?

Average annual temperature range:

_____°C (_____°F)

_____

_____

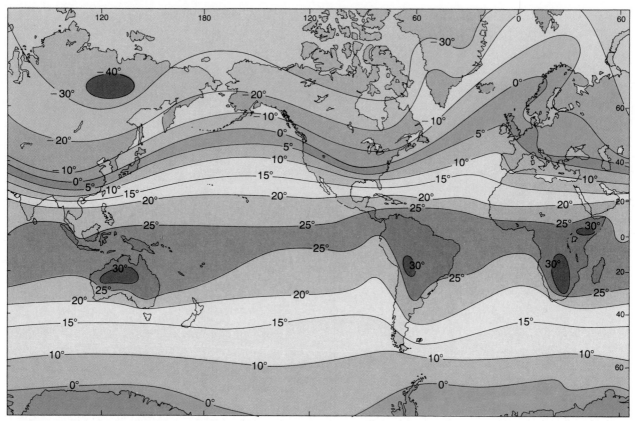

January

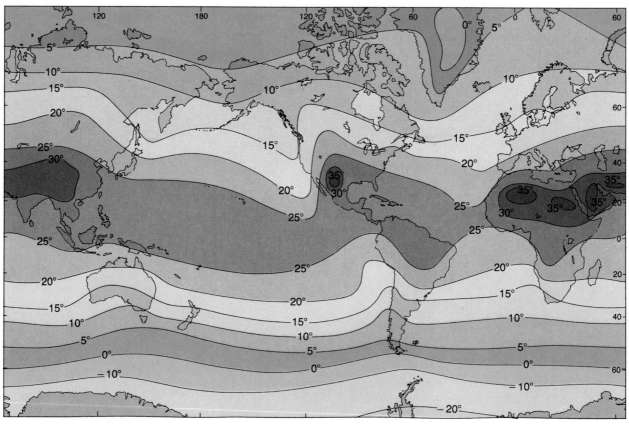

July

FIGURE 13.8 World distribution of mean surface temperatures (°C) for January and July.

51. Trace the path of the 5°C isotherm over North America in January. Explain why the isotherm deviates from a true east-west trend where it crosses from the Pacific Ocean onto the continent.

_____

_____

52. Trace the path of the 20°C isotherm over North America in July. Explain why the isotherm deviates from a true east-west trend where it crosses from the Pacific Ocean onto the continent.

_____

_____

53. Why do the isotherms in the Southern Hemisphere follow a true east-west trend more closely than those in the Northern Hemisphere?

_____

_____

54. Why does the entire pattern of isotherms shift northward from the January map to the July map?

_____

_____

Temperature Changes with Altitude. Since the primary source of heat for the lower atmosphere is the earth's surface, the normal situation found in the lower 12 kilometers of the atmosphere is a decrease in temperature with increasing altitude. This temperature decrease with altitude in the lower atmosphere is called the **environmental lapse rate.** However, at altitudes from about 12 to 45 kilometers, the atmospheric absorption of incoming solar radiation causes temperature to increase.

Using Chapter 12 of your text as a reference, answer questions 55–59.

▬▬▬▬▬

55. Label the _troposphere, mesosphere, stratosphere,_ and _thermosphere_ on the atmospheric temperature curve, Figure 13.9.
56. On Figure 13.9, mark with a line and label the _tropopause, mesopause,_ and _stratopause._
57. What is the approximate temperature of the atmosphere at each of the following altitudes?

10 km: _____°C (_____°F)

50 km: _____°C (_____°F)

80 km: _____°C (_____°F)

58. Explain the reason for each of the following.

Temperature decrease with altitude in the troposphere:

_____

_____

Temperature increase in the stratosphere:

_____

_____

Temperature increase in the thermosphere:

_____

_____

59. Of what importance is the gas ozone in the stratosphere? What will be the effect on radiation received at the earth's surface of a decrease of ozone in the stratosphere?

_____

_____

_____

The average, or normal, environmental lapse rate (temperature decrease with altitude) in the troposphere is 3.5°F per 1,000 feet (6.5°C per kilometer).

▬▬▬▬▬

60. If the surface temperature is 60°F (16°C), what would be the approximate temperature at 20,000 feet (6,000 meters)?

_____°F (_____°C)

61. If the surface temperature is 80°F (27°C), at approximately what altitude would a pilot expect to find each of the following atmospheric temperatures?

50°F: _____ feet   (10°C: _____ meters)

0°C: _____ meters   (32°F: _____ feet)

FIGURE 13.9 Atmospheric temperature curve.

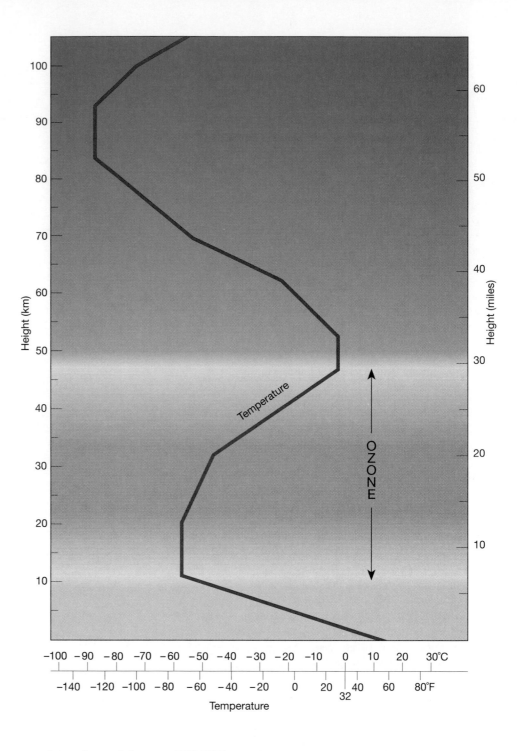

Periodically, the temperature near the surface of the earth increases with altitude. This situation, which is opposite from the normal condition, is called a **temperature inversion.** Review Chapter 16 of your text and then answer questions 62 and 63.

62. What is the cause of a surface temperature inversion?

_____

_____

63. What is the relation between a temperature inversion and atmospheric pollution?

_____

_____

WINDCHILL EQUIVALENT TEMPERATURE

Windchill equivalent temperature is the term applied to the sensation of temperature that the human body feels, in contrast to the actual temperature of the air as recorded by a thermometer. Wind cools by evaporating perspiration and carrying heat away from the body. When temperatures are cool and the wind speed increases, the body reacts as if it were being subjected to increasingly lower temperatures—a phenomenon known as _windchill._

▬▬▬▬

64. Refer to the windchill equivalent temperature chart, Table 13.4. What is the windchill equivalent temperature sensed by the human body in the following situations?

Air Temperature (°F)	Wind Speed (mph)	Windchill Equivalent Temperature (°F)
30°	10	_____
−5°	20	_____
−20°	30	_____

REVIEW

Having completed the exercise, you should know the following:

1. Long-wave radiation, which we detect as heat, is absorbed by carbon dioxide and water vapor in the atmosphere.
2. More incoming radiation from the sun is absorbed by the earth's surface than directly by the atmosphere because the atmosphere transmits most short-wave solar radiation.
3. The earth's surface releases its absorbed radiation in the form of infrared radiation, which heats the atmosphere from below.
4. Terrestrial radiation is the primary source of heat for the atmosphere.
5. The amount of radiation that is absorbed or reflected by an object is related to its color.
6. When exposed to equal amounts of radiation, land will heat up faster and reach higher temperatures than water.
7. The warmest part of the day is generally in mid-afternoon because it takes time for the earth's surface to absorb and release radiation.
8. The hottest and coldest surface temperatures on the earth are associated with continents.
9. Isotherms over continents are deflected toward the equator in winter and poleward in summer because of the unequal heating and cooling of land and water.
10. Because the Southern Hemisphere is the "water" hemisphere, temperatures are more uniform throughout the year than in the Northern Hemisphere.

TABLE 13.4 Windchill equivalent temperature (° F)

(SOURCE: NOAA, NATIONAL WEATHER SERVICE)

Air Temperature (°F)	Wind Speed (miles per hr)								
	5	10	15	20	25	30	35	40	45
40	37	28	22	18	15	13	11	10	9
35	32	22	16	11	8	5	3	2	1
30	27	16	9	4	0	−2	−4	−6	−7
25	22	10	2	−3	−7	−10	−12	−14	−15
20	16	4	−5	−10	−15	−18	−20	−22	−23
15	11	−3	−11	−17	−22	−25	−28	−29	−31
10	6	−9	−18	−25	−29	−33	−35	−37	−39
5	1	−15	−25	−32	−37	−41	−43	−45	−47
0	−5	−21	−32	−39	−44	−48	−51	−53	−55
−5	−10	−27	−38	−46	−52	−56	−59	−61	−62
−10	−15	−33	−45	−53	−59	−63	−67	−69	−70
−15	−20	−40	−52	−60	66	−71	−74	−77	−78
−20	−26	−46	−58	−67	−74	−79	−82	−85	−86
−25	−31	−52	−65	−74	−81	−86	−90	−93	−94
−30	−36	−58	−72	−82	−89	−94	−98	−101	−102

11. On an average, there is a 3.5°F/1,000 feet (6.5°C per kilometer) temperature decrease in the troposphere.
12. There is an increase in temperature in the stratosphere and thermosphere because of atmospheric absorption of incoming solar radiation.
13. A temperature inversion will occur when warmer air lies over cooler air in the lower atmosphere.
14. The body will experience cooler temperatures as the wind speed increases.

EXERCISE THIRTEEN
Atmospheric Heating

SUMMARY/REPORT PAGE

Date Due: _____

Name: _____

Date: _____

Class: _____

After you have finished Exercise Thirteen, complete the following questions. You may have to refer to the exercise for assistance or to locate specific answers. Be prepared to submit this summary/report to your instructor at the designated time.

1. Assume an average noon sun angle. What percent of the solar radiation will be absorbed by the atmosphere and what percent will be absorbed by the earth's surface?

   Atmospheric absorption: _____%

   Absorption by the earth's surface: _____%

2. What will be the atmospheric effect of each of the following?

   Less ozone in the stratosphere: _____

   _____

   More carbon dioxide in the atmosphere: _____

   _____

   A surface with a high albedo: _____

   _____

3. Briefly explain how the earth's atmosphere is heated.

   _____

   _____

   _____

4. What were the starting and ending temperatures you obtained for the black and silver containers in the albedo experiment?

	Starting Temperature	Ending Temperature
Black container:	_____	_____
Silver container:	_____	_____

5. Summarize the effect of color on the heating of an object.

   _____

   _____

6. What were the starting and ending temperatures you obtained for the water and dry sand in the land and water heating experiment?

	Starting Temperature	Ending Temperature
Water:	_____	_____
Dry sand:	_____	_____

7. Summarize the effects that equal amounts of radiation have on the heating of land and water.

   _____

   _____

8. Where are the highest and lowest average monthly temperatures located on the earth?

   Highest average monthly temperature: _____

   _____

   Lowest average monthly temperature: _____

   _____

9. Why does the Northern Hemisphere experience a greater annual range of temperature than the Southern Hemisphere?

_____

_____

10. Define each of the following:

Environmental lapse rate: _____

_____

Windchill equivalent temperature: _____

_____

Troposphere: _____

_____

Atmospheric Moisture, Pressure, and Wind

By observing, recording, and analyzing weather conditions, meteorologists attempt to define the principles that control the complex interactions that occur in the atmosphere (Figure 14.1). One important element, temperature, has already been examined in Exercises 12 and 13. However, no analysis of the atmosphere is complete without an investigation of the remaining variables—humidity, precipitation, pressure, and wind.

This exercise examines the changes of state of water, how the water vapor content of the air is measured, and the sequence of events necessary to cause cloud formation. Global patterns of precipitation, pressure, and wind are also reviewed. Although the elements are presented separately, keep in mind that all are very much interrelated. A change in any one element often brings about changes in the others.

FIGURE 14.1 Storm clouds near sunset. (Photo by E.J. Tarbuck)

OBJECTIVES

After you have completed this exercise, you should be able to

1. Explain the processes involved when water changes state.
2. Use a psychrometer or hygrometer and appropriate tables to determine the relative humidity and dew-point temperature of air.
3. Explain the adiabatic process and its effect on cooling and warming the air.
4. Calculate the temperature and relative humidity changes that take place in air as the result of adiabatic cooling.
5. Describe the relation between pressure and wind.
6. Describe the global patterns of surface pressure and wind.

TEXTBOOK REFERENCE

Chapters 13 and 14

MATERIALS

calculator

Materials Supplied by Your Instructor

psychrometer or hygrometer
beaker, ice, thermometer
barometer
atlas

TERMS

water vapor	relative humidity	psychrometer/
evaporation	saturated	hygrometer
precipitation	dew-point	condensation
latent heat	temperature	nuclei

dry adiabatic rate
wet adiabatic rate
atmospheric
  pressure
barometer
isobar

equatorial low
subtropical high
subpolar low
anticyclone
cyclone
wind

Coriolis effect
trade winds
westerlies
polar easterlies
monsoon

ATMOSPHERIC MOISTURE AND PRECIPITATION

Introduction. **Water vapor,** an odorless, colorless gas produced by the **evaporation** of water, comprises only a small percent of the lower atmosphere. However, it is an important atmospheric gas because it is the source of all **precipitation,** aids in the heating of the atmosphere by absorbing radiation, and is the source of **latent heat** (hidden or stored heat).

Changes of State. The temperatures and pressures that occur at and near the earth's surface allow water to change readily from one state of matter to another. The fact that water can exist as a gas, liquid, or solid within the atmosphere makes it one of the most unique substances on earth. Review Chapter 13 of your text and then answer questions 1–4 using Figure 14.2.

1. To help visualize the processes and heat requirements for changing the state of matter of water, write the name of the process involved (choose from the list), whether heat is absorbed or released by the process, and the amount of heat, in calories per gram, absorbed or released by the process at the indicated locations by each arrow in Figure 14.2.

Processes

Freezing	Evaporation	Deposition
Sublimation	Melting	Condensation

2. (More, Less) heat energy is required to melt a gram of ice than to evaporate a gram of water. Circle your answer.
3. The process of condensation releases (more, less) latent heat energy than freezing.
4. The energy requirement for the process of deposition is the (same as, less than) the total energy required to condense water vapor and then freeze the water. Circle your answer.

Water Vapor Capacity of Air. Any measure of water vapor in the air is referred to as *humidity*. The water vapor capacity of air is limited by, and directly related to, its temperature.

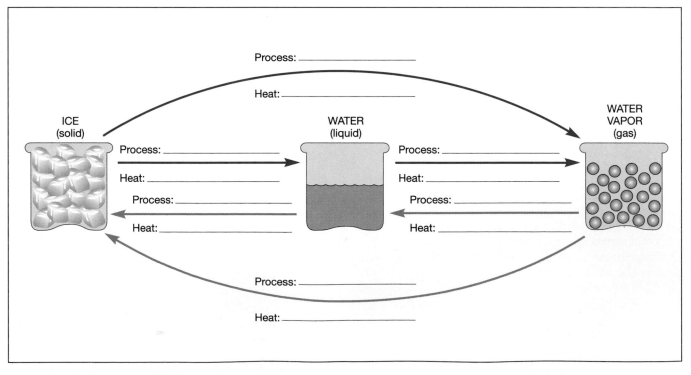

FIGURE 14.2 Changes of state of water.

Table 14.1 presents the water vapor capacity of a kilogram of air at various temperatures. Use Table 14.1 to answer questions 5–8.

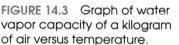

5. To illustrate the relation between water vapor capacity and air temperature, prepare a graph by plotting the data from Table 14.1 on Figure 14.3.

6. From Table 14.1 and/or Figure 14.3, what is the water vapor capacity of a kilogram of air at each of the following temperatures?

40°C: _____ grams/kilogram

20°C: _____ grams/kilogram

0°C: _____ grams/kilogram

-20°C: _____ grams/kilogram

7. Raising the air temperature of a kilogram of air 5°C, from 10°C to 15°C, (increases, decreases) its water vapor capacity by (3, 6) grams. However, raising the temperature from 35°C to 40°C (increases, decreases) the capacity by (8, 12) grams. Circle your answers.

TABLE 14.1 Water vapor capacity of a kilogram of air (at average sea level pressure)

Temperature (°C)	Grams/kg
−40	0.1
−30	0.3
−20	0.75
−10	2
0	3.5
5	5
10	7
15	10
20	14
25	20
30	26.5
35	35
40	47

8. Using Table 14.1 and/or Figure 14.3, write a statement that relates the water vapor capacity of air to the temperature of the air.

_____

_____

FIGURE 14.3 Graph of water vapor capacity of a kilogram of air versus temperature.

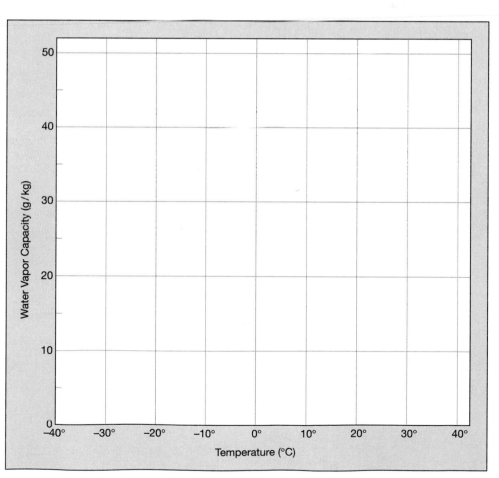

Measuring Humidity. **Relative humidity** is the most common measurement used to describe water vapor in the air. In general, it expresses how close the air is to reaching its water vapor capacity. Relative humidity is the *ratio* of the air's water vapor content (amount actually in the air) to its water vapor capacity at a given temperature, expressed as a percent. The general formula is

$$\text{Relative humidity(\%)} = \frac{\text{Water vapor content}}{\text{Water vapor capacity}} \times 100$$

For example, from Table 14.1, the water vapor capacity of a kilogram of air at 25°C would be 20 grams per kilogram. If the actual amount of water vapor in the air was 5 grams per kilogram (the water vapor content), the relative humidity of the air would be calculated as follows:

$$\text{Relative humidity(\%)} = \frac{5 \text{ g/kg}}{20 \text{ g/kg}} \times 100 = 25\%$$

9. Use Table 14.1 and the formula for relative humidity to determine the relative humidity for each of the following situations of identical temperature.

Air Temp.	Water Vapor Content	Capacity	Relative Humidity
15°C	2 g/kg	___ g/kg	___ %
15°C	5 g/kg	___ g/kg	___ %
15°C	7 g/kg	___ g/kg	___ %

10. From question 9, if the temperature of air remains constant, adding water vapor will (raise, lower) the relative humidity, while removing water vapor will (raise, lower) the relative humidity. Circle your answers.

11. Use Table 14.1 and the formula for relative humidity to determine the relative humidity for each of the following situations of identical water vapor content.

Air Temp.	Water Vapor Content	Capacity	Relative Humidity
25°C	5 g/kg	___ g/kg	___ %
15°C	5 g/kg	___ g/kg	___ %
5°C	5 g/kg	___ g/kg	___ %

12. From question 11, if the amount of water vapor in the air remains constant, cooling the air will (raise, lower) the relative humidity, while warming the air will (raise, lower) the relative humidity. Circle your answers.

13. In the winter, air is heated in homes. What effect does heating the air have on the relative humidity inside the home? What can be done to lessen this effect?

_____

_____

14. Explain why a cool basement is humid (damp) in the summer.

_____

_____

15. Write brief statements describing each of the two ways that the relative humidity of air can be changed.

1) _____

2) _____

One of the misconceptions concerning relative humidity is that it alone gives an accurate indication of the amount of water vapor in the air. For example, on a cold winter day if you hear on the radio that the relative humidity is 90 percent, can you conclude that the air contains more moisture than on a hot summer day which records a 40 percent relative humidity? Completing question 16 will help you find the answer.

16. Use Table 14.1 to determine the water vapor content for each of the following situations. As you do the calculations, keep in mind the definition of relative humidity.

Summer	Winter
Air temperature = 25°C	Air temperature = 5°C
Capacity = ___ g/kg	Capacity = ___ g/kg
Relative humidity = 40%	Relative humidity = 90%
Content = ___ g/kg	Content = ___ g/kg

17. Explain why relative humidity does *not* give an accurate indication of the amount of water vapor in the air.

_____

_____

_____

Dew-Point Temperature.
Air is **saturated** when it has reached its water vapor capacity and contains all the water vapor that it can hold at a particular temperature. In saturated air, the water vapor content equals its capacity. The temperature at which air is saturated is called the **dew-point temperature.** Put another way, the dew point is the temperature at which the relative humidity of the air is 100%.

Previously in question 11 you determined that a kilogram of air at 25°C, containing 5 grams of water vapor, had a relative humidity of 25% and was not saturated. However, when the temperature was lowered to 5°C, the air had a relative humidity of 100% and was saturated. Therefore, 5°C is the dew-point temperature of the air in that example.

18. By referring to Table 14.1, what is the dew-point temperature of a kilogram of air that contains 7 grams of water vapor?

    Dew-point temperature = _____°C

19. What is the relative humidity and dew-point temperature of a kilogram of 25°C air that contains 10 grams of water vapor?

    Relative humidity = _____%

    Dew-point temperature = _____°C

Using a Psychrometer or Hygrometer.
The relative humidity and dew-point temperature of air can be determined by using a **psychrometer** (Figure 14.4) or **hygrometer** and appropriate charts. The psychrometer consists of two thermometers mounted side by side. One of the thermometers, the wet-bulb thermometer, has a piece of wet cloth wrapped around it. As the psychrometer is spun for approximately one minute, water on the wet-bulb thermometer evaporates and cooling results. In dry air, the rate of evaporation will be high, and a low wet-bulb temperature will be recorded. The other thermometer, the dry-bulb thermometer, measures the air temperature. After using the instrument and recording both the wet- and dry-bulb temperatures, the relative humidity and dew-point temperature are determined using Table 14.2, "Relative humidity (percent)" and Table 14.3, "Dew-point temperature." With a hygrometer, relative humidity can be read directly, without the use of tables.

20. Use Table 14.2 to determine the relative humidity for each of the psychrometer readings.

	Reading 1	Reading 2
Dry-bulb temperature:	20°C	32°C
Wet-bulb temperature:	18°C	25°C
Difference between dry- and wet-bulb temperatures:	_____	_____
Relative humidity:	_____%	_____%

21. From question 20, what is the relation between the *difference* in the dry-bulb and wet-bulb temperatures and the relative humidity of the air?

    _____

    _____

22. Use Table 14.3 to determine the dew-point temperature for each of the psychrometer readings.

	Reading 1	Reading 2
Dry-bulb temperature:	8°C	30°C
Wet-bulb temperature:	6°C	24°C
Difference between dry- and wet-bulb temperatures:	_____	_____
Dew-point temperature:	_____°C	_____°C

FIGURE 14.4 Sling psychrometer. The sling psychrometer is one instrument that is used to determine relative humidity and dew–point temperature.

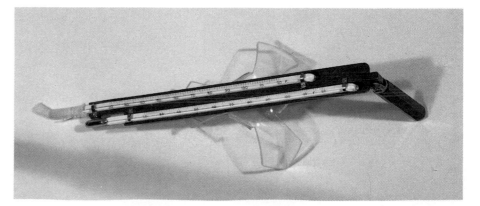

TABLE 14.2 Relative humidity (percent)*

Dry-bulb Temperature (°C)	Dry-bulb Temperature − Wet-bulb Temperature = Depression of the Wet Bulb																					
	1	2	3	4	5	6	7	8	9	10	11	12	13	14	15	16	17	18	19	20	21	22
−20	28																					
−18	40																					
−16	48	0																				
−14	55	11																				
−12	61	23																				
−10	66	33	0																			
−8	71	41	13																			
−6	73	48	20	0																		
−4	77	54	43	11																		
−2	79	58	37	20	1																	
0	81	63	45	28	11																	
2	83	67	51	36	20	6																
4	85	70	56	42	27	14																
6	86	72	59	46	35	22	10	0														
8	87	74	62	51	39	28	17	6														
10	88	76	65	54	43	33	24	13	4													
12	88	78	67	57	48	38	28	19	10	2												
14	89	79	69	60	50	41	33	25	16	8	1											
16	90	80	71	62	54	45	37	29	21	14	7	1										
18	91	81	72	64	56	48	40	33	26	19	12	6	0									
20	91	82	74	66	58	51	44	36	30	23	17	11	5	0								
22	92	83	75	68	60	53	46	40	33	27	21	15	10	4	0							
24	92	84	76	69	62	55	49	42	36	30	25	20	14	9	4	0						
26	92	85	77	70	64	57	51	45	39	34	28	23	18	13	9	5						
28	93	86	78	71	65	59	53	47	42	36	31	26	21	17	12	8	2					
30	93	86	79	72	66	61	55	49	44	39	34	29	25	20	16	12	8	4				
32	93	86	80	73	68	62	56	51	46	41	36	32	27	22	19	14	11	8	4			
34	93	86	81	74	69	63	58	52	48	43	38	34	30	26	22	18	14	11	8	5		
36	94	87	81	75	69	64	59	54	50	44	40	36	32	28	24	21	17	13	10	7	4	
38	94	87	82	76	70	66	60	55	51	46	42	38	34	30	26	23	20	16	13	10	7	5
40	94	89	82	76	71	67	61	57	52	48	44	40	36	33	29	25	22	19	16	13	10	7

*To determine the relative humidity and dew point, find the air (dry-bulb) temperature on the vertical axis (far left) and the depression of the wet bulb on the horizontal axis (top). Where the two meet, the relative humidity or dew point is found. For example, use a dry-bulb temperature of 20°C and a wet-bulb temperature of 14°C. From Table 14.2, the relative humidity is 51 percent, and from Table 14.3, the dew point is 10°C.

If a psychrometer or hygrometer is available in the laboratory, your instructor will explain the procedure for using the instrument. FOLLOW THE SPECIFIC DIRECTIONS OF YOUR INSTRUCTOR to complete question 23.

23. Use the psychrometer (or hygrometer) to determine the relative humidity and dew-point temperature of the air in the room and outside the building. If you use a psychrometer, record your information in the following space.

	Room	Outside
Dry-bulb temperature:	_____	_____
Wet-bulb temperature:	_____	_____
Difference between dry- and wet-bulb temperatures:	_____	_____
Relative humidity:	_____%	_____%
Dew-point temperature:	_____	_____

Condensation. If air is cooled below the dew-point temperature, water will condense (change from vapor

TABLE 14.3 Dew-point temperature (°C)*

Dry-bulb Temperature (°C)	Dry-bulb Temperature − Wet-bulb Temperature = Depression of the Wet Bulb																					
	1	2	3	4	5	6	7	8	9	10	11	12	13	14	15	16	17	18	19	20	21	22
−20	−33																					
−18	−28																					
−16	−24																					
−14	−21	−36																				
−12	−18	−28																				
−10	−14	−22																				
−8	−12	−18	−29																			
−6	−10	−14	−22																			
−4	−7	−22	−17	−29																		
−2	−5	−8	−13	−20																		
0	−3	−6	−9	−15	−24																	
2	−1	−3	−6	−11	−17																	
4	1	−1	−4	−7	−11	−19																
6	4	1	−1	−4	−7	−13	−21															
8	6	3	1	−2	−5	−9	−14															
10	8	6	4	1	−2	−5	−9	−14														
12	10	8	6	4	1	−2	−5	−9	−16													
14	12	11	9	6	4	1	−2	−5	−10	−17												
16	14	13	11	9	7	4	1	−1	−6	−10	−17											
18	16	15	13	11	9	7	4	2	−2	−5	−10	−19										
20	19	17	15	14	12	10	7	4	2	−2	−5	−10	−19									
22	21	19	17	16	14	12	10	8	5	3	−1	−5	−10	−19								
24	23	21	20	18	16	14	12	10	8	6	2	−1	−5	−10	−18							
26	25	23	22	20	18	17	15	13	11	9	6	3	0	−4	−9	−18						
28	27	25	24	22	21	19	17	16	14	11	9	7	4	1	−3	−9	−16					
30	29	27	26	24	23	21	19	18	16	14	12	10	8	5	1	−2	−8	−15				
32	31	29	28	27	25	24	22	21	19	17	15	13	11	8	5	2	−2	−7	−14			
34	33	31	30	29	27	26	24	23	21	20	18	16	14	12	9	6	3	1	5	12	29	
36	35	33	32	31	29	28	27	25	24	22	20	19	17	15	13	10	7	4	0	−4	−10	
38	37	35	34	33	32	30	29	28	26	25	23	21	19	17	15	13	11	8	5	1	−3	−9
40	39	37	36	35	34	32	31	30	28	27	25	24	22	20	18	16	14	12	9	6	2	−2

*See footnote to Table 14.2

to liquid) on available surfaces. In the atmosphere, the particles on which water condenses are called **condensation nuclei.** Condensation may result in the formation of dew or frost on the ground and clouds or fog in the atmosphere.

24. Examine the process of condensation by gradually adding ice to a beaker approximately one-third full of water. As you add the ice, stir the water-ice mixture gently with a thermometer. Note the temperature at the moment water begins to condense on the outside surface of the beaker. After you complete your observations, answer questions 24a and 24b.

a. The temperature at which water began condensing on the outside surface of the beaker

was _____.

b. How does the temperature at which water began to condense compare to the dew-point temperature of the air in the room you determined using the psychrometer (or hygrometer)?

_____

25. Refer to Table 14.1. How many grams of water vapor will condense on a surface if a kilogram of 10°C air with a relative humidity of 100% is cooled to 5°C?

_____ grams of water will condense

26. Assume a kilogram of 25°C air that contains 10 grams of water vapor. Use Table 14.1. How many grams of water will condense if the air's temperature is lowered to each of the following temperatures?

5°C: _____ grams of condensed water

−10°C: _____ grams of condensed water

27. What relation exists between the altitude of the base of a cloud, which consists of very small droplets of water, and the dew-point temperature of the air at that altitude?

_____

Daily Temperature and Relative Humidity. Figure 14.5 shows the typical daily variations in air tem-perature, relative humidity, and dew-point temperature during two consecutive spring days at a middle latitude city. Use Figure 14.5 to answer questions 28–32.

▬▬▬

28. In Figure 14.5, relative humidity is at its maximum at (6 A.M., 3 P.M.) on day (1, 2). Circle your answers.
29. The lowest temperature over the two-day period occurs at (6 A.M., noon, 3 P.M.) on day (1, 2).
30. The lowest relative humidity occurs at (6 A.M., noon, 4 P.M.) on day (1, 2).
31. Write a general statement describing the relation between air temperature and relative humidity throughout the time period shown in Figure 14.5.

_____

_____

FIGURE 14.5 Typical variations in air temperature, relative humidity, and dew-point temperature during two consecutive spring days at a middle latitude city.

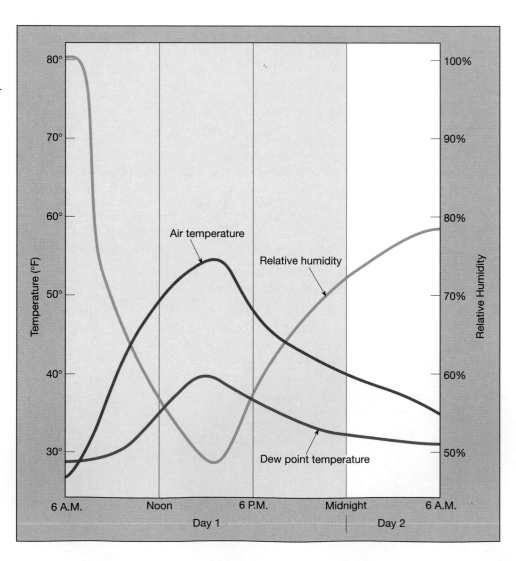

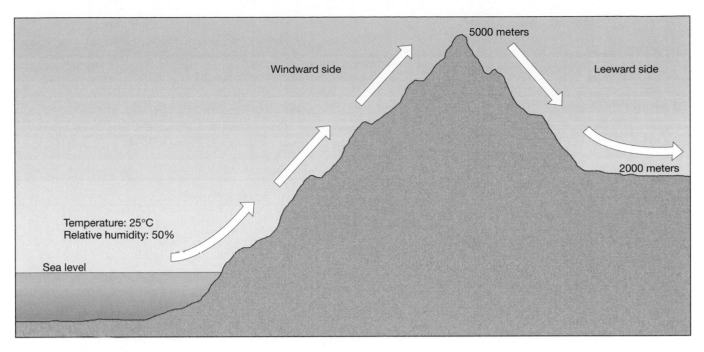

FIGURE 14.6 Adiabatic processes associated with a mountain barrier.

32. Did a dew or frost occur on either of the two days represented in Figure 14.5? If so, list when and explain how you arrived at your answer.

_____

_____

Adiabatic Processes. As you have seen, the key to causing water vapor to condense, which is necessary before precipitation can occur, is to cool the air to its dew-point temperature. In nature, when air rises, it encounters less pressure, expands, and cools. The reverse is also true. Air that is compressed will warm. Temperature changes brought about solely by expansion or compression are called *adiabatic temperature changes.* Air with a temperature above its dew point (unsaturated air) cools by expansion or warms by compression at a rate of 1°C per 100 meters of changing altitude—the **dry adiabatic rate.** After the dew-point temperature is reached, and as condensation occurs, latent heat that has been stored in the water vapor will be liberated. The heat being released by the condensing water slows down the rate of cooling of the air. Rising saturated air will continue to cool by expansion, but at a lesser rate of about 0.5°C per 100 meters of changing altitude—the **wet adiabatic rate.**

Figure 14.6 illustrates a kilogram of air at sea level with a temperature of 25°C and a relative humidity of 50%. The air is forced to rise over a 5,000 meter mountain and descend to a plateau 2,000 meters above sea level on the opposite (leeward) side. To help understand the adiabatic process, answer questions 33–45 by referring to Figure 14.6.

33. What is the water vapor capacity, content, and dew–point temperature of the air at sea level?

Capacity: _____ g/kg of air

Content: _____ g/kg of air

Dew-point temperature: _____°C

34. The air at sea level is (saturated, unsaturated). Circle your answer.

35. The air will initially (warm, cool) as it rises over the windward side of the mountain at the (wet, dry) adiabatic rate, which is (1, 0.5)°C per 100 meters. Circle the correct responses.

36. What will be the air's temperature at 500 meters?

_____°C at 500 meters

37. Condensation (will, will not) take place at 500 meters.

38. The rising air will reach its dew-point temperature at _____ meters and water vapor will begin to (condense, evaporate).

39. From the altitude where condensation begins to occur, to the summit of the mountain, the rising air will continue to expand and will (warm, cool) at the (wet, dry) adiabatic rate of about _____ °C per 100 meters.

40. The temperature of the rising air at the summit of

the mountain will be _____ °C.

41. Assume the air begins to descend on the leeward side of the mountain, it will be compressed and its temperature will (increase, decrease).

42. Assume the relative humidity of the air is below 100% during its entire descent to the plateau. The air will be (saturated, unsaturated) and will warm at

the (wet, dry) adiabatic rate of about _____ °C per 100 meters.

43. As the air descends and warms on the leeward side of the mountain, its relative humidity will (increase, decrease).

44. The air's temperature when it reaches the plateau

at 2,000 meters will be _____ °C.

45. Explain why mountains might cause dry conditions on their leeward sides.

_____

_____

Precipitation. Condensation does not always result in precipitation. This is best illustrated by the fact that it does not rain from *every* cloud. For precipitation to occur, other conditions are also necessary.

46. After you review Chapter 13 of your text, summarize what must happen to cloud droplets in order to form precipitation.

_____

_____

47. Briefly describe the atmospheric temperature conditions that could result in each of the following forms of precipitation.

Rain: _____

Snow: _____

Sleet: _____

Global Patterns of Precipitation. Figure 14.7 shows the average annual precipitation for the world. Use Figure 14.7 to answer questions 48–51.

48. Which areas of the world receive the greatest average annual precipitation?

_____

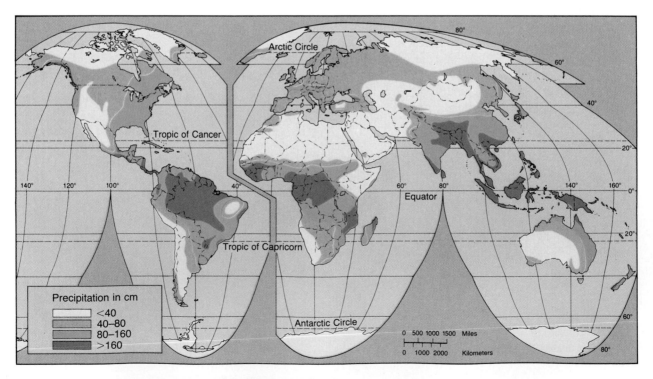

FIGURE 14.7 Average annual precipitation in centimeters.

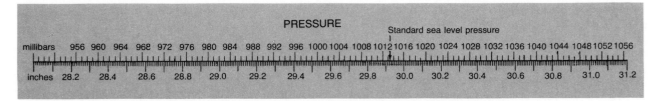

PRESSURE

Standard sea level pressure

| millibars | 956 | 960 | 964 | 968 | 972 | 976 | 980 | 984 | 988 | 992 | 996 | 1000 | 1004 | 1008 | 1012 | 1016 | 1020 | 1024 | 1028 | 1032 | 1036 | 1040 | 1044 | 1048 | 1052 | 1056 |

| inches | 28.2 | 28.4 | 28.6 | 28.8 | 29.0 | 29.2 | 29.4 | 29.6 | 29.8 | 30.0 | 30.2 | 30.4 | 30.6 | 30.8 | 31.0 | 31.2 |

FIGURE 14.8 Scale for comparing pressure readings in millibars and inches of mercury. (After NOAA)

49. The polar regions of the earth have (high, low) average annual precipitation. Circle your answer.

50. What is the average annual precipitation at your location?

_____ centimeters per year, which is equivalent to _____ inches per year.

51. Describe the pattern of average annual precipitation in North America.

_____

_____

_____

PRESSURE AND WIND

Introduction. Atmospheric pressure and wind are two elements of weather that are closely interrelated. Although pressure is the element least noticed by people in a weather report, it is pressure differences in the atmosphere that drive the winds that often bring changes in temperature and moisture.

Atmospheric Pressure. **Atmospheric pressure** is the force exerted by the weight of the atmosphere. It varies over the face of the earth, primarily because of temperature differences. Typically, warm air is less dense than cool air and, therefore, exerts less pressure. Cold, dense air is often associated with higher pressures. Pressure also changes with altitude.

The instrument used to determine atmospheric pressure is the **barometer.** Two units that can be used to measure air pressure are *inches of mercury* and *millibars.* Inches of mercury refers to the height to which a column of mercury will rise in a glass tube that has been inverted into a reservoir of mercury. Heavy air pushes with a greater force on the reservoir than lighter air, thereby causing the height of the mercury column to rise. The millibar is a unit that measures the actual force of the atmosphere pushing down on a surface. Standard pressure at sea level is 29.92 inches of mercury or 1,013.2 millibars. A pressure greater than 29.92

inches or 1,013.2 millibars is called *high pressure*. A pressure less than the standards is called *low pressure*.

52. If a barometer is located in the laboratory, record the current atmospheric pressure in both inches of mercury and millibars. If necessary, use Figure 14.8 to convert the units.

Inches of mercury: _____ inches of mercury

Millibars: _____ millibars

Figure 14.9 illustrates generalized atmospheric pressure variations with altitude. Use Figure 14.9 to answer questions 53 and 54.

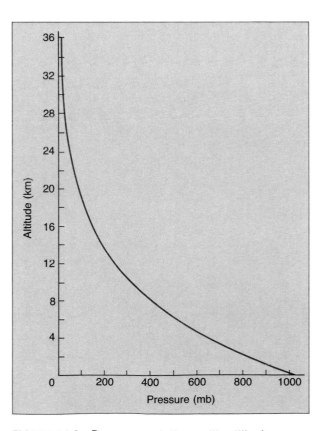

FIGURE 14.9 Pressure variations with altitude.

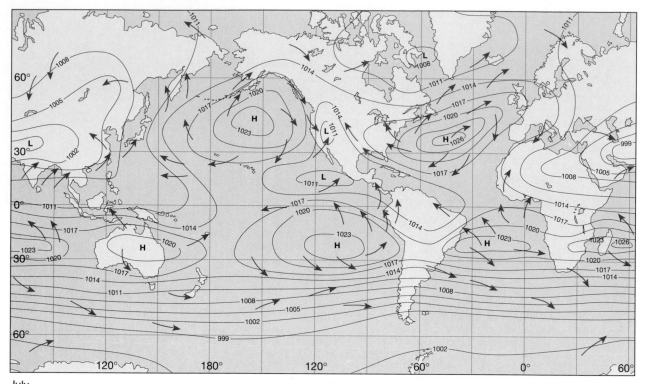

July

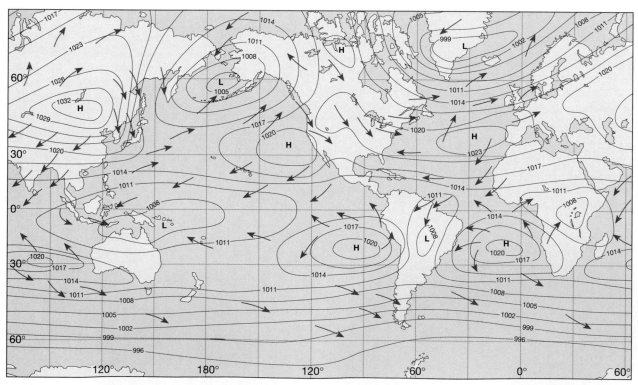

January

FIGURE 14.10 Average surface barometric pressure in millibars for July and January with associated winds.

53. Atmospheric pressure (increases, decreases) with an increase in altitude because there is (more, less) atmosphere above to exert a force. Circle your answers.

54. Pressure changes with altitude (most, least) rapidly near the earth's surface.

Since surface elevations vary, barometric readings are adjusted to indicate what the pressure would be if the barometer was located at sea level. This provides a common standard for mapping pressure, regardless of elevation.

55. A city that is 200 meters above sea level would (add, subtract) units to its barometric reading in order to correct its pressure to sea level.

Observe the average global surface pressure maps with associated winds for July and January in Figure 14.10. The lines shown are **isobars,** which connect points of equal surface barometric pressure, adjusted to sea level. Isobars can be used to identify the principal pressure zones on the earth, which include the **equatorial lows, subtropical highs,** and **subpolar lows.** Answer questions 56–62 using Figure 14.10.

56. The units used on the maps to indicate pressure are (inches of mercury, millibars). Circle your answer.

57. By writing the word "HIGH" or "LOW," indicate on the maps the general pressure at each of the following latitudes: 60°N, 30°N, 0°, 30°S, 60°S.

58. After referring to Chapter 14 of the text, write on the maps the names (equatorial low, subtropical high, or subpolar low) of each of the pressure zones you identified in question 57.

59. During the summer months, January in the Southern Hemisphere and July in the Northern Hemisphere, (high, low) pressure is more common over land. Circle your answer.

60. (High, Low) pressure is most associated with the land in the winter months.

61. Considering what you know about the unequal heating of land and water and the influence of air temperature on pressure, why does the pressure over continents change with the seasons?

_____

_____

62. Why does the air over the oceans maintain a more uniform pressure throughout the year?

_____

_____

The differential heating and cooling of land and water over most of the earth causes the zones of pressure to be broken into cells of pressure. Pressure cells are shown on a map as a system of closed, concentric isobars. High pressure cells, called **anticyclones,** are typically associated with descending (subsiding) air. Low pressure cells, called **cyclones,** have rising air in their centers. Locate and examine some of the pressure cells on the maps in Figure 14.10. Then answer questions 63 and 64.

63. In an anticyclone, the (highest, lowest) pressure occurs at the center of the cell. In a cyclone, the (highest, lowest) pressure occurs at the center. Circle your answers.

64. With which pressure cell, anticyclone or cyclone, would the vertical movement of air be most favorable for cloud formation and precipitation? Explain your answer with reference to the adiabatic process.

_____

_____

Wind. The horizontal movement of air is called **wind.** Wind is initiated because of horizontal pressure differences. Air flows from areas of higher pressure to areas of lower pressure. The direction of the wind is influenced by the **Coriolis effect** and friction between the air and earth's surface. The velocity of the wind is controlled by the difference in pressure between two areas. Examine the global wind pattern (shown with arrows) that is associated with the global pattern of pressure in Figure 14.10. The wind arrows can be used to identify the global wind belts, which include the **trade winds, westerlies,** and **polar easterlies.** Then answer questions 65–68.

65. Examine the pressure cells in Figure 14.10. Then, on Diagram 14.1, complete the diagrams of the indicated pressure cells for each hemisphere. Label the isobars with appropriate pressures *and* use arrows to indicate the surface air movement in each pressure cell.

Northern Hemisphere

High (Anticyclone)   **Low (Cyclone)**

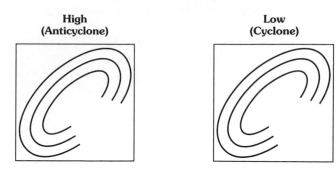

Southern Hemisphere

High (Anticyclone)   **Low (Cyclone)**

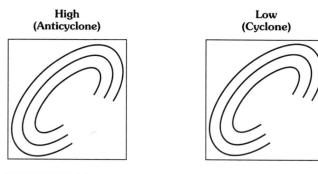

DIAGRAM 14.1

66. In the following spaces, indicate the movements of air in high and low pressure cells for each hemisphere. Write one of the two choices given in italics for each blank.

	Northern Hemisphere		Southern Hemisphere	
	High	**Low**	**High**	**Low**
Surface air moves *into* or *out of*:	___	___	___	___
Surface air will *rise* or *subside* in the center:	___	___	___	___
Surface air motion is *clockwise* or *counterclockwise*:	___	___	___	___

67. Write a brief statement that describes the difference in surface air movement between a Northern Hemisphere and Southern Hemisphere anticyclone.

_____

_____

_____

68. After reviewing Chapter 14 of the text, locate and write the name of each global wind belt (trade winds, westerlies, or polar easterlies) at the appropriate location on the maps in Figure 14.10. Also indicate on the maps the area of the *polar front*.

In Figure 14.10, notice the seasonal changes in wind direction, called **monsoons,** over continents. Then answer questions 69–72.

69. During the (summer, winter) season, the air is moving from the continent to the ocean. Circle your answer.

70. During the (summer, winter) season, the air is moving from the ocean to the continent.

71. In what way is the seasonal change in wind direction over continents related to pressure?

_____

_____

72. What effect will the seasonal reversal of wind have on moisture in the air and the potential for precipitation over the continents during the following seasons?

Summer season: _____

_____

Winter season: _____

_____

By examining Figure 14.10, you should notice that the systems of pressure belts and global winds change latitude with the seasons.

73. In what manner is the seasonal shift in pressure belts and global winds related to the movement of the overhead noon sun throughout the year?

_____

_____

REVIEW

Having completed the exercise, you should know the following:

1. The processes by which water changes its state are evaporation, melting, sublimation, condensation, freezing, and deposition. The last three processes release heat.
2. Relative humidity is the water vapor content of the air compared to the air's water vapor capacity, expressed as a percent.
3. The dew-point temperature of air is the temperature to which air has to be cooled to cause saturation. At the dew-point temperature the air has a 100% relative humidity.
4. Relative humidity and dew-point temperature can be determined using a hygrometer or psychrometer and charts.
5. Adiabatic processes cause changes in air temperature. Cooling by expansion may result in condensation.
6. Cloud formation is the result of a sequence of events that take place in the atmosphere.
7. The various forms of precipitation are each produced by specific atmospheric conditions.
8. Atmospheric pressure does not decrease at a constant rate with increasing altitude.
9. Each global pressure zone has a specific name and location.
10. Each global wind belt has a specific name and location.
11. High pressure and low pressure cells in each hemisphere have characteristic movements of air.
12. Seasonal changes in pressure over continents cause monsoons.

EXERCISE FOURTEEN
Atmospheric Moisture, Pressure, and Wind

SUMMARY/REPORT PAGE

Date Due: _____

Name: _____

Date: _____

Class: _____

After you have finished Exercise Fourteen, complete the following questions. You may have to refer to the exercise for assistance or to locate specific answers. Be prepared to submit this summary/report to your instructor at the designated time.

1. Answer the following by circling the correct response.
   a. Liquid water changes to water vapor by the process called (condensation, evaporation, deposition).
   b. (Warm, Cold) air has the greatest water vapor capacity.
   c. Lowering the air temperature will (increase, decrease) the relative humidity.
   d. At the dew-point temperature, the relative humidity is (25%, 50%, 75%, 100%).
   e. When condensation occurs, heat is (absorbed, released) by water vapor.
   f. Rising air (warms, cools) by (expansion, compression).
   g. In the early morning hours when the daily air temperature is often coolest, relative humidity is generally at its (lowest, highest).

2. What is the dew-point temperature of a kilogram of air when a psychrometer measures an 8°C dry-bulb temperature and a 6°C wet-bulb reading?

   Dew-point temperature = _____°C

3. Explain the principle that governs the operation of a psychrometer for determining relative humidity.

   _____

   _____

   _____

4. Describe the adiabatic process and how it is responsible for causing condensation in the atmosphere.

   _____

   _____

   _____

5. In the section of the exercise "Adiabatic Processes," question 38, what was the altitude where condensation occurred as the air was rising over the mountain?

   _____ meters

6. Place each of the following statements in proper sequence (1 for the first) leading to the development of clouds.

   _____: dew-point temperature reached

   _____: air begins to rise

   _____: condensation occurs

   _____: adiabatic cooling

7. Refer to Diagram 14.2. Use the indicated parallels of latitude as a guide to write the names of the global pressure zones and wind belts at their proper locations.

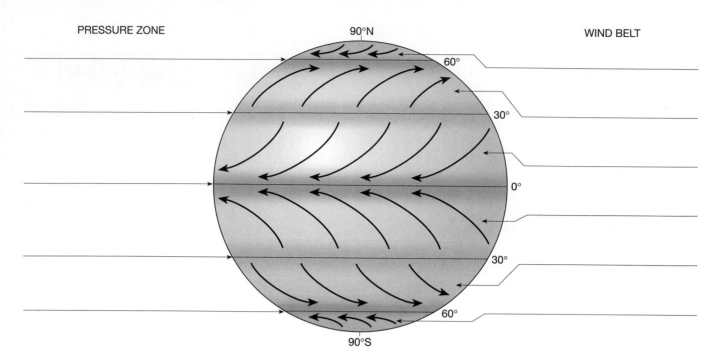

PRESSURE ZONE

WIND BELT

90°N

60°

30°

0°

30°

60°

90°S

DIAGRAM 14.2

8. On Diagram 14.3, illustrate, by labeling the isobars, a typical Northern Hemisphere cyclone and anticyclone. Use arrows to indicate the movement of surface air associated with each pressure cell.

Cyclone

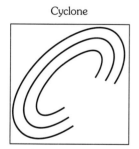

Anticyclone

DIAGRAM 14.3

Air Masses, the Middle-Latitude Cyclone, and Weather Maps

For many people living in the middle latitudes, weather patterns are the result of the movements of large bodies of air and the associated interactions among the weather elements. Exercise Fifteen investigates those atmospheric phenomena that most often influence our day-to-day weather—air masses, fronts, and traveling middle-latitude cyclones. Using the standard techniques for plotting weather station data, the exercise concludes with the preparation and analysis of a typical December surface weather map.

OBJECTIVES

After you have completed this exercise, you should be able to

1. Discuss the characteristics, movements, and source regions of North American air masses.
2. Define and draw a profile of a typical warm front.
3. Define and draw a profile of a typical cold front.
4. Diagram and label all parts of an idealized, mature, middle-latitude cyclone.
5. Interpret the data presented on a surface weather map.
6. Prepare a simple surface weather map using standard techniques.
7. Use a surface weather map to forecast the weather for a city.

TEXTBOOK REFERENCE

Chapter 15

MATERIALS

colored pencils

Materials Supplied by Your Instructor

United States map or atlas

TERMS

front	cold front	instability
air mass	occluded front	anticyclone
source region	adiabatic cooling	middle-latitude cyclone
warm front	polar front	wave cyclone

INTRODUCTION

Changes in daily weather are frequently associated with moving masses of air with contrasting temperature and moisture characteristics. One important aspect of weather forecasting involves recognizing and predicting the paths of these bodies of air.

Of particular importance is the boundary between contrasting air masses, called a **front.** As fronts move over an area, they often bring precipitation that is followed by a change in air mass type. Although there are many causes for the day-to-day variations in the weather, the most obvious changes are those associated with the passage of fronts.

AIR MASSES

An **air mass** is a large body of air that has relatively uniform temperature and moisture characteristics. The area where an air mass acquires its traits is called a **source region.** For example, air with a source region over cool ocean water tends to become cool and moist, while air that stagnates over the American Southwest in summer becomes hot and dry.

Air masses are set into motion by passing high and low pressure cells. When the air mass moves out of its source region, its temperature and moisture conditions are carried with it.

Review the section on air masses in Chapter 15. Then answer questions 1–8.

1. Air masses are classified according to their source region. Explain the meaning of each of the following air mass classification letters.

c: _____   P: _____

m: _____   T: _____

Figure 15.1 shows the source regions and directions of movement of the air masses that play an important role in the weather of North America. Use the figure to answer questions 2–8.

2. Label each of the following North American air masses on Figure 15.1. Then list the location of the source region of each in the following space.

Source Region

cP: _____

cT: _____

mP: _____

mT: _____

3. What would be the winter temperature and moisture characteristics of each of the following air masses?

Temperature   **Moisture**

cP: _____   _____

mP: _____   _____

mT: _____   _____

Notice the paths of air masses indicated by the arrows on Figure 15.1.

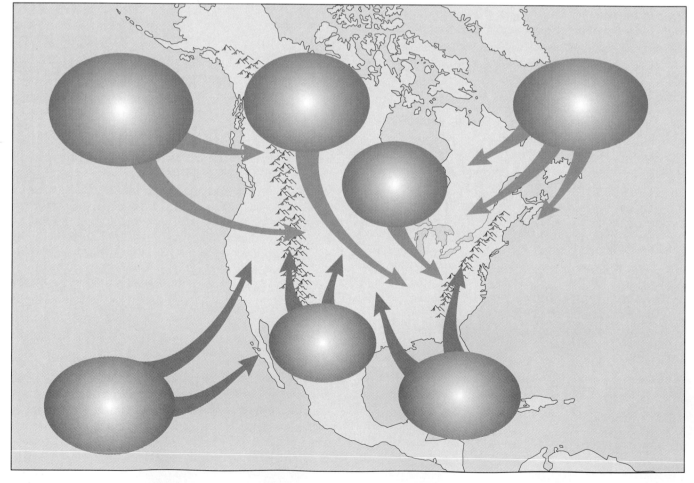

FIGURE 15.1 Source regions of North American air masses.

4. The general movement of air masses across North America is (east to west, west to east). Circle your answer.

5. How does the movement of air masses across North America correspond to the global flow of wind over the continent?

_____

6. Which air masses would have the greatest influence on the weather east of the Rocky Mountains?

_____

7. A (cP, mT) air mass would supply the greatest amount of moisture east of the Rocky Mountains. Circle your answer.

8. A (cP, mP) air mass has the greatest influence on the weather along the northwest Pacific coast. Circle your answer.

FRONTS

A front is a surface of contact between air masses of different densities. One air mass is often warmer, less dense, and higher in moisture content than the other. There is little mixing of air across a front, and each air mass retains its basic characteristics. A **warm front,** shown on a weather map by the symbol ●●●●● , occurs where warm air occupies an area formerly covered by cooler air. A **cold front,** indicated on a map with the symbol ▲▲▲▲ , forms when cold air actively advances into a region occupied by warmer air. An **occluded front,** shown on a weather map with the symbol ●▲●▲ , develops when a cold front overtakes a warm front and warm air is wedged above cold surface air.

Fronts typically act as barriers or walls over which air must rise. When it rises, air will expand and experience **adiabatic cooling.** As a consequence, clouds and precipitation often occur along fronts.

9. See Figure 15.1. Where in North America are air masses likely to collide and fronts develop?

_____

10. In the central United States, east of the Rocky Mountains, a (cP, mT) air mass will most likely be found north of a front, and a (cP, mT) mass to the south. Circle your answers.

Figure 15.2 illustrates profiles through typical cold and warm fronts. Observe the profiles closely and then answer questions 11–17.

11. Along the (cold, warm) front, the cold air is the aggressive or "pushing" air. Circle your answer.

12. Along the (cold, warm) front, the warm air rises at the steepest angle.

13. Along which front are extensive areas of stratus clouds and periods of prolonged precipitation most probable? Explain why you expect longer periods of precipitation to be associated with this type of front.

_____ front. Explanation: _____

_____

14. Assume that the fronts are moving from left to right in Figure 15.2. A drop in temperature is most likely to occur with the passing of a (cold, warm) front. Circle your answer.

The air following a cold front is frequently cold, dense, and subsiding.

15. (Clear, Cloudy) conditions are most likely to prevail after a cold front passes. Explain the reason for your choice with reference to the adiabatic process.

_____

_____

16. Clouds of vertical development and perhaps thunderstorms are most likely to occur along a (cold, warm) front.

17. As a (cold, warm) front approaches, clouds become lower, thicker, and cover more of the sky.

MIDDLE-LATITUDE CYCLONE

Contrasting air masses frequently collide in the area of the *subpolar lows.* In this region, often called the **polar front,** warm, moist air comes in contact with cool, dry air in an area of low pressure. These conditions present an ideal situation for atmospheric **instability,** rising air, adiabatic cooling, condensation, and precipitation. In contrast, in areas of high pressure, called **anticyclones,** the air typically is subsiding.

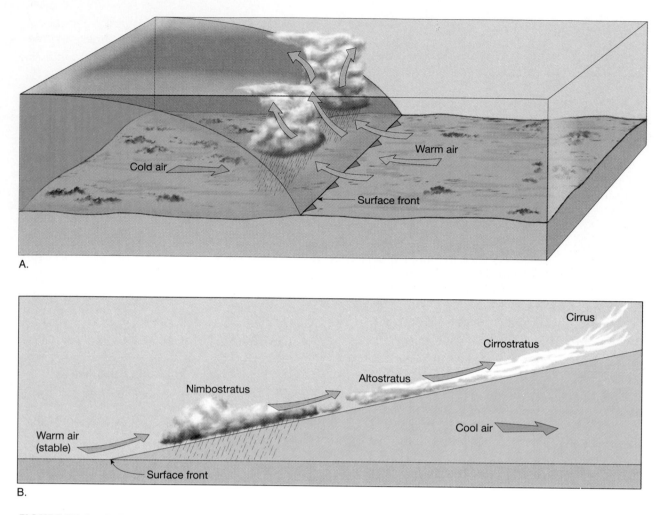

FIGURE 15.2 **A.** Typical cold front profile. **B.** Typical warm front profile.

In the Northern Hemisphere, the westerly winds to the south of the polar front and the easterly winds to the north cause a wave with counterclockwise (cyclonic) rotation to form along the frontal surface. As the low-pressure system called a **middle–latitude** (or **wave**) **cyclone** evolves, it follows a general eastward path across the United States, bringing a sequence of passing fronts and changing weather.

Figure 15.3 illustrates an idealized, mature, middle-latitude cyclone. After you review Chapter 15 of your text, use Figure 15.3 to complete questions 18–28.

18. On Figure 15.3:

a. Label the cold front, warm front, and occluded front.
b. Draw arrows showing the surface wind directions at points A, C, E, F, and G.

c. Label the sectors most likely experiencing precipitation with the word "precipitation."

19. The surface winds in the cyclone are (converging, diverging). Circle your answer.

20. The air in the center of the cyclone will be (subsiding, rising). What effect will this have on the potential for condensation and precipitation? Explain your answer.

_____

_____

_____

21. As the middle-latitude cyclone moves eastward, the barometric pressure at point A will be (rising, falling). Circle your answer.

22. After the warm front passes, the wind at point B will be from the (south, north).

23. Describe the changes in wind direction and barometric pressure that will likely occur at point D after the cold front passes.

_____

_____

24. Considering the typical air mass types and their locations in a middle-latitude cyclone, the amount of water vapor in the air will most likely (increase, decrease) at A after the warm front passes.

25. The quantity of moisture in the air at point D will most likely (increase, decrease) after the cold front passes.

26. Use Figure 15.3 to describe the sequence of weather conditions—barometric changes, wind directions, humidity, precipitation, etc.—expected for a city as the cyclone moves and the city's relative position changes from location A to B, and then to C, D, and E.

_____

_____

_____

_____

27. Near the center of the low, a/an (warm, cold, occluded) front has formed where the cold front has

overtaken the warm front. Circle your answer. Then answer questions 27a and 27b.

a. What happens to the warm mT air in this type of front?

_____

b. With reference to the adiabatic process, why is there a good chance for precipitation with this type of front?

_____

_____

After the entire wave cyclone passes, pressure will rise.

28. Describe the general weather often associated with a high pressure cell, called an anticyclone.

_____

_____

As mentioned previously, middle-latitude cyclones form in the belt of subpolar lows. After you have reviewed the subpolar lows in your notes, text, and Exercise 14, answer the following question by circling the correct responses.

FIGURE 15.3 Mature, middle-latitude cyclone (idealized).

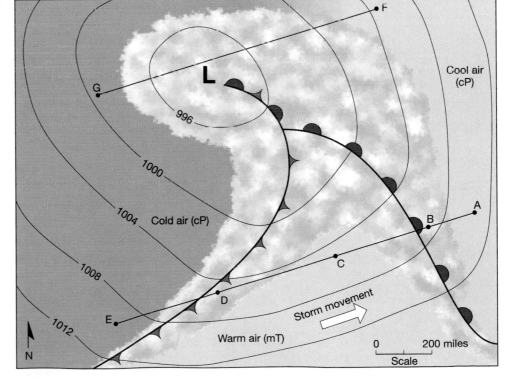

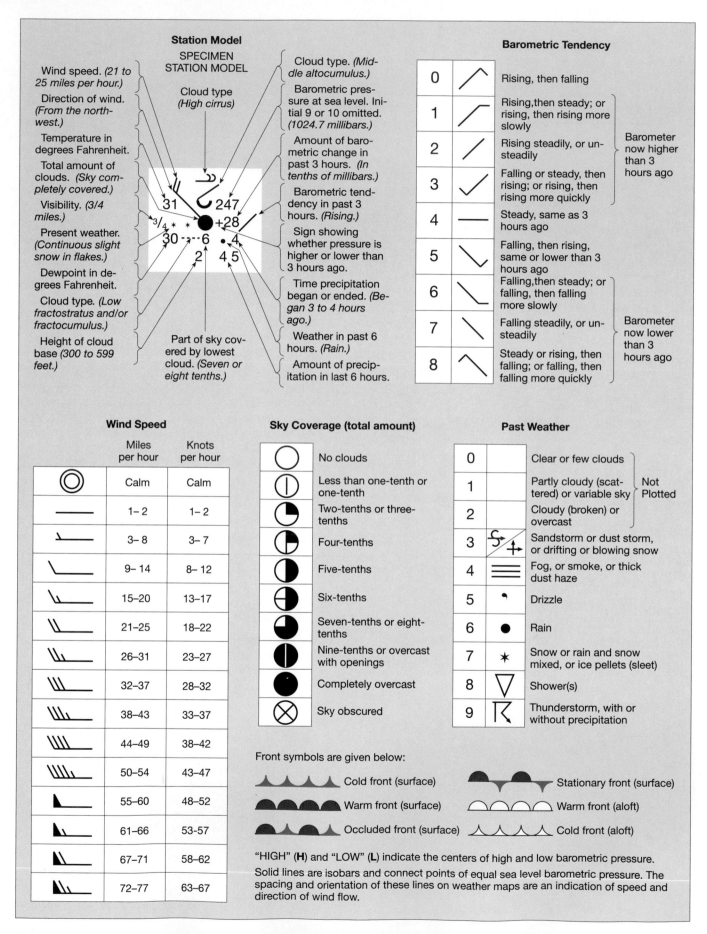

FIGURE 15.4 Specimen weather station model and standard symbols. (Source: Daily Weather Maps, U.S. Department of Commerce)

29. During the (summer, winter) season the belt of subpolar lows and the polar front are farthest south in North America, and the central United States will experience a (greater, lesser) frequency of passing middle-latitude cyclones.

WEATHER STATION ANALYSIS AND FORECASTING

In order to understand, analyze, and predict the weather, observers at hundreds of weather stations throughout the United States collect and record weather data several times a day. This information is forwarded to offices of the National Weather Service where it and satellite data are computer processed and mapped. Weather maps, containing data from throughout the country, are then distributed to any interested individual or agency.

Weather Station Data. To manage the great quantity of information necessary for accurate maps, meteorologists have developed a system for coding weather data. Figure 15.4 illustrates the system and many of the symbols that are used to record data for a weather station.

Figure 15.5 is a coded weather station, shown as it would appear on a simplified surface weather map.

30. Using the specimen station model and explanations shown in Figure 15.4 as your guide, interpret the weather conditions reported at the station illustrated in Figure 15.5.

Percent of sky cover: _____ %

Wind direction: _____

Wind speed: _____ mph

Temperature: _____ °F

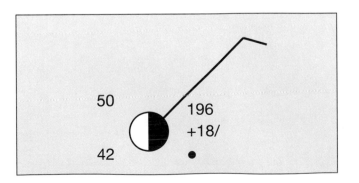

FIGURE 15.5 Coded weather station (abbreviated).

Dew-point temperature: _____ °F

Barometric pressure: _____ millibars

Barometric change in past 3 hours: _____ mb

Weather during the past 6 hours: _____

31. Encode and plot the weather conditions for the following weather station on the station symbol shown below. Use Figures 15.4 and 15.5 as your guides.

The sky is six-tenths covered by clouds. Air temperature is 82°F with a dew point of 50°F. Wind is from the south at 22 miles per hour. Barometric pressure is 1022.4 millibars and has fallen from 1023.0 millibars during the past three hours. There has been no precipitation during the past six hours.

Preparing a Weather Map and Forecast. Table 15.1 contains weather data for several cities in the central and eastern United States on a December day. Data for several of the cities have been plotted on the map, Figure 15.6. Use Table 15.1 and Figure 15.6 to complete questions 32 and 33.

32. Refer to stations where the data has already been plotted and the station model in Figure 15.4. Plot the data for the remaining stations on the map. Then complete the following steps.

Step 1. Beginning with 988 millibars, draw isobars as accurately as possible at four millibar intervals (992 mb, 996 mb, 1000 mb, etc.). (*Note:* You will have to estimate pressures

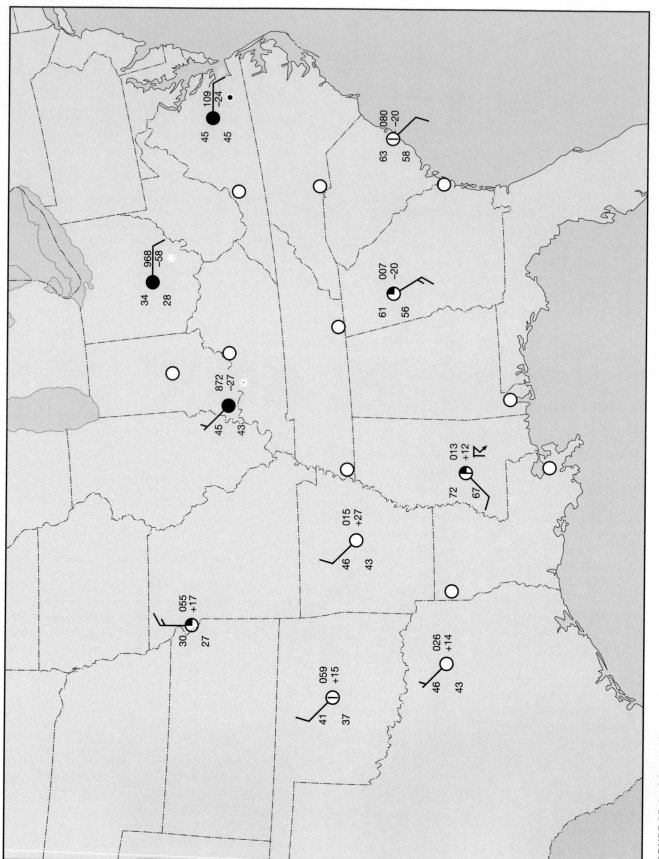

FIGURE 15.6 Map of December weather data for selected cities in the central and eastern United States.

TABLE 15.1 December surface weather data for selected cities in the central and eastern United States

Station	%Cloud Cover	Wind Direction	Wind Speed (mph)	Temp.	Dew Pt. Temp.	Pressure (mb)	Pressure 3 Hour + or –	Precip.
Atlanta, GA	20	SE	18	61	56	1000.7	–2.0	
Charleston, SC	10	SE	12	63	58	1008.0	–2.0	
Charlotte, NC	70	SW	14	54	49	1003.4	–4.4	Drizzle
Chattanooga, TN	60	SW	12	66	60	995.3	–2.5	Drizzle
Columbus, OH	100	E	10	34	28	996.8	–5.8	Snow
Evansville, IN	100	NW	7	45	43	987.2	–2.7	Snow
Fort Worth, TX	0	NW	5	46	43	1002.6	+1.4	
Indianapolis, IN	100	NE	30	34	32	993.2	–5.6	Snow
Jackson, MS	40	SW	10	72	67	1001.3	+1.2	Thunderstorm
Kansas City, MO	30	N	18	30	27	1005.5	+1.7	
Little Rock, AR	0	NW	10	46	43	1001.5	+2.7	
Louisville, KY	100	E	12	34	34	993.0	–4.2	Snow
Memphis, TN	80	NW	12	50	45	996.7	+5.8	
Mobile, AL	60	SW	10	72	68	1004.3	–.3	
New Orleans, LA	20	SW	1?	75	70	1003.9	–.1	
Oklahoma City, OK	10	NW	13	41	37	1005.9	+1.5	
Richmond, VA	100	E	1C	45	45	1010.9	–2.4	Rain
Roanoke, VA	100	SE	10	39	39	1007.5	–3.6	Snow
Savannah, GA	30	SE	7	61	55	1007.6	–2.0	
Shreveport, LA	0	NW	8	46	43	1002.6	+1.4	

between cities to determine the location of isobars. Also, it may be a good idea to first sketch the isobars lightly in pencil.)

Step 2. By observing the data plotted on the map, determine the locations of the cold and warm fronts. Using the proper symbols, label the cold and warm fronts as accurately as possible on the map.

Step 3. Label the air mass types that are most likely to be located to the northwest and to the southeast of the cold front.

Step 4. Indicate areas of precipitation by lightly shading the map with a pencil.

33. Assume that the middle-latitude cyclone illustrated on the map was centered in Oklahoma the previous day and is moving northeastward. Indicate your forecast (temperature, wind direction, probability of precipitation, cloud cover, and barometric pressure) for the next 12–24 hours at the following locations.

Chattanooga, TN: _____

_____

_____

Little Rock, AR: _____

_____

_____

Washington, D.C.: _____

_____

_____

Raleigh, NC: _____

_____

_____

REVIEW

Having completed the exercise, you should know the following:

1. Basic information regarding the source regions, characteristics, classification, and movements of the air masses that affect the weather of North America.
2. A front is a boundary separating air masses of different densities.
3. The structures of idealized cold, warm, and occluded fronts.
4. Why precipitation occurs along fronts.
5. Middle-latitude cyclones originate in the subpolar low along the polar front.
6. The structure of an idealized middle-latitude cyclone.
7. Weather station data are placed on a map in an orderly manner using standard symbols.
8. The procedure used to construct a simple surface weather map.
9. How a surface weather map is interpreted when making forecasts.

Air Masses, the Middle-Latitude Cyclone, and Weather Maps

SUMMARY/REPORT PAGE

Date Due: _____

Name: _____

Date: _____

Class: _____

After you have finished Exercise Fifteen, complete the following questions. You may have to refer to the exercise for assistance or to locate specific answers. Be prepared to submit this summary/report to your instructor at the designated time.

1. List the source region(s) and winter temperature/moisture characteristics of each of the following North American air masses.

   cP: _____

   mT: _____

   mP: _____

2. In the following space, diagram a profile (side view) of an idealized cold front. Label the cold air, warm air, and sketch the probable cloud type at the appropriate location. Draw an arrow on the diagram showing the direction of movement of the front.

Cold Front Profile

3. Indicate the type of front (cold, warm, or occluded) best described by each of the following statements.

   a. Steep wall of cold air: _____

   b. Warm air replaces cool air: _____

   c. Thunderstorms: _____

   d. Drop in temperature: _____

   e. After passing, wind comes from the south: _____

   f. Narrow belt of precipitation: _____

   g. Cold front overtakes warm front: _____

   h. Gradual rise of warm air over cool air: _____

4. Describe the sequence of weather events that a city would experience as an idealized, mature, middle-latitude cyclone that has not developed an occluded front passes over it. Assume that the center of the wave cyclone passes, west to east, 150 miles to the north of the city. Using a diagram may be helpful.

   _____

   _____

   _____

   _____

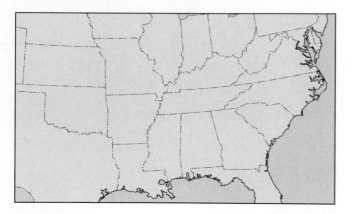

DIAGRAM 15.1

5. Refer to Diagram 15.1. Draw a sketch of the December weather map that you prepared at the end of the exercise, question 32. Show isobars and wind direction arrows. Indicate and label the fronts.

6. Based on the December weather map that you constructed at the end of the exercise, from question 33, what was your forecast for Little Rock, AR?

_____

_____

Global Climates and the Human Impact

Exercise Sixteen investigates world climates using the system of climate classification devised by Wladimir Koeppen. Climographs for several climatic types will be prepared and the global distribution of climates will be examined. The exercise concludes with an examination of the human impact on climate and weather.

OBJECTIVES

After you have completed this exercise, you should be able to

1. Read and prepare a climograph.
2. List the criteria used to define each principal climatic type.
3. Describe the general location of each principal climate group.
4. List the primary atmospheric pollutants and their sources.
5. Discuss the effect that cities have on weather elements.
6. Describe greenhouse warming and its consequences.
7. List some of the natural events that have, or could, cause climate to change.

TEXTBOOK REFERENCE

Appendix F and Chapter 16

MATERIALS

ruler
calculator

Materials Supplied by Your Instructor

world map or atlas

TERMS

climate	Koeppen system	photochemical
climatology	climograph	reaction

INTRODUCTION

No investigation of the atmosphere is complete without examining the global distribution of the major atmospheric elements and the impact that humans have on weather and climate. Each day we become more aware of the complex relations and delicate balances that exist in the environment. To help understand this complexity, scientists have devised a variety of classification systems that simplify and describe the general weather conditions that occur at various places on the earth.

CLIMATE

Climate may be defined as the synthesis, or summary, of weather conditions at a particular place over a long period of time. Climatic classification simplifies the complex distribution of the weather elements for analysis and explanation.

Climatology involves grouping those areas that have similar weather characteristics. Temperature and precipitation are the two elements most commonly used in climate classifications. However, other methods, using different criteria, have also been developed. It should be remembered that any classification is artificial, and its value depends upon the intended use.

Appendix F of your text presents the climatic classification system devised by Dr. Wladimir Koeppen (1846–1940). Since its introduction, the **Koeppen system,** with some modification, has become the best-known and most-used classification for presenting the general world pattern of climates.

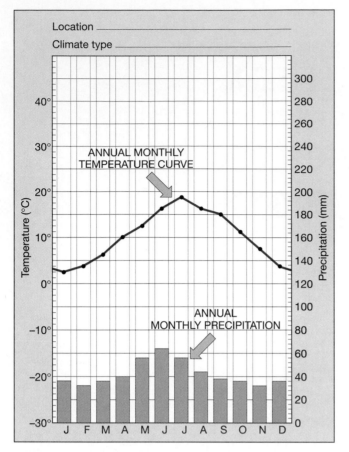

Location _____

Climate type _____

FIGURE 16.1 Typical climograph. Letters along the bottom margin represent the months, **J**anuary through **D**ecember. Average monthly temperatures (°C) are plotted with a single line using the scale on the left axis. Precipitation for each month (in mm) is plotted with a bar using the right axis.

Using a Climograph. Temperature and precipitation are presented on a **climograph** such as the one shown in Figure 16.1. Average monthly temperatures are connected with a single line and read from the temperature scale on the left axis. Average precipitation for each month is represented with a bar and read from the precipitation scale on the right axis.

Refer to the climograph, Figure 16.1, to answer questions 1–6. Circle the correct response.

1. At the place represented by the climograph, the month of (May, June, January) receives the greatest amount of precipitation.
2. The lowest temperature occurs during the month of (July, December, January).
3. The approximate average annual temperature is (0, 10, 20)°C.
4. The total annual precipitation is approximately (240, 480, 720) millimeters.
5. The place represented by the climograph is in the (Northern, Southern) Hemisphere.
6. The (summer, winter) months receive the greatest amount of precipitation.

KOEPPEN SYSTEM OF CLIMATIC CLASSIFICATION

The Koeppen system of climatic classification employs five principal climate groups. Four of the groups are defined on the basis of temperature characteristics and the fifth has precipitation as its primary criterion. Further division of the groups into climatic types allows for a more detailed climatic description. Koeppen believed that the distribution of natural vegetation was the best expression of the totality of climate. Therefore, the boundaries he chose were based largely on the limits of certain plant associations.

7. On Table 16.1, list the names and characteristics of each principal climate group next to its designated classification letter. Use Appendix F of the text as a reference.

TABLE 16.1 Characteristics of the principal climate groups

Climate Group	Name	Temperature and/or Precipitation Characteristics
A:		
B:		
C:		
D:		
E:		

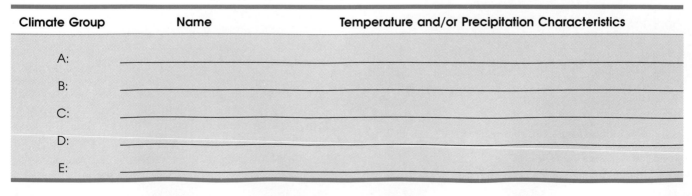

TABLE 16.2 Climatic data for representative stations

	J	F	M	A	M	J	J	A	S	O	N	D	Year
Iquitos, Peru (Af); lat. 3°39'S; 115 m													
Temp. (°C)	25.6	25.6	24.4	25.0	24.4	23.3	23.3	24.4	24.4	25.0	25.6	25.6	24.7
Precip. (mm)	259	249	310	165	254	188	168	117	221	183	213	292	2619
Rio de Janeiro, Brazil (Aw); lat. 22°50'S; 26 m													
Temp. (°C)	25.9	26.1	25.2	23.9	22.3	21.3	20.8	21.1	21.5	22.3	23.1	24.4	23.2
Precip. (mm)	137	137	143	116	73	43	43	43	53	74	97	127	1086
Faya, Chad (BWh); lat. 18°00'N; 251 m													
Temp. (°C)	20.4	22.7	27.0	30.6	33.8	34.2	33.6	32.7	32.6	30.5	25.5	21.3	28.7
Precip. (mm)	0	0	0	0	0	2	1	11	2	0	0	0	16
Salt Lake City, Utah (BSk); lat. 40°46'N; 1288 m													
Temp. (°C)	−2.1	0.9	4.7	9.9	14.7	19.4	24.7	23.6	18.3	11.5	3.4	−0.2	10.7
Precip. (mm)	34	30	40	45	36	25	15	22	13	29	33	31	353
Washington, D.C. (Cfa); lat. 38°50'N; 20 m													
Temp. (°C)	2.7	3.2	7.1	13.2	18.8	23.4	25.7	24.7	20.9	15.0	8.7	3.4	13.9
Precip. (mm)	77	63	82	80	105	82	105	124	97	78	72	71	1036
Brest, France (Cfb); lat. 48°24'N; 103 m													
Temp. (°C)	6.1	5.8	7.8	9.2	11.6	14.4	15.6	16.0	14.7	12.0	9.0	7.0	10.8
Precip. (mm)	133	96	83	69	68	56	62	80	87	104	138	150	1126
Rome, Italy (Csa); lat. 41°52'N; 3 m													
Temp. (°C)	8.0	9.0	10.9	13.7	17.5	21.6	24.4	24.2	21.5	17.2	12.7	9.5	15.9
Precip. (mm)	83	73	52	50	48	18	9	18	70	110	113	105	749
Peoria, Illinois (Dfa); lat. 40°45'N; 180 m													
Temp. (°C)	−4.4	−2.2	4.4	10.6	16.7	21.7	23.9	22.7	18.3	11.7	3.8	−2.2	10.4
Precip. (mm)	46	51	69	84	99	97	97	81	97	61	61	51	894
Verkhoyansk, Russia (Dfd); lat. 67°33'N; 137 m													
Temp. (°C)	−46.8	−43.1	−30.2	−13.5	2.7	12.9	15.7	11.4	2.7	−14.3	−35.7	−44.5	−15.2
Precip. (mm)	7	5	5	4	5	25	33	30	13	11	10	7	155
Ivigtut, Greenland (ET); lat. 61°12'N; 129 m													
Temp. (°C)	−7.2	−7.2	−4.4	−0.6	4.4	8.3	10.0	8.3	5.0	1.1	−3.3	−6.1	0.7
Precip. (mm)	84	66	86	62	89	81	79	94	150	145	117	79	1132
McMurdo Station, Antarctica (EF); lat. 77°53'S; 2 m													
Temp. (°C)	−4.4	−8.9	−15.5	−22.8	−23.9	−24.4	−26.1	−26.1	−24.4	−18.8	−10.0	−3.9	−17.4
Precip. (mm)	13	18	10	10	10	8	5	8	10	5	5	8	110

Table 16.2 contains climatic data for several stations that are representative of Koeppen climatic types. Use Appendix F of your text and the data in Table 16.2 to answer the following questions.

Humid Tropical (type A) Climates.

With the exception of the dry climates, no other climate covers as large an area on the earth as the humid tropical climates.

8. What temperature criterion is used for defining an A climate?

9. Plot the monthly temperature and precipitation data for Iquitos, Peru, an A climate, given in Table 16.2 on the climate chart, Diagram 16.1.

Use the Iquitos, Peru, climograph you prepared in question 9 to answer questions 10–13.

10. What is the *annual temperature range* (difference between highest and lowest monthly temperatures) for Iquitos?

_____ °C

11. Describe the yearly variability of temperature for A climates.

12. Notice that Iquitos receives an average of 2,619 mm of precipitation per year. How many inches of precipitation per year would this equal? (*Hint:*

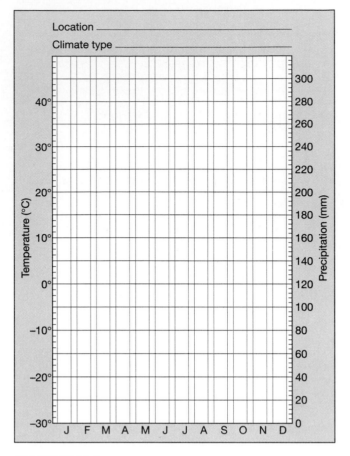

Location _____

Climate type _____

DIAGRAM 16.1

You may have to refer to the conversion tables located on the inside back cover of this manual.)

2,619 mm equals _____ inches

13. The precipitation at Iquitos is (concentrated in one season, distributed throughout the year). Circle your answer.

Use the world climate map, Figure 16.2, to answer questions 14–16.

14. In what latitude belt are A climates located?

_____

15. The most extensive areas of tropical rain forest (Af) climates are located (along coasts, in the interiors) of continents. Circle your answer.

16. Considering the locations of A climates, would weather fronts or columns of rapidly rising, hot, surface air be most likely responsible for the precipitation? Explain your answer.

_____

_____

Dry (type B) Climates. Of all the climate groups, the dry climates cover the greatest portion of the earth's surface. To be classified as a dry climate does not necessarily imply little or no precipitation, but rather indicates that the yearly precipitation is not as great as the potential loss of moisture by evaporation.

17. What are the three variables that the Koeppen classification uses to establish the boundary between dry and humid climates?

1) _____

2) _____

3) _____

18. What name is applied to the following climatic types?

BW: _____

BS: _____

19. What is the primary cause of arid climates in the tropics?

_____

20. What factors contribute to the formation of arid climates in the middle latitudes?

_____

_____

To answer questions 21 and 22, refer to the world climate map, Figure 16.2.

21. At what latitudes, North and South, are the most extensive arid areas located?

_____

22. The Sahara desert in northern Africa is the largest area in the world with a BWh climate. What are some other regions that have the same climate?

_____

Humid Middle-Latitude Climates with Mild Winters (type C).

A large percentage of the world's population is located in areas with C climates. It is a climate characterized by weather contrasts brought about by changing seasons. On the average, the regions of C climates are dominated by contrasting air masses and associated middle-latitude cyclones.

23. What temperature criterion is used to define the boundary of C climates?

_____

24. Using the data in Table 16.2, on Diagram 16.2, prepare climographs for Washington, D.C., and Rome, Italy.

Answer questions 25–28 using the climographs you have constructed for Washington, D.C., and Rome, Italy, Diagram 16.2.

25. In what manner are the temperature curves for each of the two cities similar?

_____

26. How does the annual distribution of precipitation vary between the two cities?

_____

27. What is the difference between a Cf (Washington, D.C.) climate and a Cs (Rome, Italy) climate?

_____

28. (Weather fronts, Columns of rapidly rising, hot, surface air) are most likely responsible for the winter precipitation in Cf climates. Circle your answer.

Use the world climate map, Figure 16.2, to answer questions 29–31.

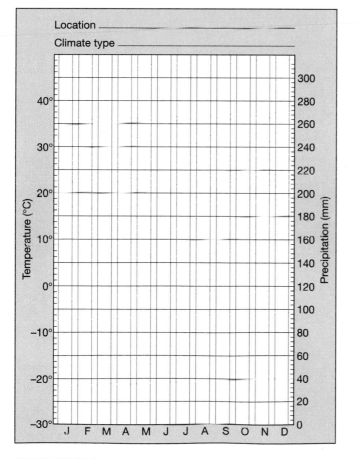

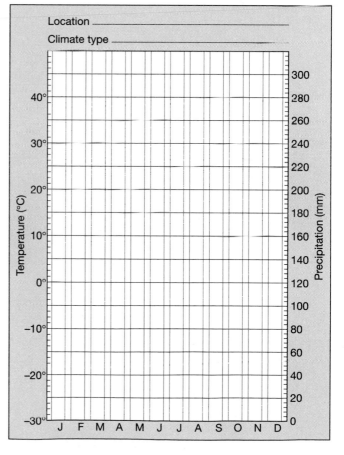

DIAGRAM 16.2

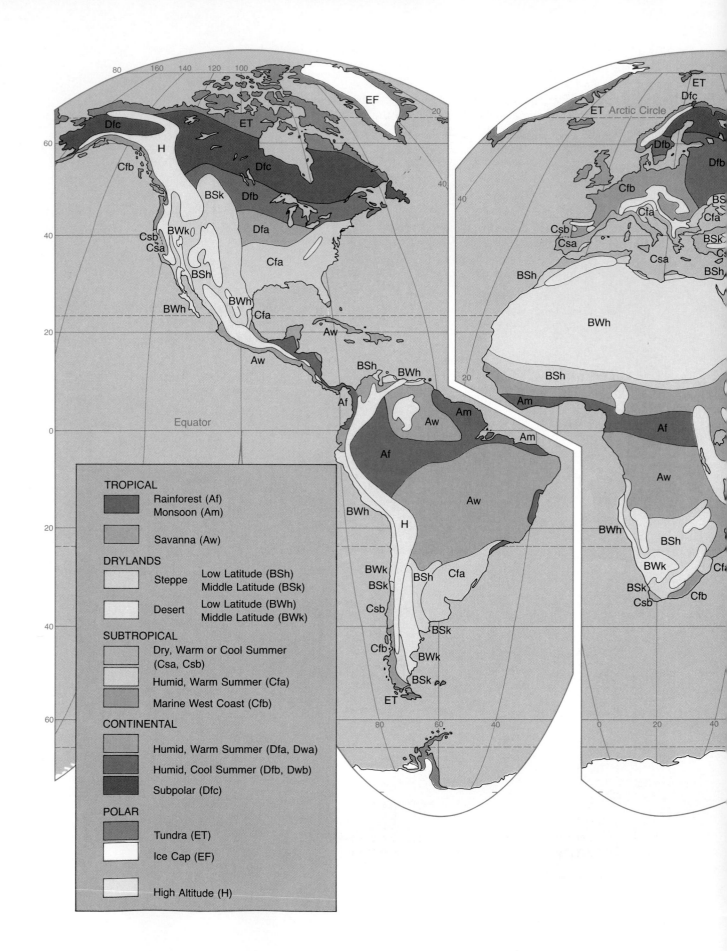

TROPICAL
Rainforest (Af)
Monsoon (Am)
Savanna (Aw)

DRYLANDS
Steppe   Low Latitude (BSh)
          Middle Latitude (BSk)
Desert   Low Latitude (BWh)
          Middle Latitude (BWk)

SUBTROPICAL
Dry, Warm or Cool Summer
(Csa, Csb)
Humid, Warm Summer (Cfa)
Marine West Coast (Cfb)

CONTINENTAL
Humid, Warm Summer (Dfa, Dwa)
Humid, Cool Summer (Dfb, Dwb)
Subpolar (Dfc)

POLAR
Tundra (ET)
Ice Cap (EF)

High Altitude (H)

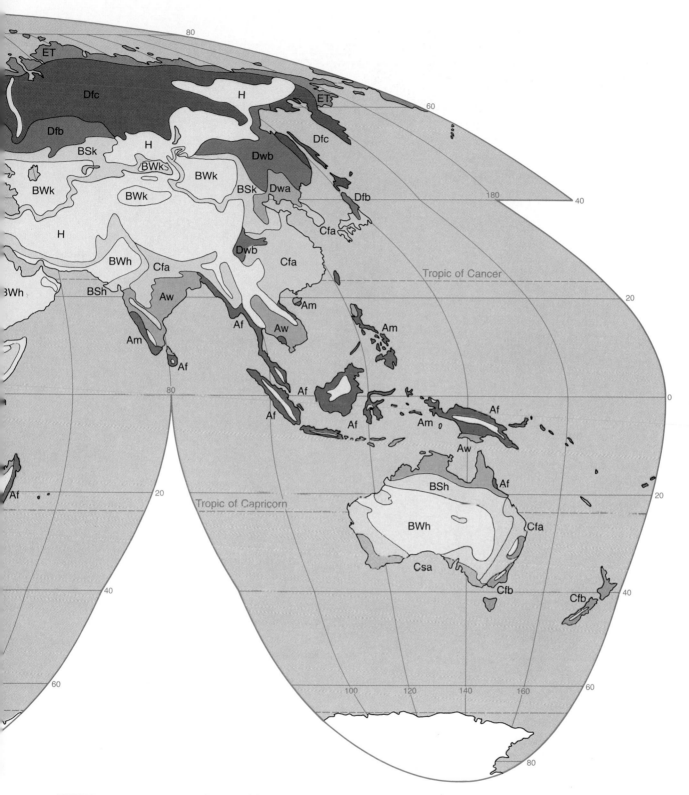

FIGURE 16.2 Climates of the world (Koeppen). (Adapted from F. Willard Miller, *Physical Geography*, Columbus, Ohio, Macmillan/Merrill, 1985. Plate 2)

29. What countries in Asia have areas of climate similar to that of Washington, D.C.?

_____

30. What Southern Hemisphere countries have climates similar to that of Washington, D.C.?

_____

31. What state has a climate similar to that of Rome, Italy?

_____

Humid Middle-Latitude Climates with Severe Winters (type D).

The harsh winters and a relatively short growing season restrict agricultural activity in much of the area of D climates. The northern portions of D climate regions are covered by coniferous forests, with lumbering being a significant economic activity.

32. What criteria are used for defining a D climate?

_____

33. Use the data from Table 16.2 to plot a climograph for Peoria, Illinois, on Diagram 16.3.

Use the climograph you have constructed for Peoria, Illinois, Diagram 16.3, to answer questions 34–36.

34. What is the annual range of temperature in Peoria, Illinois?

_____ °C

35. How does the annual range of temperature in Peoria, Illinois, compare to the temperature ranges in Iquitos, Peru, and Rome, Italy?

_____

36. During what season does Peoria receive its greatest precipitation and how does this compare with the seasonal distribution of precipitation for Rome, Italy?

_____

_____

Use the world climate map, Figure 16.2, to answer questions 37–39.

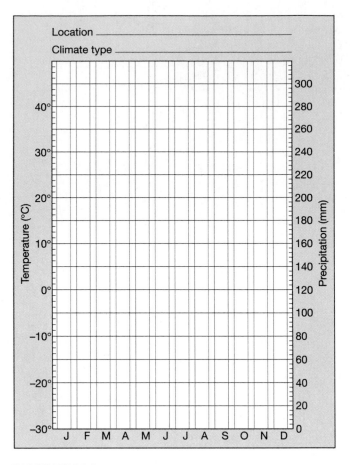

Location _____

Climate type _____

DIAGRAM 16.3

37. Which continent has the greatest continuous expanse of D climates?

_____

38. D climates are located only in the (Northern, Southern) Hemisphere. Circle your answer.

39. Suggest a reason why D climates are located in only the hemisphere you selected in question 38.

_____

Polar (type E Climates).

The polar climates, found at high latitudes and scattered high altitudes in mountains, are regions of cold temperatures and sparse population. Low evaporation rates allow these areas to be classified as humid, even though the annual precipitation is modest.

40. What criterion is used for defining E climates?

_____

41. Contrast the characteristics and locations of the two basic polar climates.

ET climates: _____

_____

EF climates: _____

_____

Climate and Altitude. Although often not included with the principal climate groups, *high-altitude,* or *highland (type H)* climates exist in all climatic regions. They are the result of the changes in radiation, temperature, humidity, precipitation, and atmospheric pressure that take place with elevation and orientation of mountain slopes.

Examine the climatic data for Quito, Ecuador, in Table 16.3. Quito is located in a region where A climates are expected. However, its altitude of 2,770 meters (9,086 feet) changes its Koeppen classification.

42. Use Table 16.3 and Table F.1 "Koeppen system of climatic classification" in Appendix F of your text to determine the climatic classification of Quito, Ecuador.

Climate of Quito, Ecuador: _____

Criteria used for the selection of the climate type

of Quito, Ecuador: _____

_____

43. Locate Quito, Ecuador, on a map or globe. Considering its location, what effect has altitude had on the climatic classification?

_____

44. Why would you expect the vegetation in the area of Quito to be different from that found at Guayaquil, Ecuador, a city located on the coast?

_____

45. From Figure 16.2, where is the greatest continuous expanse of high-altitude climate located?

_____

HUMAN IMPACT ON CLIMATE AND WEATHER

Recently the human impact on climate and weather has been receiving close attention by scientists. Accelerating urbanization, with an increasing number of factories, office buildings, roads, and homes, alters existing *microclimates* and may contribute to the modification of global climates. In this section you will investigate some of the ways that human activities are changing the atmospheric environment and the possible consequences.

Using Chapter 16 of your text as a reference, answer questions 46–60.

46. Atmospheric pollutants can be grouped into two categories—*primary* and *secondary.* Briefly describe each category.

Primary pollutants: _____

_____

Secondary pollutants: _____

_____

TABLE 16.3 Climatic data for Quito, Ecuador

Quito, Ecuador
lat. 0°, long. 79°W; 2770 m

	J	F	M	A	M	J	J	A	S	O	N	D	Year
Temp. (°C)	13	13	13	13	13	13	13	14	14	13	13	13	13
Precip. (mm)	99	110	142	175	137	43	20	30	69	113	95	93	1126

TABLE 16.4 Estimated nation-wide emissions (millions of metric tons/year)

Source	Carbon Monoxide	Particulates	Sulfur Oxides	Volatile Organics	Nitrogen Oxides	Total
Transportation	40.7	1.4	0.9	6.0	8.4	57.4
Fuel combustion in stationary sources	7.2	1.8	16.4	2.3	10.3	38.0
Industrial processes	4.7	2.5	3.1	8.3	0.6	19.2
Solid waste disposal	1.7	0.3	0.0	0.6	0.1	2.7
Miscellaneous	7.1	1.0	0.0	2.4	0.1	10.6
Total	61.4	7.0	20.4	19.6	19.5	127.9

SOURCE: National Air Pollutant Emission Estimates, 1940-1987, U.S. Environmental Protection Agency Publication No. EPA-450/4-88-022, 1989.

Table 16.4 lists many of the types, sources, and amounts of primary pollutants. Use the table to answer questions 47–49.

TABLE 16.5 Average climatic changes produced by cities

Element	Comparison with Rural Environment
Particulate matter	10 times more
Temperature	
Annual mean	0.5-1.5°C higher
Winter	1-2°C higher
Solar radiation	15-30% less
Ultraviolet, winter	30% less
Ultraviolet, summer	5% less
Precipitation	5-15% more
Thunderstorm frequency	16% more
Winter	5% more
Summer	29% more
Relative humidity	6% lower
Winter	2% lower
Summer	8% lower
Cloudiness (frequency)	5-10% more
Fog (frequency)	60% more
Winter	100% more
Summer	30% more
Wind speed	25% lower
Calms	5-20% more

SOURCE: After Landsberg, Changnon, and others.

47. What is the leading source (by weight) of primary pollutants? How many metric tons of material does this source add to the atmosphere each year?

_____; _____million metric tons/year

48. (Carbon monoxide, Sulfur oxide) is the most abundant primary pollutant. It accounts for approximately (25%, 50%, 75%) of all primary pollutants. Circle your answers.

49. What is the total weight of all primary pollutants added to the atmosphere each year?

_____ million metric tons/year

Many secondary pollutants are the result of **photochemical reactions.**

50. What triggers many photochemical reactions?

_____

51. What is one secondary pollutant produced by the photochemical process?

_____

Table 16.5 shows the average climatic changes produced by cities. Use the table to answer questions 52–55.

52. Compared to rural areas, which elements are increased by urbanization?

_____

_____

53. Compared to rural areas, which elements are decreased by urbanization?

_____

_____

54. Of all the elements listed in Table 16.5, which shows the greatest increase due to urbanization? What might be a contributing factor to the increase?

_____

55. Indicate a possible reason for each of the following effects that cities have on their weather.

Increased frequency of thunderstorms: _____

_____

Lower wind speed: _____

_____

Increased precipitation: _____

_____

One of the major concerns of atmospheric scientists is the effect that increasing levels of carbon dioxide and other gases will have on global climates and the environment in general. Answer questions 56–60 using Chapter 16 of your text as a reference.

▬▬▬▬

56. Explain what is meant by the term *greenhouse warming*.

_____

_____

57. In what way does each of the following contribute to greenhouse warming?

Burning fossil fuels: _____

_____

Removing large areas of forests: _____

_____

58. If the level of greenhouse gases continues to increase at the current rate, what is the predicted increase in global surface temperatures by the second half of the next century?

Between _____ °C and _____ °C

59. What might be the consequences of greenhouse warming on each of the following?

Sea level: _____

_____

Agriculture: _____

_____

Global pattern of precipitation: _____

_____

Paths of large-scale cyclonic storms: _____

_____

Many natural events have, or could, cause global climates to change.

▬▬▬▬

60. List two or three natural events that cause global climates to change.

_____

_____

REVIEW

Having completed the exercise, you should know the following:

1. Climate data are often presented on a climograph.
2. The Koeppen system is the most-used classification for presenting the world pattern of climates.
3. Four of the five Koeppen climate groups are defined on the basis of temperature characteristics.
4. Each of the five Koeppen climate groups has definite characteristics.
5. How the climatic classification of a place is determined.
6. The global distribution of the principal Koeppen climate groups.
7. There are two categories of atmospheric pollutants—primary and secondary.
8. Cities, when compared to rural areas, experience more precipitation, have considerably greater amounts of particulate matter in the air, and higher temperatures.
9. The level of the primary greenhouse gas, carbon dioxide, is increasing at a rate that may significantly alter the global temperature by the second half of the next century.
10. Natural events, such as volcanic activity and the movement of lithospheric plates, are responsible for changes in climate.

Global Climates and the Human Impact

SUMMARY/REPORT PAGE

Date Due: _____

Name: _____

Date: _____

Class: _____

After you have finished Exercise Sixteen, complete the following questions. You may have to refer to the exercise for assistance or to locate specific answers. Be prepared to submit this summary/report to your instructor at the designated time.

1. Explain how a climograph is constructed.

   _____

   _____

   _____

2. In your own words, provide some key words to describe the characteristics of each of the following climate groups.

   A climates: _____

   _____

   B climates: _____

   _____

   C climates: _____

   _____

   D climates: _____

   _____

   E climates: _____

   _____

   H climates: _____

   _____

3. List the general location of each of the following climate groups.

   A climates: _____

   _____

   B climates: _____

   _____

   C climates: _____

   _____

   D climates: _____

   _____

   E climates: _____

   _____

   H climates: _____

   _____

4. Indicate by name the climate group best described by each of the following statements.

   Vast areas of northern coniferous forests: _____

   _____

   Smallest annual range of temperature: _____

   _____

   The highest annual precipitation: _____

   _____

Mean temperature of the warmest month is below 10°C: _____

_____

The result of high elevation and mountain slope orientation: _____

_____

Potential evaporation exceeds precipitation: _____

_____

Very little change in the monthly precipitation and temperature throughout the year: _____

_____

Caused by the subsidence of air beneath high pressure cells: _____

_____

5. Summarize the effects that cities have on their weather and microclimate.

_____

_____

_____

6. What category represents the single greatest source of primary pollutants?

_____

7. Describe some possible long-range consequences of continued greenhouse warming.

_____

_____

_____

Astronomical Observations

Scientific inquiry often starts with the systematic collection of data from which hypotheses and general theories are developed. The study of astronomy begins by carefully observing and recording the changing positions of the sun, moon, planets, and stars. To become proficient in astronomy requires developing the skills necessary to become a keen observer, using both the unaided eye and the telescope.

In this exercise you will observe several celestial objects. Observing and recording the changing positions of the sun, moon, and stars will aid in the interpretation and understanding of their movements in future exercises.

OBJECTIVES

After you have completed this exercise, you will have

1. Records of the changing position of the sun as it rises or sets on the horizon.
2. Measurements of the angle of the sun above the horizon at noon on several days.
3. A record of the phases of the moon over a period of several weeks.
4. Data on the times that the moon rises and sets.
5. Records of the position and motion of stars.
6. An understanding of the parts of a telescope.

TEXTBOOK REFERENCE

Chapters 17 and 19; Appendix E

MATERIALS

meterstick (or yardstick)	calculator
ruler	star chart

Materials Supplied by Your Instructor

telescope(s) (optional)

TERMS

revolution
altitude
rotation

INTRODUCTION

Astronomy is an observational science. Since antiquity observers have noticed that many celestial objects appear in different parts of the sky with certain regularity. To explain these movements ancient astronomers proposed many models of the "universe."

SUN OBSERVATIONS

Many people are unaware that the sun rises and sets at different locations on the horizon each day. As the earth **revolves** about the sun, the orientation of its axis to the sun continually changes. The result is that the location of the rising and setting sun, as well as the **altitude** (angle above the horizon) of the sun at noon, changes throughout the year.

Sunset (or Sunrise) Observations. Use the procedure presented in question 1 to make several observations and recordings of the sun's location on the horizon at sunset or sunrise.

1. Following Steps 1 through 4, record at least four separate observations of the setting or rising sun.

SUNSET (Sunrise) DATA SHEETS

————————————— HORIZON —————————————

Date of observation _____ Time of observation _____

————————————— HORIZON —————————————

Date of observation _____ Time of observation _____

————————————— HORIZON —————————————

Date of observation _____ Time of observation _____

————————————— HORIZON —————————————

Date of observation _____ Time of observation _____

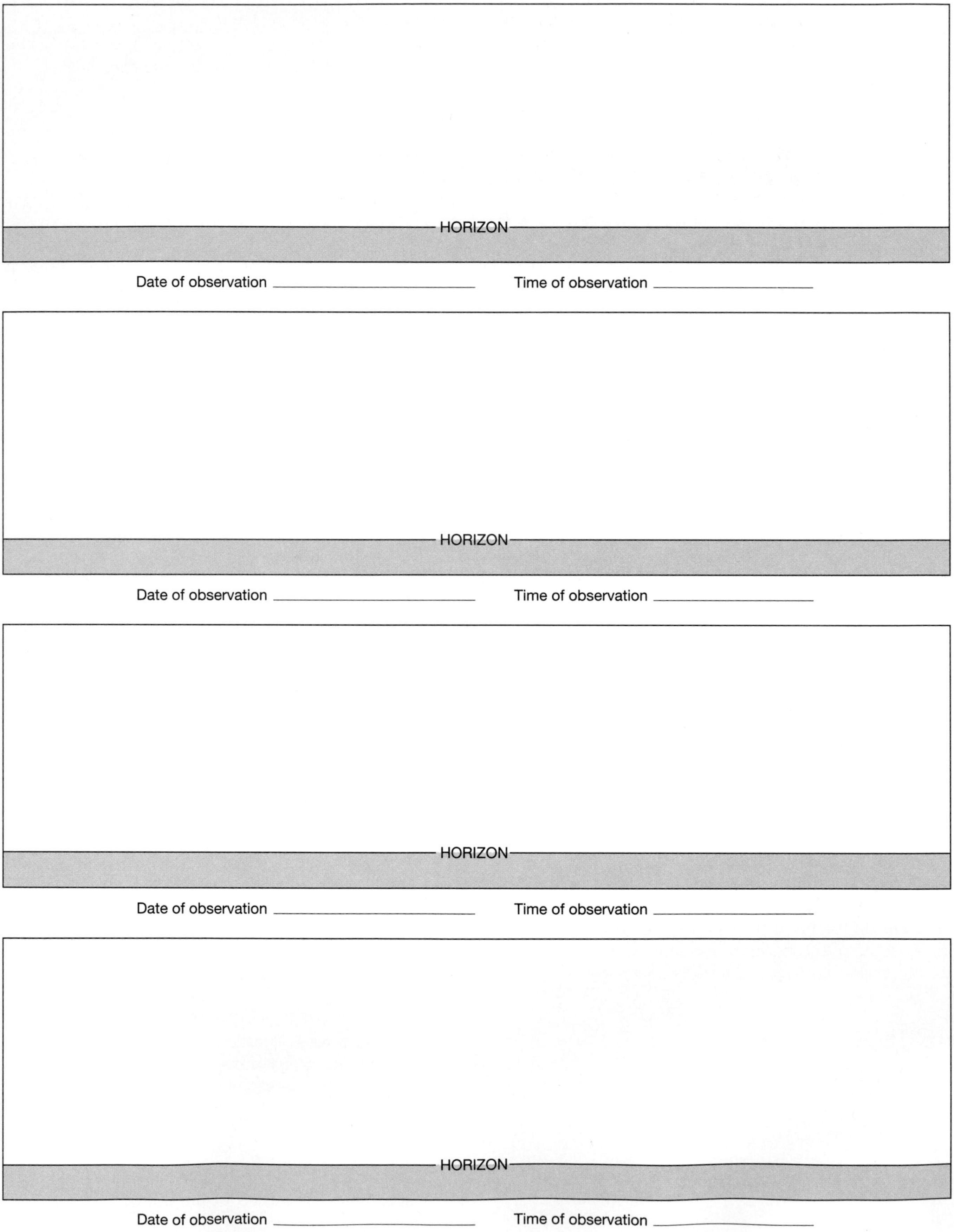

FIGURE 17.1 Sunset (sunrise) data sheets.

Gather the data over a period of several weeks; *wait four or five days between each observation.* The directions are for sunset, although some minor changes will also allow their use for sunrise.

Step 1. Several minutes prior to sunset, estimate where the sun will set on the western horizon. Draw the prominent features (buildings, trees, etc.) to the north and south of the sun's approximate setting position on a sunset data sheet in Figure 17.1 (*Note:* As you observe the sun setting in the west, south will be to your left and north to your right.)

Step 2. As the sun sets, draw its position on the data sheet relative to the fixed features on the horizon. CAUTION: NEVER LOOK DIRECTLY AT THE SUN; EYE DAMAGE MAY RESULT.

Step 3. Note the date and time of your observation on the data sheet.

Step 4. Return to the same location several days later. Repeat your observation and record the results on a new data sheet.

2. Describe the changing location of the sun at sunset, or sunrise, that you have observed over the past several weeks.

Measuring the Noon Sun Angle. Observe the method for measuring the altitude (angle) of the noon sun above the horizon illustrated in Figure 17.2. Then, following the steps listed in question 3, determine the altitude of the sun at noon on several days.

3. To determine the altitude (angle) of the sun at noon:

Step 1. Place a yardstick (a meterstick or ruler will do) vertical to the ground or a table top.

Step 2. When the sun is at its highest position in the sky (noon, standard time, or 1 P.M. daylight savings time, will be close enough), measure the length of the shadow. CAUTION: NEVER LOOK DIRECTLY AT THE SUN; EYE DAMAGE MAY RESULT.

Step 3. Divide the height of the stick by the length of the shadow.

Step 4. Consult Table 17.1 to determine the sun angle. To find the sun angle, locate the number on the table that comes closest to your answer from Step 3, and read the angle listed next to it.

Step 5. Repeat the measurement at exactly the same time on several different days over a period of four or five weeks. Record the date and results of each measurement in the following space.

FIGURE 17.2 Illustration of the method for measuring the angle of the sun above the horizon at noon.

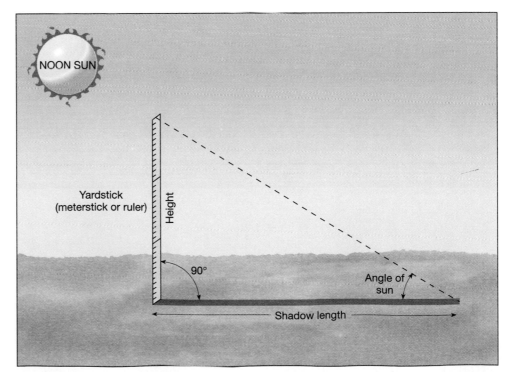

NOON SUN

Yardstick (meterstick or ruler)

Height

90°

Angle of sun

Shadow length

Date: _____ Noon sun angle: _____°

Date: _____ Noon sun angle: _____°

Date: _____ Noon sun angle: _____°

Date: _____ Noon sun angle: _____°

Answer questions 4–6 after you have completed your measurements of the altitude of the noon sun.

4. The altitude of the noon sun has (increased, decreased) over the period of the measurements. Circle your answer.
5. How many degrees has the noon sun angle changed over the period of your observations?

_____ °

6. What is the approximate average change of the noon sun angle per day?

_____ ° per day

TABLE 17.1 Data table for determining the noon sun angle. Select the nearest number to the quotient determined by dividing the height of the stick by the length of the shadow. Read the corresponding sun angle.

If $\dfrac{\text{Height of stick}}{\text{Length of shadow}}$	Then sun angle is	If $\dfrac{\text{Height of stick}}{\text{Length of shadow}}$	Then sun angle is
0.2679	15°	1.235	51°
0.2867	16°	1.280	52°
0.3057	17°	1.327	53°
0.3249	18°	1.376	54°
0.3443	19°	1.428	55°
0.3640	20°	1.483	56°
0.3839	21°	1.540	57°
0.4040	22°	1.600	58°
0.4245	23°	1.664	59°
0.4452	24°	1.732	60°
0.4663	25°	1.804	61°
0.4877	26°	1.881	62°
0.5095	27°	1.963	63°
0.5317	28°	2.050	64°
0.5543	29°	2.145	65°
0.5774	30°	2.246	66°
0.6009	31°	2.356	67°
0.6249	32°	2.475	68°
0.6494	33°	2.605	69°
0.6745	34°	2.748	70°
0.7002	35°	2.904	71°
0.7265	36°	3.078	72°
0.7536	37°	3.271	73°
0.7813	38°	3.487	74°
0.8098	39°	3.732	75°
0.8391	40°	4.011	76°
0.8693	41°	4.332	77°
0.9004	42°	4.705	78°
0.9325	43°	5.145	79°
0.9657	44°	5.671	80°
1.0000	45°	6.314	81°
1.0360	46°	7.115	82°
1.0720	47°	8.144	83°
1.1110	48°	9.514	84°
1.1500	49°	11.430	85°
1.1920	50°		

MOON OBSERVATIONS

Most people have noticed that the shape of the illuminated portion of the moon as observed from Earth changes regularly. However, few take the time to systematically record and explain these changes. Following the procedure presented in question 7, begin your study of the moon by observing its phases and recording your observations.

7. Record at least four observations of the moon by completing each of the following steps regularly at a two- or three-day interval.

   Step 1. On a moon observation data sheet provided in Figure 17.3, indicate the approximate east-west position of the moon in the sky by drawing a circle at the appropriate location. (*Note:* As you look to the south to observe the moon, east will be to your left and west to your right.)

   Step 2. By shading the circle, indicate the shape of the illuminated portion of the moon you observe.

   Step 3. Note the date and time of your observation on the data sheet.

   Step 4. Keep in mind that the approximate time between moonrise on the eastern horizon and moonset on the western horizon is twelve hours. Estimate when the moon may have risen and when it may set. Write your estimate on the data sheet.

   Step 5. Repeat your observation in several days, using a new data sheet.

Answer questions 8–11 after you have completed all your observations of the moon.

8. What happened to the size and shape of the illuminated portion of the moon over the period of your observations?

   _____

   _____

9. The moon moved further (eastward, westward) in the sky with each successive observation. Circle your answer.

10. The times of moonrise and moonset became (earlier, later) with each successive observation. Circle your answer.

11. Based upon your observations and your answers to questions 9 and 10, the moon revolves around the earth from (east to west, west to east).

STAR OBSERVATIONS

Throughout history people have been recording the nightly movement of stars that results from the earth's **rotation,** as well as the seasonal changes in the constellations as the earth revolves about the sun. Early astronomers offered many explanations for the changes before the true nature of the earth's motions was understood in the 17th century.

To best observe the stars, select a suitable dark area on a clear, moonless night. Then complete questions 12 through 19.

12. Make a list of the different colors of the stars you can observe in the sky.

   _____

Select one star that is overhead, or nearly so, and observe its movement over a period of one hour.

13. With your arm extended, approximately how many widths of your fist has the position of the star changed?

   _____ fist widths

14. The star appears to move (eastward, westward) over a period of one hour. Circle your answer.

15. How is the movement of the star you observed in question 14 related to the direction of rotation of the earth?

   _____

   _____

Use a suitable star chart (see Appendix E of your text) to locate several constellations and the North Star (Polaris).

16. Refer to Diagram 17.1. Sketch the pattern of stars for two constellations you were able to locate. List the name of each constellation by its diagram.

MOON OBSERVATION DATA SHEETS

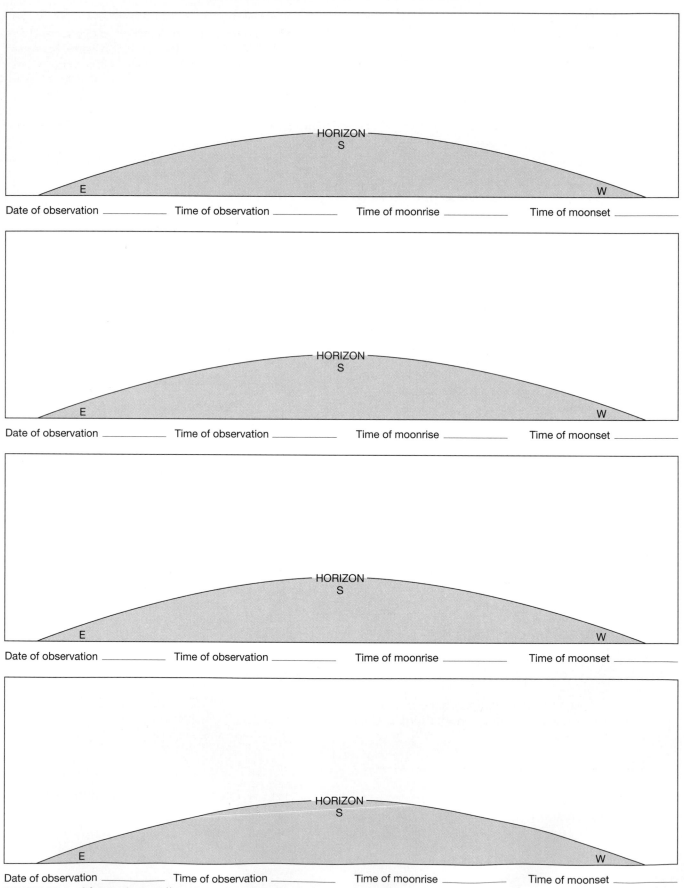

FIGURE 17.3 Moon observation data sheets.

Constellation Star Pattern

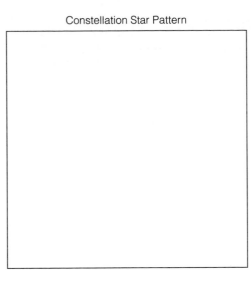

Constellation name: _____

Constellation Star Pattern

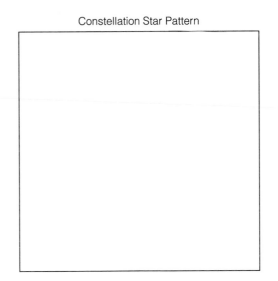

Constellation name: _____

DIAGRAM 17.1

Over a period of several hours, observe the motion of the stars in the vicinity of Polaris, the North Star.

17. Write a brief summary of the motion of the stars in the vicinity of Polaris.

_____

_____

If you can, come back to the same location at the exact same time, several weeks later.

18. The same star you observed overhead several weeks earlier (is still overhead, has moved to the east, has moved to the west). Circle your answer.

19. How is the change in position of the star you observed overhead several weeks earlier related to the revolution of the earth?

_____

_____

_____

TELESCOPIC OBSERVATION

Should time and facilities permit, your instructor will inform you of the nights that telescopes are available for your use. During a night observation session you will be instructed in the operation of a telescope and will use the instrument to observe celestial objects. Answer questions 20–26 after you have completed your observation session.

20. Did you use a *refracting* or *reflecting* telescope for your observations?

_____

21. What is the difference between a refracting and a reflecting telescope?

_____

_____

22. How are the sizes of telescopes given? For example, a six inch telescope means. . . .

_____

23. How do you increase or decrease the *magnification* of a telescope?

_____

24. What is the advantage of having an electric drive mechanism on the telescope mounting?

_____

25. What is the function of a *finderscope* on a telescope?

_____

26. Refer to Diagram 17.2. Prepare sketches and describe two of the objects (moon, planet, nebula, etc.) you were able to locate and observe in the telescope.

Sketch of Object Observed

Object description:

_____

_____

_____

Sketch of Object Observed

Object description:

_____

_____

_____

DIAGRAM 17.2

REVIEW

Although it may have taken several weeks to complete the exercise, you should know the following:

1. The sun does not set or rise at the same location on the horizon each day.
2. The angle of the sun at noon at any one location does not remain constant over a period of weeks.
3. The illuminated portion of the moon that you can observe from the earth changes with regularity.
4. The time that the moon rises and sets changes each day.
5. Stars are not all the same color.
6. As the earth rotates, stars appear to move westward in the night sky.
7. The name and purpose of many parts of a telescope.
8. The correct procedure for using a telescope to observe celestial objects.

Astronomical Observations

SUMMARY/REPORT PAGE

Date Due: _____

Name: _____

Date: _____

Class: _____

After you have finished Exercise Seventeen, complete the following questions. You may have to refer to the exercise for assistance or to locate specific answers. Be prepared to submit this summary/report to your instructor at the designated time.

1. On Diagram 17.3, prepare a single sketch illustrating your observed positions of the setting (or rising) sun on the horizon during the past several weeks. Show the reference features you used on the horizon. Label each position of the sun with the date of the observation. Write a brief summary of your observations below the diagram.

2. From question 3, step 5, in the exercise, list the noon sun angle that you calculated for the first and last day of your measurements.

Noon sun angle on the first day: _____ °

Date of observation: _____

Noon sun angle on the last day: _____ °

Date of observation: _____

3. Draw two sketches of the moon—the first illustrating the moon as you saw it on your first lunar observation, the second as you saw it on your last observation. Label the date and time of each observation.

First Moon Observation

Date: _____

Time: _____

Last Moon Observation

Date: _____

Time: _____

DIAGRAM 17.3 Sunset (Sunrise) Observations

Horizon

Summary: _____

_____

4. Did the moon rise earlier or later each night that you observed it?

_____

5. List the different colors of stars that you observed.

_____

6. Approximately how many widths of your fist, with your arm extended, will a star appear to move in one hour? Toward which direction do the stars appear to move throughout the night and what is the reason for the motion?

_____

_____

_____

7. Refer to Diagram 17.4. Sketch the pattern of stars for any constellation you have been able to locate in the sky. What is the name of the constellation?

Constellation Star Pattern

Constellation name: _____

DIAGRAM 17.4

Patterns in the Solar System

Although composed of many diverse objects, the solar system (Figure 18.1) exhibits various degrees of order and several regular patterns. To simplify the investigation of planetary sizes, masses, etc., the planets can be arranged into two distinct groups, with the members of each displaying similar attributes. This exercise examines the physical properties and motions of the planets with the goal of summarizing these characteristics in a few general, easily remembered statements.

OBJECTIVES

After you have completed this exercise, you should be able to

1. Describe the appearance of the solar system when it is viewed along the plane of the ecliptic.
2. Summarize the distances and spacing of the planets in the solar system.

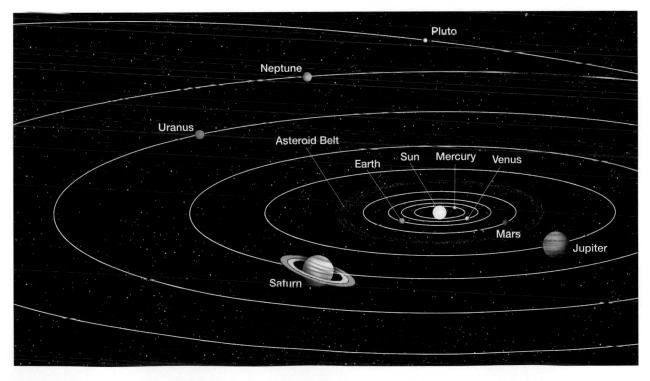

FIGURE 18.1   The solar system showing the orbits of the planets to scale. A different scale has been used for the sizes of the sun and planets. Therefore, the diagram is not a true scale model representation of the solar system.

3. Summarize and compare the physical characteristics of the terrestrial and Jovian planets.
4. Describe the motions of the planets in the solar system.

TEXTBOOK REFERENCE

Chapters 17 and 18

MATERIALS

ruler
colored pencils
calculator

Materials Supplied by Your Instructor

4-meter length of adding machine paper
meterstick

TERMS

nebula	mass	revolution
terrestrial planets	density	Kepler's laws
Jovian planets	weight	astronomical unit
plane of the ecliptic	rotation	

INTRODUCTION

The order that exists within the solar system is directly related to the laws of physics that governed its formation. Astronomers have determined that the sun and planets originated approximately 4.6 billion years ago from an enormous cloud of dust and gas. As this **nebula** contracted, it began to rotate and flatten. Eventually the temperature and pressure in the center of the cloud was great enough to initiate nuclear fusion and form the sun.

Near the center of the nebula, the planets Mercury, Venus, Earth, and Mars evolved under nearly the same conditions and consequently exhibit similar physical properties. Because these planets are rocky objects with solid surfaces, they are collectively called the **terrestrial** (earthlike) **planets.**

The outer planets, Jupiter, Saturn, Uranus, and Neptune, being farther from the sun than the terrestrial planets, formed under much colder conditions and are gaseous objects with central cores of ices and rock. Since the four planets are very similar, they are often grouped together and called the **Jovian** (Jupiterlike) **planets.** The planet Pluto is not included in either group since its great distance and small size make determining its complete characteristics impossible at the present time.

Table 18.1 illustrates many of the individual characteristics of the planets in the solar system.

————

1. Examine the data in Table 18.1; then,
   a. Draw lines on the upper and lower parts of Table 18.1 that separate the terrestrial planets from the Jovian planets. Label the lines, "Belt of Asteroids."
   b. On both parts of the table write the word "terrestrial" next to Mercury, Venus, Earth, and Mars and the word "Jovian" next to Jupiter, Saturn, Uranus, and Neptune.

THE "SHAPE" OF THE SOLAR SYSTEM

When the solar system is viewed from the side, the orbits of the planets all lie in nearly the same plane, called the **plane of the ecliptic** (see Figure 18.1). The column labeled "Inclination of Orbit" in Table 18.1 lists how many degrees the orbit of each planet is inclined from the plane. Answer questions 2–4 by referring to Table 18.1.

————

2. The orbits of which two planets have the greatest inclination to the plane of the ecliptic?

_____

3. With the exception of the two planets indicated in question 2, the orbits of the remaining planets all lie within (4, 6, 10) degrees of the plane. Circle your answer.

4. Consider the nebular origin of the solar system, and suggest a reason why the orbits of the planets are nearly all in the same plane.

_____

_____

DISTANCE AND SPACING OF THE PLANETS

An examination of any scale model solar system reveals that the distances from the sun and the spacing between the planets appear to follow a regular pattern. Although many early astronomers were concerned with planetary distances and spacing, it was not until 1766 that J. Titius, a German astronomer, found a simple mathematical relation that described the arrangement of the planets known at the time. A few years later, Johann

TABLE 18.1 Planetary data.

Planet	Symbol	Mean Distance from Sun: AU	Mean Distance from Sun: Millions of Miles	Mean Distance from Sun: Millions of Kilometers	Period of Revolution	Inclination of Ecliptic	Orbital Velocity: mi/s	Orbital Velocity: km/s
Mercury	☿	0.387	36	58	88^{d}	7°00′	29.5	47.5
Venus	♀	0.723	67	108	224.7^{d}	3°24′	21.8	35.0
Earth	⊕	1.000	93	150	365.25^{d}	0°00′	18.5	29.8
Mars	♂	1.524	142	228	687^{d}	1°51′	14.9	24.1
Jupiter	♃	5.203	483	778	11.86^{yr}	1°19′	8.1	13.1
Saturn	♄	9.539	886	1427	29.46^{yr}	2°30′	6.0	9.6
Uranus	⛢	19.180	1780	2866	84^{yr}	0°46′	4.2	6.8
Neptune	♆	30.060	2790	4492	165^{yr}	1°46′	3.3	5.3
Pluto	♇	39.440	3670	5909	248^{yr}	17°12′	2.9	4.7

Planet	Period of Rotation	Diameter: Miles	Diameter: Kilometers	Relative Mass (Earth = 1)	Average Density (g/cm^3)	Polar Flat-tening (%)	Eccen-tricity	Number of Known Satellites
Mercury	59^{d}	3015	4854	0.056	5.1	0.0	0.206	0
Venus	243^{d}	7526	12,112	0.82	5.3	0.0	0.007	0
Earth	$23^{h}56^{m}04^{s}$	7920	12,751	1.00	5.52	0.3	0.017	1
Mars	$24^{h}37^{m}23^{s}$	4216	6788	0.108	3.94	0.5	0.093	2
Jupiter	$\sim 9^{h}50^{m}$	88,700	143,000	318.000	1.34	6.5	0.048	16
Saturn	$\sim 10^{h}25^{m}$	75,000	121,000	95.200	0.70	10.5	0.056	17
Uranus	$10^{h}45^{m}$	29,000	47,000	14.600	1.55	7.0	0.047	15
Neptune	$\sim 16^{h}$	28,900	46,529	17.300	1.64	2.5	0.008	8
Pluto	6.4^{d}	~1500	~2400	~0.01(?)	~1.5(?)	?	0.250	1

Bode, also a German, popularized the mathematical scheme.

A Scale Model of Planetary Distances. Perhaps the best way to examine distance and spacing of the planets in the solar system is to use a scale model.

5. Prepare a distance scale model of the solar system according to the following steps.

Step 1. Obtain a 4-meter length of adding machine paper and a meterstick from your instructor.

Step 2. Draw an "X" about 10 centimeters from one end of the adding machine paper and label it "sun."

Step 3. Using the mean distances of the planets from the sun in miles presented in Table 18.1 and the following scale, draw a small circle for each planet at its proper scale mile distance from the sun. Use a different colored pencil for the terrestrial and Jovian planets and write the name of the planet next to its position.

Scale

1 millimeter = 1 million miles
1 centimeter = 10 million miles
1 meter = 1,000 million miles

Step 4. Write the word "asteroids" 258 million scale miles from the sun.

Answer questions 6–9 using the distance scale model you constructed in question 5.

6. What feature of the solar system separates the terrestrial planets from the Jovian planets?

____

7. Observe the scale model diagram and summarize the spacing for each of the two groups of planets.

Spacing of the terrestrial planets:_____

_____

Spacing of the Jovian planets:_____

_____

8. Write a brief statement that describes the spacing of the planets in the solar system.

_____

_____

9. Which planet(s) vary the most from the general pattern of spacing?

_____

COMPARING THE TERRESTRIAL AND JOVIAN PLANETS

The physical characteristics such as diameter, density, and mass of the terrestrial planets are very similar and can be summarized in a few statements. Likewise, the characteristics exhibited by the Jovian planets as a group can also be generalized.

To gain an understanding of the similarities of the planets within each of the two groups and the contrasts between the two groups, complete the following sections using the planetary data presented in Table 18.1. Because of the lack of sufficient data, Pluto will not be used in determining group characteristics or for general comparisons between the groups.

Size of the Planets. The similarities in the diameters of the planets within each of the two groups and the contrast between the groups are perhaps the most obvious patterns in the solar system. The diameter of each planet is given in both miles and kilometers in Table 18.1.

10. To visually compare the relative sizes of the planets and sun, complete the following steps using the unmarked side of your 4-meter length of adding machine paper.

Step 1. Determine the radius of each planet in kilometers by dividing its diameter (in kilometers) by 2. List your answers in the "Radius" column of Table 18.2.

Step 2. Use a scale of 1 cm = 2000 km. Determine the scale model radius of each planet and list your answer in the "Scale Model Radius" column of Table 18.2.

TABLE 18.2 Planetary radii with scale model equivalents.

Planet	Radius (in kilometers)	Scale Model Radius
Mercury	_____	_____ cm
Venus	_____	_____ cm
Earth	_____	_____ cm
Mars	_____	_____ cm
Jupiter	_____	_____ cm
Saturn	_____	_____ cm
Uranus	_____	_____ cm
Neptune	_____	_____ cm

Step 3. Draw an "X" about 10 cm from one end of the adding machine paper and label it "Starting point."

Step 4. Using the scale model radius in Table 18.2, begin at the starting point and mark the radius of each planet with a line on the paper. Use a different colored pencil for the terrestrial and Jovian planets. Label each line with the planet's name.

Step 5. The diameter of the sun is approximately 1,350,000 kilometers. Using the same scale as you used for the planets (1 cm = 2000 km), determine the scale model radius of the sun. Mark the sun's radius on the adding machine paper using a different colored pencil from the two planet groups. Label the line "sun."

Answer questions 11–17 using both Table 18.1 and the scale model radius diagram you constructed in question 10.

11. Which is the largest of the terrestrial planets and what is its diameter?

_____, _____ miles

12. Which is the smallest Jovian planet and what is its diameter?

_____, _____ miles

13. Complete the following statement.

The smallest Jovian planet, _____ , is

_____ times larger than the largest terrestrial planet.

14. Summarize the sizes of the planets within each group.

The diameters of the terrestrial planets:_____

_____

The diameters of the Jovian planets:_____

_____

15. Write a general statement that compares the sizes of the terrestrial planets to those of the Jovian planets.

_____

_____

16. Complete the following statement.

The sun is _____ times larger than Earth and

_____ times larger than Jupiter.

17. Refer to Table 18.1. The diameter of Pluto is most like the diameters of the (terrestrial, Jovian) planets. Circle your answer.

Mass and Density of the Planets. **Mass** is a measure of the quantity of matter an object contains. In Table 18.1 the masses of the planets are given in relation to the mass of Earth. For example, the mass of Mercury is given as 0.056, which means that it consists of only a small fraction of the quantity of matter that Earth contains. On the other hand, the Jovian planets all contain several times more matter than Earth.

Density is the mass per unit volume of a substance. In Table 18.1 the average densities of the planets are expressed in grams per cubic centimeter (g/cm^3). As a reference, the density of water is approximately one gram per cubic centimeter.

Using the relative masses of the planets given in Table 18.1, answer questions 18–22.

▬▬▬▬

18. Complete the following statements:

a. The planet _____ is the most massive planet in the solar system. It is _____ times more massive than Earth.

b. The least massive planet (excluding Pluto) is

_____ , which contains only _____ as much mass as Earth.

The gravitational attraction of a planet is directly related to its mass.

▬▬▬▬

19. Which planet exerts the greatest pull of gravity? Explain your answer.

_____

_____

Your **weight** is a function of the gravitational attraction of an object on your mass.

▬▬▬▬

20. On which planet (excluding Pluto) would you weigh the least? Explain your answer.

_____

_____

21. Which of the two groups of planets would have the greatest ability to hold large quantities of gas as part of their compositions? Explain your answer.

_____

_____

22. Write a general statement comparing the masses of the terrestrial planets to the masses of the Jovian planets.

_____

_____

Diameter vs. Density. To visually compare the diameters and densities of the planets, use the data in Table 18.1 to complete the diameter vs. density graph, Figure 18.2, according to the procedure in question 23.

▬▬▬▬

23. Plot a point on the diameter vs. density graph, Figure 18.2, for each planet, (excluding Pluto) where its diameter intersects its density. Label each point with the planet's name. Use a different colored pencil for the terrestrial and Jovian planets.

Answer questions 24–34 using Table 18.1 and the diameter vs. density graph, Figure 18.2, you constructed in question 23.

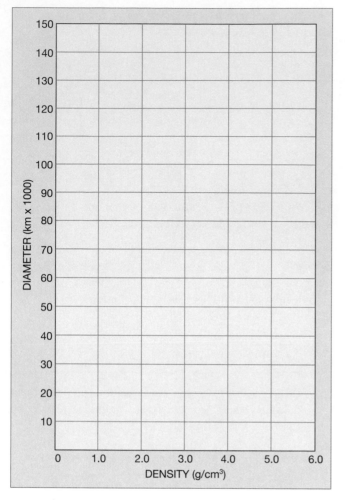

FIGURE 18.2   Diameter vs. density graph.

24. What general relation exists between a planet's size and its density?

_____

25. Consider the fact that the densities of the two rocks that form the majority of Earth's surface, the igneous rocks granite and basalt, are each about 3.0 g/cm³. Therefore, the average density of the terrestrial planets is (greater, less than) the density of Earth's surface. Circle your answer.

26. The term (rocky, gaseous) best describes the terrestrial planets.

27. The average density of Earth is about 5.5 g/cm³. Considering that the densities of the surface rocks are much less than the average, what does this suggest about the density of Earth's interior?

_____

28. Which of the planets has a density less than water and therefore would "float"?

29. Write a brief statement comparing the densities of the Jovian planets to the density of water.

_____

_____

30. The Jovian planets can be best described as (rocky, frozen ice and gas) worlds. Circle your answer.

31. Explain why Jupiter can be such a massive object and yet have such a low density.

_____

32. Write a general statement comparing the densities of the terrestrial planets to the Jovian planets.

_____

_____

33. Why are the densities of the terrestrial and Jovian planets so different?

_____

_____

34. Examine the estimated mass and density of Pluto presented in Table 18.1 and then complete the following statement by circling the correct responses.

The mass of Pluto is most like the masses of the (terrestrial, Jovian) planets, while its density is similar to the (terrestrial, Jovian) planets. This suggests that Pluto is a (small, large) planet made of (rocky, ice and frozen gas) material.

Number of Moons of the Planets.   The column labeled "Number of Known Satellites" in Table 18.1 indicates the number of known moons orbiting each planet.

35. Write a brief statement comparing the number of known moons of the terrestrial planets to the number orbiting the Jovian planets.

_____

_____

36. What is the general relation between the number of moons a planet has compared to its mass?

_____

_____

Rotation and Revolution of the Planets.

Rotation is the turning of a planet about its axis that is responsible for day and night. When the solar system is viewed from above the Northern Hemisphere of Earth, the planets, with the exception of Venus, rotate in a counterclockwise direction. Venus exhibits a very slow clockwise rotation. The time that it takes for a planet to complete one 360° rotation on its axis is called the *period of rotation*. The units used to measure a planet's period of rotation are Earth hours and/or days.

Revolution is the motion of a planet around the sun. The time that it takes a planet to complete one revolution about the sun is the length of its year, called the *period of revolution*. The units used to measure a planet's period of revolution are Earth days and/or years. Without exception, the direction of revolution of the planets is counterclockwise around the sun when the solar system is viewed from above the Northern Hemisphere of Earth.

Use the planetary data in Table 18.1 to answer questions 37–46.

----

37. If you could live on Venus or Jupiter, approximately how long would you have to wait between sunrises?

On Venus a sunrise would occur every _____ days.
On Jupiter a sunrise would occur every _____ hours.

38. Write a statement comparing the periods of rotation of the terrestrial planets to those of the Jovian planets.

_____

_____

The giant planet Jupiter rotates once on its axis approximately every 10 hours. If an object were on the equator of the planet and rotating with it, it would travel approximately 280,000 miles (the equatorial circumference or distance around the equator) in about 10 hours.

----

39. Calculate the equatorial rotational velocity of Jupiter using the following formula.

$$\text{Velocity} = \frac{\text{Distance}}{\text{Time}} = \frac{\text{mi}}{\text{hr}} =$$

_____ mi/hr

40. The equatorial circumference of the earth is about 24,000 miles. What is the approximate equatorial rotational velocity of Earth?

_____ miles/hour

41. How many times faster is Jupiter's equatorial rotational velocity than Earth's?

_____ times faster

42. Compare the planets' periods of rotation to their periods of revolution and then complete the following statement by circling the correct responses.

The terrestrial planets all have (long, short) days and (long, short) years, while the Jovian planets all have (long, short) days and (short, long) years.

43. In one Earth year, how many revolutions will the planet Mercury complete and what fraction of a revolution will Neptune accomplish?

Mercury: _____ revolutions in one earth year

Neptune: _____ of a revolution in one earth year

44. On Venus, how many days (sunrises) would there be in each of its years?

_____ day(s) per year

45. How many days (rotations) will Mercury complete in one of its years?

Mercury: _____ Mercury days in one Mercury year

46. Explain the relation between a planet's period of rotation and period of revolution that would cause one side of a planet to face the sun throughout its year.

_____

_____

In the early 1600s Johannes Kepler set forth three laws of planetary motion. Use Chapter 17 of your text as a reference to answer questions 47–49.

----

47. Use your own words to summarize each of Kepler's laws in the following spaces.

Kepler's first law: _____

_____

Kepler's second law: _____

_____

Kepler's third law: _____

_____

48. Considering Kepler's first and second laws, the distance a planet is from the sun as well as the planet's orbital velocity both (remain constant, change) throughout its year. Circle your answer.

According to Kepler's third law, the period of revolution of a planet, measured in Earth years, is related to its distance from the sun in astronomical units (one **astronomical unit (AU)** is defined as the average distance from the sun to Earth—93 million miles or 150 million kilometers).

▬▬▬

49. Applying Kepler's third law, what would be the period of revolution of a planet that is 4 AUs from the sun? Show your calculation in the following space.

_____

_____

_____

REVIEW

Having completed the exercise, you should know the following:

1. The orbits of the planets all lie nearly in a plane, called the plane of the ecliptic. The two planets whose orbits are most inclined to the plane are Pluto and Mercury.
2. In general, the space between planets increases the farther planets are from the sun.
3. As a group, the terrestrial planets are small, dense, and rocky planets with few moons and long days.
4. As a group, the Jovian planets are large, have low densities, and are gas objects with ice and rocky interiors. The Jovian planets have short days and multiple moon systems.
5. With few exceptions, motion in the solar system is counterclockwise when objects are viewed from above the Northern Hemisphere of Earth.
6. Kepler's three laws describe the shapes of planetary orbits, velocities of the planets in their orbits, and the mathematical relation between a planet's distance from the sun and its period of revolution.

Patterns in the Solar System

SUMMARY/REPORT PAGE

Date Due: _____

Name: _____
Date: _____
Class: _____

After you have finished Exercise Eighteen, complete the following questions. You may have to refer to the exercise for assistance or to locate specific answers. Be prepared to submit this summary/report to your instructor at the designated time.

1. On Diagram 18.1, prepare a sketch illustrating the planets Mercury, Venus, Earth, and Mars at their approximate proportional distance from the sun. View the solar system from above the Northern Hemisphere of the earth. Draw arrows around each planet to illustrate its direction of rotation. Also, draw an arrow in the orbit of each planet that shows the direction of revolution.

2. Briefly describe the spacing of the planets in the solar system.

_____

_____

_____

3. Define the following terms:

Terrestrial planets:_____

_____

Jovian planets:_____

_____

Plane of the ecliptic:_____

_____

Rotation:_____

_____

Mass:_____

_____

Astronomical unit:_____

_____

4. Referring to the nebular origin of the solar system, describe and explain the direction of revolution of the planets.

_____

_____

DIAGRAM 18.1

(SUN)

5. Write a brief statement for each of the following characteristics that compares the terrestrial to the Jovian planets.

Diameter:_____

_____

Density:_____

_____

Period of rotation:_____

_____

Number of moons:_____

_____

Mass:_____

_____

6. If you knew the distance of a planet from the sun, explain how you would calculate its period of revolution.

_____

_____

_____

EXERCISE NINETEEN

Planet Positions

The ability to recognize a planet in the night sky is one of the first skills developed by any astronomer. Locating a planet begins with a knowledge of the motions of celestial objects and the proficient use of astronomical charts. This exercise examines a "working" scale model of the solar system. Using diagrams, you will investigate the movement of the planets in their orbits.

OBJECTIVES

After you have completed this exercise, you should be able to

1. Explain the observed motion of a planet when it is viewed from Earth.
2. Give the position of a planet by listing the constellation in which it is located.
3. Prepare a scale model of the solar system for a specified date showing the positions of the five planets that can be viewed from Earth without a telescope.
4. Explain the conditions that determine whether or not a planet can be seen on a specified date.
5. Use a diagram of the solar system to estimate the time that a planet will rise and set.

TEXTBOOK REFERENCE

Chapters 17 and 18

MATERIALS

ruler
colored pencils
calculator

TERMS

period of revolution
constellations of the zodiac
retrograde motion

INTRODUCTION

Generations of professional and amateur astronomers have noticed that certain celestial objects change position in the sky relative to the background of stars. These "wanderers" include the moon, Mercury, Venus, Mars, Jupiter, Saturn, and the sun; these objects, with no recorded discoverer, have been observed since antiquity and have given their names, through mythology, to the seven days of the week.

A WORKING SCALE MODEL OF THE SOLAR SYSTEM

Most scale models of the solar system, such as the one you may have constructed in Exercise Eighteen, simply illustrate the planets in a line and/or do not incorporate the element of planetary motion. To accurately portray the solar system involves both showing the planets at their correct scale model distances from the sun as well as placing them in their proper position around the sun.

Figures 19.1, 19.2, and 19.3 present the December 31, 1957, position of the five planets that can be seen from Earth without a telescope. The solar system is viewed from above the Northern Hemisphere of the earth. The dot in the center of each figure represents the center of the sun. Surrounding each figure is a circle, marked in degrees, which is used to reference the locations of the planets.

The zero point on each planet's orbit specifies its position on December 31, 1957. For Mercury, Venus,

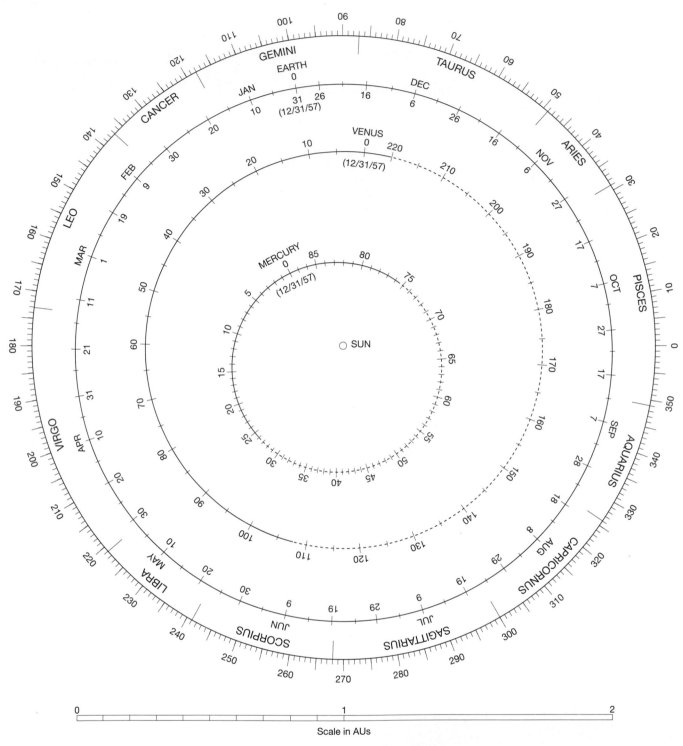

FIGURE 19.1 Positions of Mercury, Venus, and Earth on December 31, 1957. (Adapted from *Field Guide to the Stars and Planets* by Donald Menzel. Copyright 1964 by Donald Menzel. By permission of Houghton Mifflin Company)

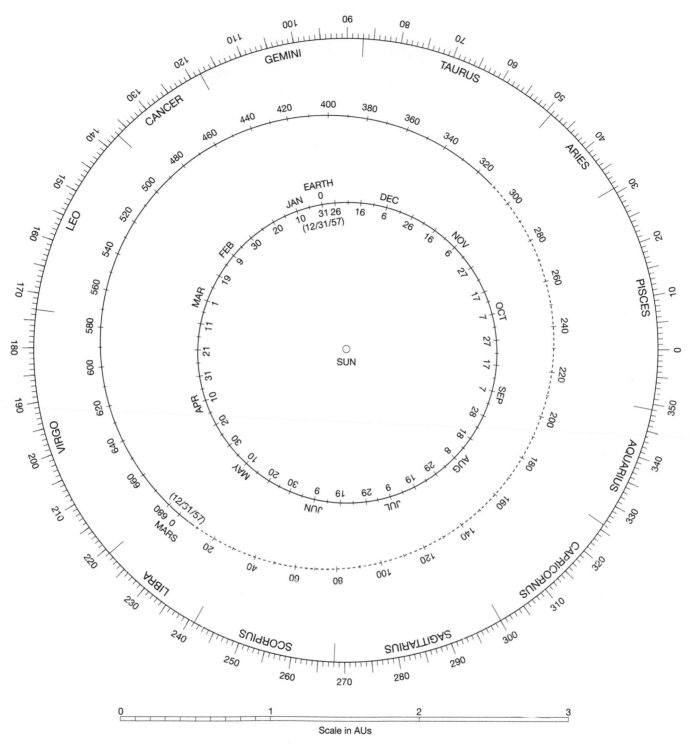

FIGURE 19.2 Positions of Earth and Mars on December 31, 1957. (Adapted from *Field Guide to the Stars and Planets* by Donald Menzel. Copyright 1964 by Donald Menzel. By permission of Houghton Mifflin Company)

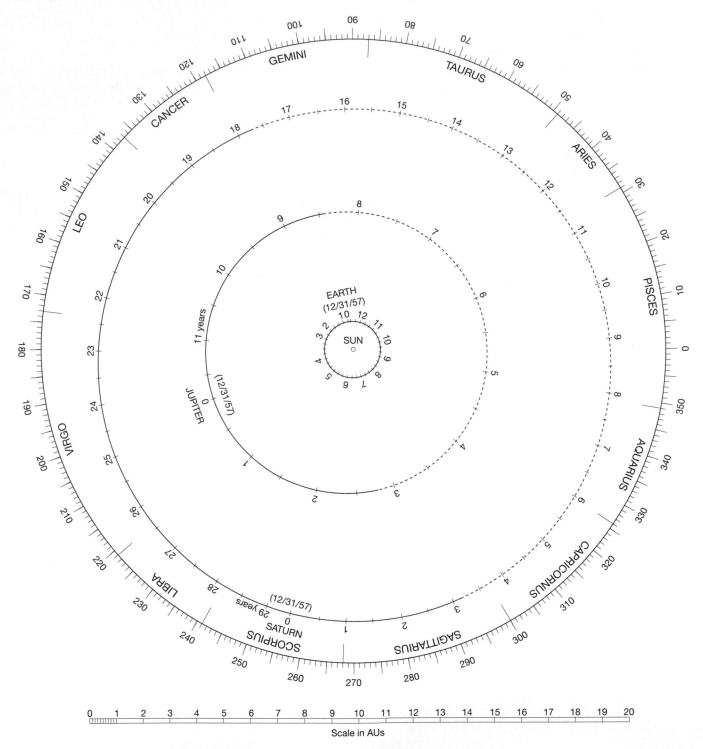

FIGURE 19.3 Positions of Earth, Jupiter, and Saturn on December 31, 1957. (Adapted from *Field Guide to the Stars and Planets* by Donald Menzel. Copyright 1964 by Donald Menzel. By permission of Houghton Mifflin Company)

and Mars (Figures 19.1 and 19.2), the numbers marked on each orbit indicate the planet's position on successive days during its revolution. Since Jupiter and Saturn (Figure 19.3) have relatively long **periods of revolution,** their orbits have been divided into years. Earth's orbit has been divided into months and select days of the month to simplify its location and allow for direct positioning.

The planets, moon, and sun lie in nearly the same plane. Therefore, when observed from Earth, they move along the same region of the sky, called the *zodiac* ("zone of animals"). Indicated on the outer reference circle of each figure are the twelve **constellations of the zodiac,** which form the background of stars.

Revolution of the Planets. The period of revolution of a planet is directly related to its distance from the sun. The direction of revolution around the sun is the same for all planets.

Examine each of the orbits of the planets shown in Figures 19.1–19.3. Then answer questions 1–4.

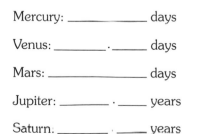

1. On Figures 19.1–19.3, draw an arrow on each planet's orbit showing the direction of revolution of the planet.
2. As shown on Figures 19.1–19.3, the direction of revolution of the planets in the solar system, when viewed from above the Northern Hemisphere of the earth, is (clockwise, counterclockwise). Circle your answer.
3. Using Figures 19.1, 19.2, and 19.3, estimate the periods of revolution of Mercury, Mars, and Saturn.

   Mercury: _____ days

   Mars: _____ days

   Saturn: _____ years

4. Refer to Table 18.1, "Planetary Data," in Exercise Eighteen, and record the *exact* period of revolution for the planets listed below. Also, write each planet's exact period of revolution by the zero point (December 31, 1957) on its orbit in Figures 19.1–19.3.

Period of Revolution

   Mercury: _____ days

   Venus: _____._____ days

   Mars: _____ days

   Jupiter: _____._____ years

   Saturn: _____._____ years

Determining the Location of a Planet Using the Reference Circle. The reference circle surrounding Figures 19.1–19.3 is used to locate a planet's position

around the sun. The following steps are used to determine a planet's position on the reference circle.

Step 1. Draw a straight line that connects the center of the sun (the dot at the center of each figure) with the planet in its orbital position.
Step 2. Extend the line until it intersects the reference circle and note the degree.

5. By using the procedure for locating a planet's position on the reference circle, accurately determine the degree location of the following planets for December 31, 1957 (the zero location of each planet on the charts). As an example, Mercury has already been done.

Planet	Degree Location (on December 31, 1957)
Mercury (Figure 19.1)	(125°)
Venus (Figure 19.1)	_____
Earth (Figure 19.1)	_____
Mars (Figure 19.2)	_____
Jupiter (Figure 19.3)	_____
Saturn (Figure 19.3)	_____

Determining in Which Constellation a Planet Is Located. A planet is often referred to as being "in" a constellation. The reference is to the particular constellation that is located directly behind the planet when that planet is viewed from Earth on that same date. The following steps are used to determine in which constellation a planet is located.

Step 1. Position Earth at the appropriate date in its orbit.
Step 2. Draw a line beginning at the position of Earth through the position of the planet on the same date.
Step 3. Extend the line until it intersects the surrounding reference circle and note the constellation.

6. By following the procedure for determining in which constellation a planet is located, determine the constellation in which each of the planets was

observed on December 31, 1957. As examples, Mercury and Jupiter have already been done.

Planet	Constellation (on December 31, 1957)
Mercury (Figure 19.1)	(Scorpius)
Venus (Figure 19.1)	
Mars (Figure 19.2)	
Jupiter (Figure 19.3)	(Virgo)
Saturn (Figure 19.3)	

Relative Movement of the Planets. The planets that are farthest from the sun take longer to complete one revolution than those that are nearest to the sun (Kepler's third law). Therefore, given the same number of days, planets that are near the sun will move further around their orbits than planets that are at a great distance.

To aid in understanding the effects that different periods of revolution have on the location of planets in the sky, complete questions 7–11.

-----

7. On Figures 19.1 and 19.3, advance Mercury, Earth, and Jupiter ninety days (0.25 year) beyond the December 31, 1957, position in their orbits. Place a dot in the orbit of each planet at the new position. Use the new position to determine the constellations in which Mercury and Jupiter are located.

**Constellation
(90 days after December 31, 1957)**

Mercury: _____

Jupiter: _____

8. Compare the constellation locations of Mercury and Jupiter ninety days after December 31, 1957, to their constellation locations on December 31, 1957. What effect has ninety days of motion had on where each planet is seen from Earth?

_____

_____

9. How many revolutions will Mercury complete in one Earth year (365.25 days)?

_____ revolutions

10. What fraction of a revolution will Jupiter complete in one Earth year?

_____ of a revolution

11. Relative to the background of stars, throughout the year the positions of Mercury and Venus change (slightly, considerably), while the positions of Jupiter and Saturn change (slightly, considerably). Circle the correct responses.

Retrograde Motion. The fact that Earth periodically overtakes the more distant planets in their orbits makes the motion of the outer planets *appear*, on occasion, to move backward. The apparent backward or westward movement of a planet when viewed from Earth is called **retrograde motion.** (*Note*: A similar observation is made when one vehicle passes another going the same direction on a highway.)

Figure 19.4 illustrates five equally successive positions of Earth and Mars in their orbits over a period of approximately six months. Completing questions 12–16 will help explain retrograde motion.

-----

12. To illustrate retrograde motion, complete the following steps using Figure 19.4.

Step 1. On Figure 19.4, accurately draw a line connecting Earth to Mars at each of the five positions in the orbits. Extend each line from Mars to the background of stars shown on the figure. Mark the place where each line intersects the background of stars and label it with the number of the position. (*Note*: Positions 1 and 2 have already been done and can serve as guides.)

Step 2. Using a continuous line, connect the five numbered positions on the background of stars in order, from 1 to 5.

Use the diagram you have constructed in Figure 19.4 to answer questions 13–16.

-----

13. Describe the apparent motion of Mars, as viewed from Earth, relative to the background stars.

_____

_____

14. At any time did Mars actually move backward in its orbit?

_____

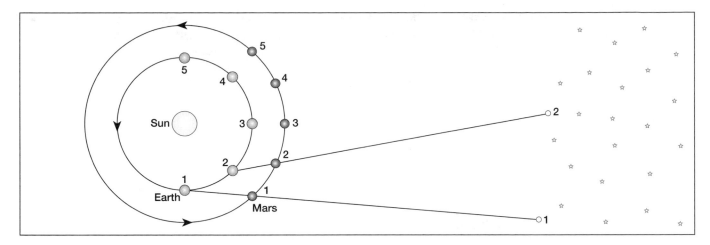

FIGURE 19.4 Diagram of Sun, Earth, and Mars for illustrating retrograde motion. (Diagram not drawn to scale)

15. Explain the reason for the observed motion of Mars, relative to the background of stars.

_____

_____

16. Assuming, as Ptolemy erroneously did in A.D. 141, that Earth does *not* move in an orbit, how must the observed motion of Mars be explained?

_____

_____

Adjusting the Planets' Positions on the Orbital Charts. As you have seen, given the same number of days or years, each planet will move a varying amount in its orbit. Therefore, to use the orbital charts (Figures 19.1–19.3) for any date other than December 31, 1957, requires that the charts be adjusted.

The following steps are used to update the locations of the planets on Figures 19.1, 19.2, and 19.3 for any date.

Step 1. Accurately determine the total amount of time (in days and in years) that has elapsed between December 31, 1957 (the date that the current figures represent), and the date you are seeking. (*Note:* One year equals 365.25 days.)

Step 2. Advance each planet the number of days (for Mercury, Venus, and Mars) or years, including fractions (for Jupiter and Saturn), that have elapsed between December 31, 1957, and the new date. Mark the new position with a dot on the planet's orbit. Label the position with the new date.

Step 3. For Earth, locate the date on the orbit that corresponds to the date you are seeking and mark the new position. (The time that has elapsed since the figures were drawn can be disregarded because in any year Earth will be located at the specified date in its orbit.)

Step 4. Determine the position of the planet on the 360° reference circle by drawing a line from the sun through the new position. To determine the constellation in which a planet is located, draw a line from the new position of Earth through the new position of the planet.

Using the four steps for adjusting the planets' positions on the charts for any date, complete question 16.

---

16. Determine the positions of the planets for today's date (or the date specified by your instructor). After calculating the number of days and years that have elapsed since the charts were drawn and entering your results below, complete Table 19.1 by listing the new location of each planet in degrees on the reference circle. Also, determine in which constellation the planet is located when viewed from Earth on the specified date and record your answer in Table 19.1.

Date of new positions of the planets: _____

Number of elapsed years since December 31,

1957 = _____ . _____ years

TABLE 19.1 Planet positions on today's date (or the date specified by your instructor).

	Location in Degrees	In the Constellation
Mercury:	_____	_____
Venus:	_____	_____
Earth:	_____	not applicable
Mars:	_____	_____
Jupiter:	_____	_____
Saturn:	_____	_____

Date of planet positions specified in the table: _____

Number of elapsed days since December 31,

1957 = _____ days

Viewing a Planet from Earth. Earth rotates on its axis in a counterclockwise direction when viewed from above the Northern Hemisphere, the reference point of Figures 19.1–19.3. To see a planet, you must be in darkness and on the same side of Earth as the planet is located. Also, a planet that is opposite the sun from Earth cannot be seen.

17. Label the positions of noon, midnight, sunrise, and sunset on Earth at its location for today's date (or the date specified by your instructor) in Figures 19.1, 19.2, and 19.3.

18. Using the locations of the planets for today's date (or the date specified by your instructor) in Figures 19.1–19.3, write a brief description of each planet's visibility on that same date. Indicate in which constellation the planet is located. Also, by examining the times marked on Earth, include, when appropriate, at what time the planet can first be seen (the time it rises) and at what time it sets.

Visibility from Earth of:

Mercury: _____

_____

_____

Venus: _____

_____

_____

Mars: _____

_____

_____

Jupiter: _____

_____

_____

Saturn: _____

_____

_____

REVIEW

Having completed the exercise, you should know the following:

1. The planets each have different periods of revolution.
2. Planets closer to the sun than Earth change position in the sky frequently during the course of a year.
3. The farthest planets from the sun change position in the sky only slightly during a year.
4. Planets all revolve in the same direction around the sun.
5. When viewed from Earth, planets appear to be located in constellations.
6. The reason that planets periodically exhibit retrograde motion.
7. How to determine where the planets are located around the sun for a specific date.
8. How to estimate at what times a planet will rise and set.

Planet Positions

SUMMARY/REPORT PAGE

Date Due: _____

Name: _____

Date: _____

Class: _____

After you have finished Exercise Nineteen, complete the following questions. You may have to refer to the exercise for assistance or to locate specific answers. Be prepared to submit this summary/report to your instructor at the designated time.

1. In the following space prepare a diagram of the solar system showing the relative positions of the planets Mercury through Saturn for today's date (or the date specified by your instructor). View the solar system from above the Northern Hemisphere of Earth. Show the planets' locations as accurately as possible and label each.

Date illustrated on the diagram: _____

2. Which, if any, planet(s) is/are visible from Earth on the diagram you prepared in question 1 of this Summary/Report page? At what times do each of the visible planets rise and set?

_____

_____

_____

3. What is the reason that a planet will periodically exhibit retrograde motion?

_____

_____

4. Describe what is meant when a planet is said to be "in" a particular constellation.

_____

_____

5. When viewed from Earth, why does the position of Saturn change only slightly from year to year?

_____

_____

6. Describe the relative positions of Earth and Mars when Mars is visible from Earth.

_____

_____

7. The following diagram illustrates the sun, Earth, and Venus viewed from above the Northern Hemisphere of Earth. Venus moves to the opposite side of the sun in its orbit approximately every 112 days. The two positions of Venus represent the planet in its orbit on either side of the sun.

Draw an arrow around the earth showing its direction of rotation. Label the location of noon, sunset, midnight, and sunrise on the earth. Indicate, by circling your answers, in which position Venus will be visible from Earth prior to sunrise and in which position it will be visible from Earth after sunset.

VENUS   **SUN**   VENUS

○   ○

(morning visible)   (morning visible)

(evening visible)   (evening visible)

○

EARTH

EXERCISE TWENTY

The Moon and Sun

The moon and sun are two of the nearest celestial objects to Earth, and each has inspired generations of observers to wonder about the nature of the cosmos (Figure 20.1). Today, lunar and solar studies are providing astronomers with information that helps to answer questions about the evolution of the solar system and the source of energy of stars. This exercise investigates the motions and features of the moon, as well as the structure and dynamic surface of our "nearest star," the sun.

FIGURE 20.1 Telescopic view of the lunar surface. (Copyright UC Regents; UCO/Lick Observatory image)

Mare Imbrium (Sea of Rains)

Copernicus

Mare Tranquillitatus

0   600 km
Scale

OBJECTIVES

After you have completed this exercise, you should be able to

1. Recognize and name each of the phases of the moon.
2. Diagram the Earth, moon, and sun in their proper relation for each of the phases of the moon.
3. Explain the difference between the synodic and sidereal cycles of the moon.
4. Diagram the Earth, moon, and sun in their proper relation during a solar and lunar eclipse.
5. Discuss the difference between lunar terrae and maria and be able to recognize each on a lunar map or photograph.
6. Determine the relative ages of lunar features.
7. Recognize and name the different types of lunar craters.
8. Diagram and label the parts of the sun.
9. Describe several of the features found on the solar surface.

TEXTBOOK REFERENCE

Chapters 17, 18, and 19

MATERIALS

metric ruler
calculator

Materials Supplied by Your Instructor

stereoscope

TERMS

synodic month	maria	solar interior
sidereal month	crater	prominence
solar eclipse	corona	sunspot
lunar eclipse	chromosphere	
terrae	photosphere	

INTRODUCTION

The two celestial objects that have received the greatest attention throughout history are the moon and sun. Many ancient civilizations revered these objects and devised methods for keeping track of their changing positions in the sky. Today, the precise movements of the Earth-moon-sun system are still being monitored and astronomers are continuously refining their measurements.

The cratered surface of the moon and the dynamic surface of the sun have intrigued modern observers ever since Galileo first turned his telescope skyward. Using a variety of instruments, as well as the knowledge gained from manned lunar landings, modern astronomers have gathered a vast amount of information about the age, surface, and internal structure of the moon. Furthermore, several solar telescopes around the world keep the sun under constant surveillance and provide data concerning many phenomena including its surface features, temperatures, and composition.

THE MOON

At an average distance of 384,401 kilometers (238,329 miles), Earth's nearest celestial neighbor and only natural satellite is the moon. The monthly counterclockwise revolution around the earth and associated Earth-moon revolution about the sun change the moon's position relative to the sun, producing the phases viewed by earthbound observers. Furthermore, as the moon moves in its orbit, its slow counterclockwise rotation results in the same side continuously facing earthward.

The moon has no atmosphere, and therefore no landforms produced by wind or water erosion. Instead, the vast majority of lunar surface features are the result of ancient volcanic eruptions and impact cratering by *meteoroids*. Many features that formed on the moon's surface over 3 billion years ago are still discernible and have been only slightly modified by the continuous bombardment of tiny *micrometeorites*.

Phases of the Moon. The changing phases of the moon have been recorded throughout history and were among the first astronomical phenomena to be understood. The lunar phases observed from Earth are the result of the motion of the moon and sunlight that is reflected from its surface. The half of the moon facing directly toward the sun is illuminated at all times. However, to an earthbound observer the percentage of the bright side that is visible depends on the location of the moon with respect to the sun and Earth. As illustrated in Figure 20.2, when the moon lies between the sun and Earth, none of the bright side can be seen by an earthbound observer—a phase called *new-moon* ("no–moon"). Conversely, when the moon lies on the side of Earth opposite the sun, all of its bright side is visible, producing the *full-moon* phase. At any position between these extremes, only a fraction of the moon's illuminated half is visible.

Figure 20.2 is a view of the Earth-moon-sun system from above the Northern Hemisphere of Earth. Eight positions of the moon during its monthly journey around the Earth are illustrated. Notice on the figure that the illuminated half of the moon is always directly

FIGURE 20.2 The lunar cycle as viewed from above the Northern Hemisphere of Earth, including phases of the moon as observed from Earth (A–H).

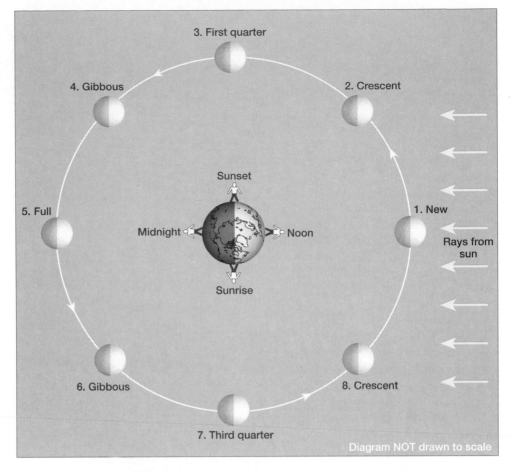

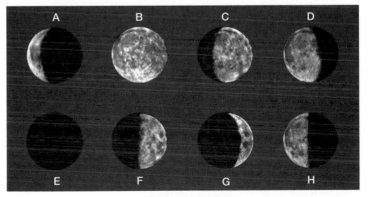

PHASES FROM EARTH

toward the sun. However, the portion of the illuminated half visible from Earth changes during the lunar cycle. To help you understand the lunar phases, complete questions 1–13 using Figure 20.2. Use Chapter 17 of your text as a reference.

1. For each of the eight positions of the moon in Figure 20.2, draw a line through the moon which separates the half of the moon that is *visible from*

Earth (the half directly toward Earth) from the half which cannot be seen.

2. As the moon journeys from position 1 to position 5, the proportion of its *illuminated side visible from Earth* is (increasing, decreasing). Circle your answer.

3. As the moon journeys from position 5 to position 8, the proportion of its *illuminated side visible from Earth* is (increasing, decreasing).

4. Complete the following by matching each of the eight drawings of the lunar phases as seen from

Earth in Figure 20.2 (A–H) with the moon's appropriate position (1–8) on the figure.

Position Number	Letter of Phase Seen from Earth	Name of Phase
1.	_____	_____
2.	_____	_____
3.	_____	_____
4.	_____	_____
5.	_____	_____
6.	_____	_____
7.	_____	_____
8.	_____	_____

Observe the time of day (noon, sunset, etc.) represented by the four positions of the earthbound observer on Figure 20.2. Note that an observer can see the moon approximately 90° on either side of his or her location—that is, about 1/4 of a circle in both directions. Using the 8 positions of the moon in its orbit and the times represented on Earth, answer questions 5–13.

5. The new moon is highest in the sky to an earthbound observer at (noon, sunset, midnight, sunrise). Circle your answer.

6. The full moon appears highest in the sky to an earthbound observer at (noon, sunset, midnight, sunrise). Circle your answer.

7. Throughout the lunar cycle, the moon moves further (eastward, westward) in the sky each day. Therefore, to an earthbound observer, the time of day when the moon is highest in the sky becomes progressively (earlier, later). Circle the correct responses.

8. Can a full moon be observed from Earth by an observer positioned at noon? Explain the reason for your answer.

_____

_____

9. A full moon first becomes visible to an earthbound observer positioned at (noon, sunset, midnight, sunrise), and she or he must look (eastward, westward) to see the rising moon. Circle the correct responses.

10. Can the first- and third-quarter lunar phases be observed during the daylight hours? Explain the reason for your answer.

_____

_____

11. At approximately what times will the first-quarter moon rise and set?

Rise: _____   Set: _____

12. At approximately what times will the third-quarter moon rise and set?

Rise: _____   Set: _____

13. Assume a crescent-phase moon is observed in the early evening in the western sky. During the next few days the moon will be rising (earlier, later) and the visible, illuminated portion of the moon will become progressively (larger, smaller). Circle your answers.

Synodic and Sidereal Months. The time interval required for the moon to complete a full cycle of phases is 29.5 days, a period of time called the **synodic month.** This complete cycle of the phases of the moon (i.e. new-moon to the next new-moon) is the basis of the word "month" (or "moonth"). Although the cycle of phases requires 29.5 days, the true period of the moon's 360° revolution around Earth takes only 27.3 days and is known as the **sidereal month.** The difference of approximately 2 days results from the fact that as the moon revolves around Earth, the Earth-moon system also is moving around the sun.

Figure 20.3 illustrates an exaggerated month of motion of the Earth-moon system around the sun. Refer to the figure to answer questions 14–22.

14. On Month I of Figure 20.3, indicate the dark half of the moon on each of the eight lunar positions by shading the appropriate area with a pencil.

15. Select from the eight lunar positions in Month I, and indicate which represents the following lunar phases.

Phase	Lunar Position (Month I)
New-moon	_____
Third quarter	_____
Full-moon	_____
First quarter	_____

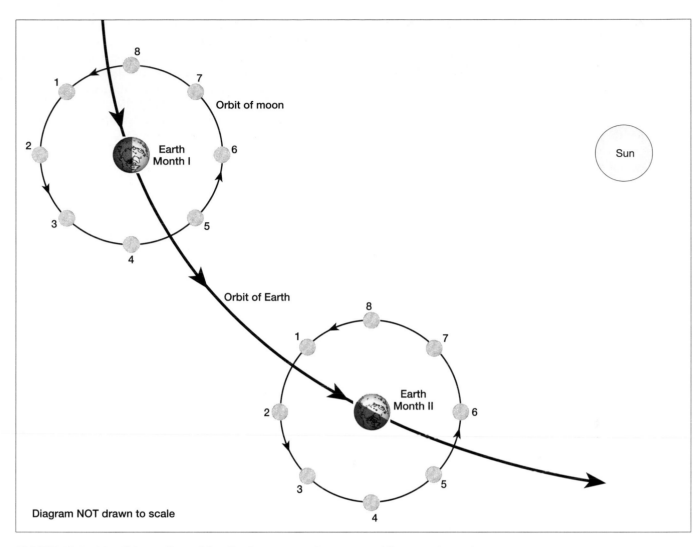

FIGURE 20.3 Monthly motion of the Earth-moon system around the sun viewed from above the Northern Hemisphere.

16. On Month I, label the position of the new-moon phase with the words "new moon."
17. In Month II of Figure 20.3, lunar position number (1, 3, 5, 7) represents the new-moon phase. Circle your answer and label the position of the new-moon phase on Month II with the words "new moon."

   Begin with the position of the new-moon phase in Month I, and imagine revolving the moon 360° around Earth while at the same time moving it to Month II.

18. After a 360° revolution beginning at the new-moon phase in Month I, the moon is located at position (6, 7, 8) in Month II. Circle your answer.

19. A complete 360° revolution of the moon around Earth is called a _____ month, which takes _____ days.
20. The position of the moon you determined in question 18 occurs (before, after) the moon completes a full cycle of phases from Month I to Month II. Circle your answer.
21. In Month II, when the moon moves the additional distance in its orbit, from position 6 to 7, and again is at the new-moon phase, it will have completed

    a _____ month, which takes _____ days.
22. In your own words, explain the difference between a sidereal and synodic month.

    _____

    _____

Eclipses. Eclipses occur when the sun, moon, and Earth are in a direct line. An eclipse can be either a **solar eclipse** or a **lunar eclipse.** Using Chapter 17 of your text as a reference, answer questions 23–27.

23. Refer to Diagram 20.1. Describe a solar eclipse and prepare a "side view" diagram showing the relation of the sun, Earth, and moon during the eclipse.

Description: _____

_____

DIAGRAM 20.1   Solar eclipse diagram

24. A solar eclipse occurs during the (new-moon, first–quarter, full-moon) phase of the moon. Circle your answer.
25. Refer to Diagram 20.2. Describe a lunar eclipse and prepare a "side view" diagram showing the relation of the sun, Earth, and moon during the eclipse.

Description: _____

_____

DIAGRAM 20.2   Lunar eclipse diagram

26. A lunar eclipse occurs during the (new-moon, third quarter, full-moon) phase of the moon. Circle your answer.
27. Why does Earth not experience a solar and lunar eclipse during each synodic month?

_____

_____

The Lunar Surface. Since the moon lacks an atmosphere, its landscape has been shaped primarily by meteoroid impacts and ancient volcanic processes. In general, the moon's surface can be classified as one of two types. As illustrated in Figure 20.4, **terrae** are lunar highlands, which are the bright areas of the moon seen from Earth. The dark areas of the moon, called **maria** (plural for mare, the Latin word for sea), are flat lowland regions. Together, the arrangement of terrae and maria on the lunar surface result in the well-known "man in the moon."

The most obvious features on the lunar surface are **craters.** Many different types of craters exist (see Figure 20.4), with most of them produced when rapidly moving debris impacted the lunar surface.

Figure 20.5 is a near-side, enhanced photograph of the moon with many of the major features labeled. Lunar latitude and longitude are indicated along the edges of the photo. A graphic scale for determining distances is also included. Use Figures 20.4 and 20.5 to answer questions 28–43.

28. Using Chapter 18 of your text as a reference, describe the origin of the lunar maria.

_____

_____

29. By examining Figure 20.5 (also see Figure 20.1), approximately (20, 40, 70) percent of the near side of the moon consists of lunar maria. Circle your answer.
30. What is the name and approximate width of the mare located at 16°N latitude and 59°E longitude?

Mare _____ , _____ km wide

31. What is the lunar latitude and longitude of the crater named Copernicus?

Latitude: _____ , longitude: _____

32. Locate the following lunar features on Figure 20.5 and give the lunar latitude and longitude of each. Also, use Figure 20.4 as a guide to indicate the type of feature represented by each.

	Location	**Type of Feature**
Sinus Iridum:	_____ ,	_____
Humboldt:	_____ ,	_____
Mare Orientale:	_____ ,	_____
Rupes Altai:	_____ ,	_____
Kepler:	_____ ,	_____

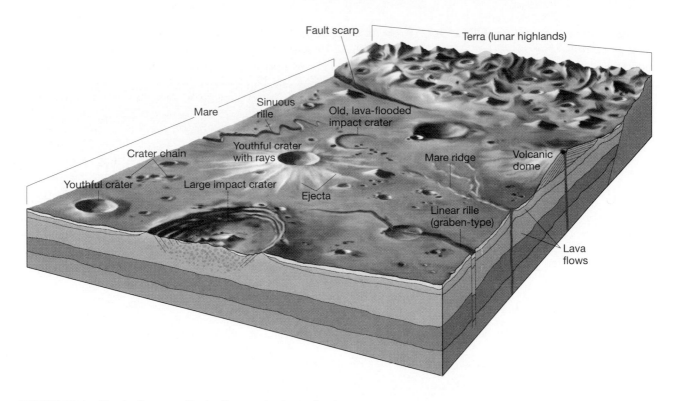

FIGURE 20.4 Block diagram illustrating major lunar features.

The number of impact craters within an area can be used to determine the age of a surface. In general, the more craters that are present, the longer the surface has been in existence.

33. Examine the frequency of craters on the lunar highlands compared to those on the maria. Lunar (terrae, maria) have formed most recently. Circle your answer.

34. Rocks brought back from the lunar maria during the manned Apollo landings are about 3.2–3.8 billion years old. Therefore, lunar highlands are (older, younger) than 3.2–3.8 billion years. Circle your answer.

35. Compare the crater frequency on the floor of Mare Smythii (2°S, 87°E) with that of Sinus Iridum (45°N, 31°W). Sinus Iridum appears to be (older, younger) than Mare Smythii.

36. Mare Smythii appears to be (older, younger) than Mare Crisium (16°N, 59°E).

When craters overlap, the rim of the most recent crater will cut through the rim of the older.

37. Using a stereoscope, examine the stereogram of the overlapping lunar craters shown in Figure 20.6 and label the most recent crater with the word "youngest."

38. Observe the crater Gasserdi (17°S, 39°W) and its relation to Mare Humorum in Figure 20.5. Crater Gasserdi is (older, younger) than Mare Humorum.

39. Locate craters Mee, Hainzel, and the unnamed crater northwest of Hainzel at approximately latitude 42°S and longitude 35°W in Figure 20.5. List the craters in order, from youngest to oldest.

Youngest: _____

_____

Oldest: _____

Most crater rims become rounded after long periods of bombardment by sand-sized particles (micrometeorites). Therefore, the "sharpness" of a crater is an indication of its age.

40. Locate the crater Copernicus on the lunar map, Figure 20.5, and compare the "sharpness" of its rim to other lunar craters. What conclusion can you make about the age of Copernicus?

_____

41. What conclusion can be made about the age of

FIGURE 20.5 Near side image of the moon. ((c) Mitchell Beazley Publishers, 1981, distributed in the U.S.A. by Rand McNally and Company. Used with permission)

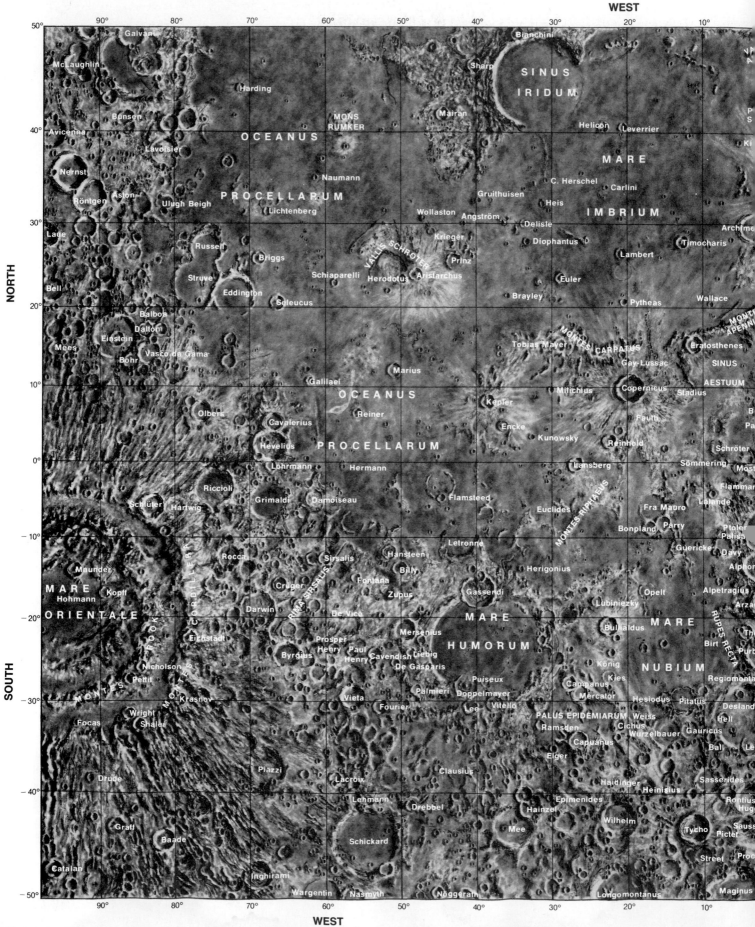

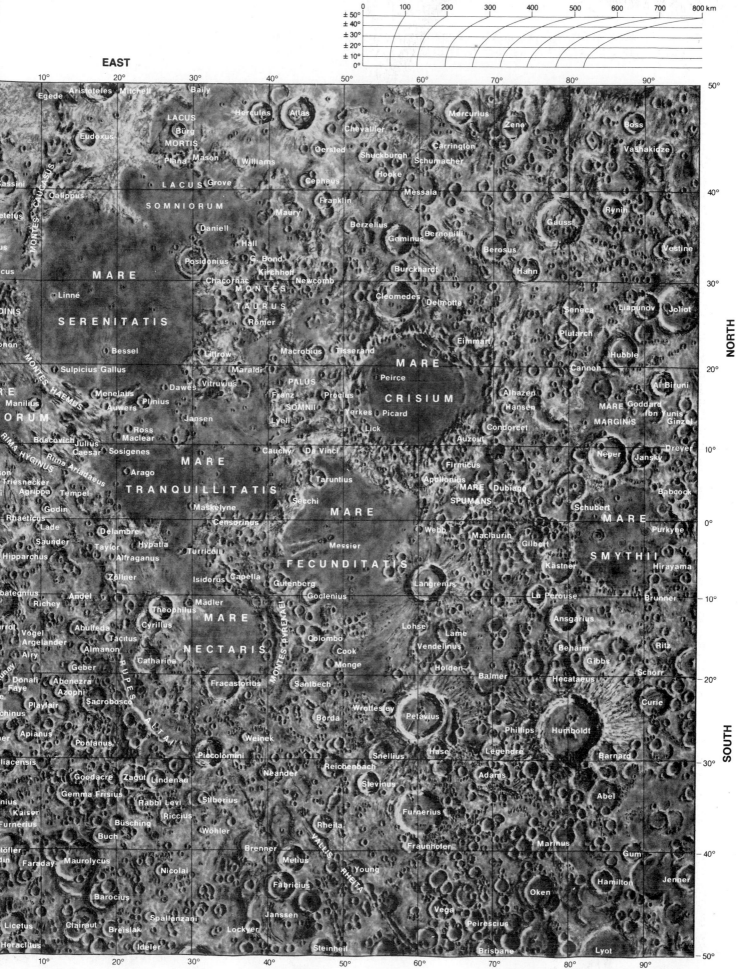

Scale (top right):
0   100   200   300   400   500   600   700   800 km
± 50°
± 40°
± 30°
± 20°
± 10°
0°

10° 20° 30° 40° 50° 60° 70° 80° 90°

50°

Egede Aristoteles Mitchell Baily
Eudoxus Hercules Atlas Chevallier Mercurius Zeno Boss
LACUS Bürg Oersted Shuckburgh Carrington Vashakidze
MORTIS Plana Mason Williams Schumacher
Daniell Grove Maury Hooke Messala Gauss Rynin
LACUS Franklin Berzelius Bernoulli Berosus Vestine
SOMNIORUM Geminus Hahn
Calippus Hall G. Bond Kirchhoff Newcomb Burckhardt
MARE Posidonius Chacornac Cleomedes Delmotte
Linné MONTES Eimmart Seneca Liapunov Joliot
SERENITATIS TAURUS Römer Plutarch
Bessel Littrow Macrobius Tisserand MARE Hubble
Sulpicius Gallus Maraldi CRISIUM Cannon
MONTES Vitruvius PALUS Peirce Alhazen Al Biruni
Manilius HAEMUS Menelaus Dawes Franz Proclus Hansen MARE Goddard
PORUM Auwers Plinius SOMNII Yerkes Picard Condorcet MARGINIS Ibn Yunis Ginzel
Boscovich Julius Ross Jansen Lyell Lick Auzout Dreyer
RIMA HYGINUS Caesar Maclear Cauchy Da Vinci Firmicus Neper Jansky
Rima Ariadaeus Sosigenes MARE Taruntius Apollonius MARE Babcock
Triesnecker Arago Secchi SPUMANS Dubiago Schubert
Agrippa Tempel TRANQUILLITATIS Maskelyne MARE Webb Maclaurin MARE
Godin Censorinus Messier Gilbert SMYTHII
Rhaeticus Lade Delambre MARE Kästner Purkyne
Saunder Taylor Hypatia Turricelli FECUNDITATIS Langrenus La Perouse Hirayama
Hipparchus Alfraganus Isidorus Capella Gutenberg Brunner
Andel Zöllner Goclenius Ansgarius
Richey Mädler MARE Lohse Lame Behaim Ritz
Abulfeda Theophilus Colombo Vendelinus Gibbs Schorr
Vogel Tacitus Cyrillus NECTARIS Cook Holden
Argelander Almanon Catharina Monge Balmer Hecataeus
Airy Geber MONTES PYRENAEI Fracastorius Santbech Curie
Donati Abenezra Wrottesley Petavius Phillips Humboldt
Faye Azophi Borda Holden Barnard
Playfair Sacrobosco Weinek Snellius Hase Legendre
Apianus Pontanus RUPES ALTAI Piccolomini Reichenbach Adams Abel
Aliacensis Santbech Neander Slevinus
Goodacre Zagut Lindenau Furnerius
Gemma Frisius Rabbi Levi Stiborius Marinus Gum
Kaiser Büsching Riccius Rheita Fraunhofer Oken Hamilton Jenner
Furnerius Buch Wöhler VALLIS Young
Stöfler Brenner RHEITA Vega
Faraday Maurolycus Metius Peirescius Lyot
Barocius Nicolai Fabricius
Licetus Clairaut Spallanzani Janssen Brisbane
Heraclitus Breislak Ideler Lockyer Steinheil

NORTH
SOUTH

10° 20° 30° 40° 50° 60° 70° 80° 90°

FIGURE 20.6 Stereogram of lunar craters. (Photo courtesy of NASA)

crater Mee compared to crater Tycho, directly east of Mee?

_____

Examine the crater Copernicus and the area around it closely in Figure 20.5.

▬▬▬▬

42. What type of crater is Copernicus? You may find Figure 20.4 useful.

_____

43. What is the origin of the bright rays that radiate outward from Copernicus?

_____

_____

44. It is likely that the earth was bombarded with meteoroids early in its history at least as frequently as the moon. If so, why are there so few craters visible on the earth's surface today?

_____

_____

THE SUN

At an average distance of 150 million kilometers (93 million miles), the sun is the nearest star to Earth. When compared to the other billions of stars in our galaxy, our sun is considered only "average." However, it is not only important to us as our primary source of energy, but also to astronomers, since it is the only star whose surface can be observed in detail.

Structure of the Sun. For convenience, the sun can be divided into four parts; the **corona, chromosphere, photosphere,** and **solar interior.** Of the four, only the first three are observable.

Use Chapter 19 of your text as a reference to answer questions 45–53.

▬▬▬▬

45. On Diagram 20.3, prepare a cross-sectional illustration of the sun showing its four parts in their proper positions. Label each of the parts.

46. Briefly summarize the appearance, temperature, and composition of each of the following parts of the sun.

DIAGRAM 20.3   Cross-sectional diagram of the sun

Photosphere: _____

_____

Chromosphere: _____

_____

Corona: _____

Solar Features.   As the sun rotates, it does so differentially, taking fewer days to complete one rotation on its equator than near the poles. The unequal period of rotation causes variations in the sun's magnetic field, which in turn influence many of its surface features.

Figure 20.7 illustrates several features of the active sun when the solar disk is photographed in hydrogen alpha light. The actual diameter of the solar image in the figure is approximately 870,000 miles. Use Figure 20.7 and Chapter 19 of your text to answer questions 47–53.

47. Using a metric ruler, measure the diameter of the solar image in millimeters. Then determine the scale of the photograph in miles per millimeter and write the scale on the photograph.

$$\text{Scale} = \frac{870{,}000 \text{ miles}}{\underline{\hspace{2cm}} \text{ mm}} = \underline{\hspace{2cm}} \text{ miles/millimeter}$$

Notice the "granulated" appearance of the solar image on the photograph.

48. What is the cause of the irregular, grainy appear-

FIGURE 20.7   The solar disk photographed in hydrogen alpha light. (This composite courtesy of Hale Observatories and National Solar Observatory, Sacramento Peak)

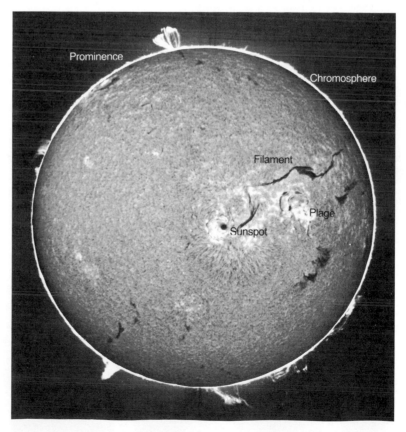

ance of the solar surface? Use Chapter 19 of your text as a reference.

_____

_____

Locate the large solar **prominence** near the top of the photograph.

▬▬▬

49. Using the scale you prepared in question 47, approximately how many miles does the prominence extend above the surface of the sun?

_____ miles

50. Describe the appearance and apparent cause of prominences.

_____

_____

Examine the large **sunspot** near the center of the solar disk.

▬▬▬

51. What is the approximate diameter of the sunspot?

_____ miles

52. On Diagram 20.4, draw a typical sunspot showing its *umbra* and *penumbra*.

DIAGRAM 20.4   Sunspot umbra and penumbra

53. What is the approximate temperature of sunspots compared to the solar surface? Why do sunspots appear as dark areas on the sun?

_____

_____

REVIEW

Having completed the exercise, you should know the following:

1. Lunar phases are the result of the moon's motions causing changes in the proportion of the illuminated half of the moon that is visible.
2. The moon takes 29.5 days to complete a cycle of phases, a period of time called the synodic month.
3. A 360° revolution of the moon around Earth takes 27.3 days, a period of time called the sidereal month.
4. The difference of about two days between the synodic and sidereal month is the result of the motion of the Earth-moon system around the sun.
5. Lunar and solar eclipses occur only when the Earth, sun, and moon lie along the same line.
6. The lunar surface can be divided into terrae (highlands) and maria (lowlands).
7. Lunar maria are more recent than terrae because the density of impact craters is less.
8. Rayed craters with sharp rims, such as Copernicus, are among the most recent features on the moon.
9. The four general parts to the sun are solar interior, photosphere, chromosphere, and corona.
10. Most of the features of the sun are related to the sun's magnetic field.
11. A solar prominence occurs when coronal material condenses along lines of magnetic force above the chromosphere.
12. Sunspots appear as dark spots on the sun because they are 1500K cooler than the photosphere.

The Moon and Sun

SUMMARY/REPORT PAGE

Date Due: _____

Name: _____

Date: _____

Class: _____

After you have finished Exercise Twenty, complete the following questions. You may have to refer to the exercise for assistance or to locate specific answers. Be prepared to submit this summary/report to your instructor at the designated time.

1. Write a brief paragraph explaining the reasons for the changes that occur in the phases of the moon as observed from Earth during a full lunar cycle of 29.5 days.

_____

_____

_____

2. Each of the four photographs in Figure 20.8 was taken from the earth when the moon was at its highest position in the sky. In the space provided below each photo, write the name of the phase represented and the time of day when the picture was taken.

3. Diagram 20.5 illustrates the Earth-moon system viewed from above the Northern Hemisphere. If you have completed Exercise Seventeen, "Astronomical Observations," draw a circle representing the moon in the proper position in its orbit for each of the lunar phases you observed and recorded in the exercise. Indicate the dark half of the moon by shading each circle. Label each position with the date and time of your observation. (If you have not completed Exercise Seventeen, complete the diagram by illustrating the moon at its proper position during full moon, new moon, first quarter, and third quarter. Label each position with the name of the phase.)

4. Explain the difference between a sidereal and synodic month.

_____

_____

_____

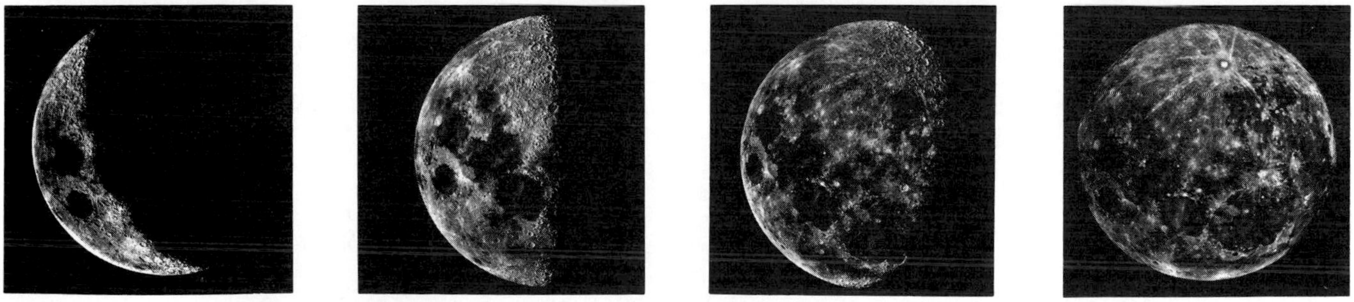

Phase: _____

Time: _____

Phase: _____

Time: _____

Phase: _____

Time: _____

Phase: _____

Time: _____

FIGURE 20.8 Lunar photos. (Copyright UC Regents; UCO/Lick Observatory image)

DIAGRAM 20.5 Lunar phases
diagram

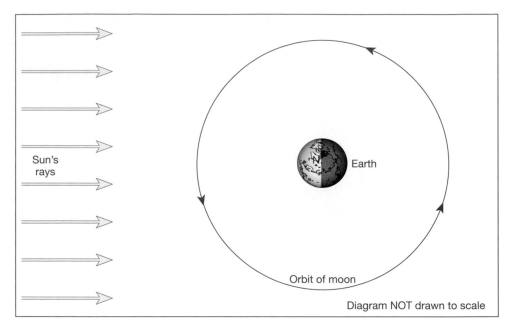

5. What are the most obvious differences in appearance between terrae and maria on a lunar photograph or map?

_____

_____

6. Questions 6a–6d refer to Figure 20.9, a photograph of a 20-kilometer-wide lunar crater.

a. The crater is located in a (terra, mare) region of the moon. Circle your answer.

b. What evidence suggests that the crater is of comparatively recent origin?

_____

c. The large crater in the photograph is of what type?

_____

FIGURE 20.9   A 20 km-wide lunar crater. (Photo courtesy of NASA)

d. What is the origin of the rays that extend from the crater rim?

_____

_____

7. When two craters overlap, how can you determine which is the most recent?

_____

_____

8. Define each of the following terms.

Chromosphere: _____

_____

Solar eclipse: _____

_____

Sunspot: _____

_____

Lunar terrae: _____

_____

EXERCISE TWENTY-ONE

Location and Distance on the Earth

The ability to find places and features on the earth's surface using maps and globes is an essential skill required of all earth scientists. This exercise introduces the system used for determining location on the earth. Using the system as a foundation, you will examine ways to measure distance on the surface of the earth.

OBJECTIVES

After you have completed this exercise, you should be able to

1. Explain the earth's grid system used for locating places and features.
2. Use the earth's grid system to accurately locate a place or feature.
3. Explain the relation between latitude and the angle of the North Star (Polaris) above the horizon.
4. Explain the relation between longitude and solar time.
5. Determine the shortest route and distance between any two places on the earth's surface.

TEXTBOOK REFERENCE

Appendix C

MATERIALS

ruler   calculator
protractor

Materials Supplied by Your Instructor

globe   atlas
world wall map 50–80 cm length of string

TERMS

earth's grid	South Pole	solar time
latitude	longitude	standard time
parallel of latitude	meridian of longitude	great circle
equator	prime meridian	small circle
North Pole	hemisphere	

INTRODUCTION

Globes and maps each have a system of north-south and east-west lines, called the **earth's grid,** that forms the basis for locating points on the earth. The grid is, in effect, much like a large sheet of graph paper that has been laid over the surface of the earth. Using the system is very similar to using a graph; that is, the position of a point is determined by the intersection of two lines.

Latitude is north-south distance on the earth (Figure 21.1). The lines (circles) of the grid that extend around the earth in an east-west direction are called **parallels of latitude** (see Figure 21.1). *Parallels of latitude mark north and south distance from the equator* on the earth's surface. As their name implies, these circles are parallel to one another. Two places on the earth, the **North Pole** and **South Pole,** are exceptions; they are points of latitude rather than lines.

Longitude is east-west distance on the earth (see Figure 21.1). **Meridians of longitude** are each halves of circles that extend from the North Pole to the South Pole on one side of the earth. *Meridians of longitude mark east and west distance from the* **prime meridian** *on the earth's surface.* Adjacent meridians are farthest apart on the equator and converge (come together) toward the poles.

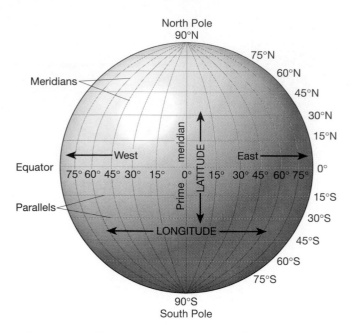

FIGURE 21.1 The earth's grid system.

The intersection of a parallel of latitude with a meridian of longitude determines the location of a point on the earth's surface.

The earth's shape is nearly spherical. Since parallels and meridians mark distances on a sphere, their designation, like distance around a circle, is given in *degrees* (°). For more precise location, a degree can be subdivided into sixty equal parts, called *minutes* ('), and a minute of angle can be divided into sixty parts, called *seconds* ("). Thus, 31°10'20" means 31 degrees, 10 minutes, and 20 seconds.

The type of map or globe used determines the accuracy to which a place may be located. On detailed maps it is often possible to estimate latitude and longitude to the nearest degree, minute, and second. On the other hand, when using a world map or globe, it may only be possible to estimate latitude and longitude to the nearest whole degree or two.

In addition to showing location on the earth, latitude and longitude can be used to determine distance. Knowing the shape and size of the earth, the distance in miles and kilometers covered by a degree of latitude or longitude has been calculated. These measurements provide the foundation for navigation.

DETERMINING LATITUDE AND LONGITUDE

Latitude. The equator is a circle drawn on a globe that is equally distant from both the North Pole and South Pole. It divides the globe into two equal halves,

called **hemispheres.** The equator serves as the beginning point for determining latitude and is assigned the value 0°00'00" latitude.

Latitude is distance north and south of the equator, measured as an angle in degrees from the center of the earth (Figure 21.2).

Latitude begins at the equator, extends north to the North Pole, designated 90°00'00"N latitude (a 90° angle measured north from the equator), and also extends south to the South Pole, designated 90°00'00"S latitude. *The poles and all parallels of latitude, with the exception of the equator, are designated either N (if they are north of the equator) or S (if they are south of the equator).*

1. Locate the equator on a globe. Diagram 21.1 represents the earth, with point B its center. Sketch and label the equator on the diagram. Also label the Northern Hemisphere and Southern Hemisphere on the diagram.

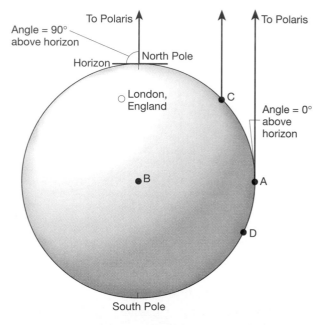

DIAGRAM 21.1

2. On Diagram 21.1, make an angle by drawing a line from point A on the equator to point B (the center of the earth). Then extend the line from point B to point C in the Northern Hemisphere. The angle you have drawn (∠ABC) is 45°. Therefore, by definition of latitude, point C is at 45°N latitude.

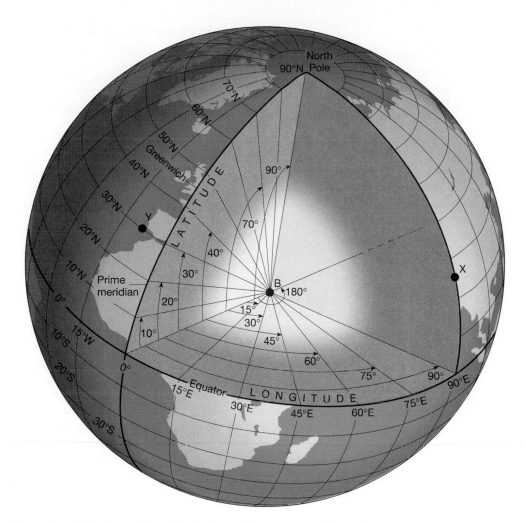

FIGURE 21.2 Measuring latitude and longitude. The angle measured from the equator to the center of the earth (B), and then northward to the parallel where point Y is located is 30°. Therefore, the latitude of point Y is 30°N. All points on the same parallel as Y are designated 30°N latitude.

The angle measured from the prime meridian where it crosses the equator to the center of the earth (B), and then eastward to the meridian where point X is located, is 90°. Therefore, the longitude of point X is 90°E. All points on the same meridian as X are designated 90°E longitude.

3. Draw a line on Diagram 21.1 parallel to the equator that also goes through point C. All points on this line are at 45°N latitude.

4. Using a protractor, measure ∠ABD on Diagram 21.1. Then draw a line parallel to the equator that also goes through point D. Label the line with its proper latitude.

On a map or globe, parallels may be drawn at any interval.

5. How many degrees of latitude separate the parallels on the globe you are using?

_____ degrees of latitude between each parallel

6. Keep in mind that the lines of latitude are parallel to the equator and each other. Locate some other parallels on the globe. Sketch and label a few of these parallels on Diagram 21.1.

7. Use the diagram that illustrates parallels of latitude, Diagram 21.2, to answer questions 7a and 7b.

   a. Accurately draw and label the following additional parallels of latitude on Diagram 21.2.

   5°N latitude

   10°S latitude

   25°N latitude

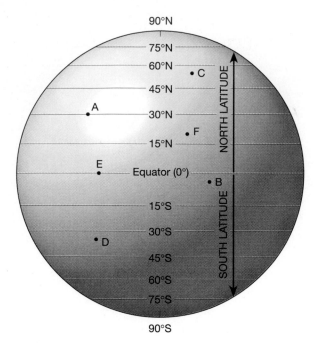

DIAGRAM 21.2

b. Refer to Diagram 21.2. Write out the latitude for each designated point as was done for points A and B. Remember to indicate whether the point is north or south of the equator by writing an N or S and include the word "latitude."

Point A: (30°N latitude)   Point D: _____

Point B: (5°S latitude)   Point E: _____

Point C: _____ Point F: _____

8. Use a globe or atlas to locate the cities listed below and give their latitude to the nearest degree. Indicate N or S and include the word "latitude."

Moscow, Russia: _____

Durban, South Africa: _____

Your home city: _____

Your college campus city: _____

9. By using a globe or atlas, give the name of a city or feature that is equally as far south of the equator as your home city is north.

_____

10. The farthest one can be from the equator is (45, 90, 180) degrees of latitude. Circle your answer.

11. The two places on the earth that are farthest from

the equator to the north and to the south are called the

_____ and _____

There are four special parallels of latitude marked and named on most globes.

▬▬▬

12. Use a globe or atlas to locate the following special parallels and indicate the name given to each.

Name of Parallel

66°30'00"N latitude: _____

23°30'00"N latitude: _____

23°30'00"S latitude: _____

66°30'00"S latitude: _____

Latitude and the North Star. Today most ships use navigational satellites to determine their location. However, early explorers were well aware of the concept of latitude and could use the angle of the North Star (a star named Polaris) above the horizon to determine their north-south position in the Northern Hemisphere. As shown on Diagram 21.1, someone standing at the North Pole would look overhead (90° angle above the horizon) to see Polaris. Their latitude is 90°00'00"N. On the other hand, someone standing on the equator, 0°00'00" latitude, would observe Polaris on the horizon (0° angle above the horizon).

Use Diagram 21.1 as a guide to answer questions 13–15.

▬▬▬

13. The angle of Polaris above the horizon for someone standing at point C in Diagram 21.1 would be (45°, 90°, 180°). Circle your answer.

14. What is the relation between a particular latitude and the angle of Polaris above the horizon at that latitude?

_____

15. What is the angle of Polaris above the horizon at the following cities?

**Angle of Polaris
Above the Horizon**

Fairbanks, AK: _____ degrees

St. Paul, MN: _____ degrees

New Orleans, LA: _____ degrees

Your home city: _____ degrees

Your college campus city: _____ degrees

Longitude. Meridians are the north-south lines on the globe that converge at the poles and are farthest apart on the equator. They are used to determine longitude, which is distance east and west on the earth (see Figure 21.1). Each meridian is a half circle on one side of the globe.

Notice on the globe that all meridians are alike. The choice of a zero, or beginning, meridian was arbitrary. The meridian that was chosen by international agreement in 1884 to be 0°00'00" longitude passes through the Royal Astronomical Observatory at Greenwich, England, located near London. This internationally accepted reference for longitude is named the prime meridian.

> Longitude is distance, measured as an angle in degrees east and west of the prime meridian (see Figure 21.2).

Longitude begins at the prime meridian (0°00'00" longitude) and extends to the east and to the west, half way around the earth to the 180°00'00" meridian, which is directly opposite the prime meridian. *All meridians, with the exception of the prime meridian and the 180° meridian, are designated either E (if they are east of the prime meridian) or W (if they are west of the prime meridian).*

16. Locate the prime meridian on a globe. Sketch and label it on the diagram of the earth, Diagram 21.1.
17. Label the Eastern Hemisphere, that half of the globe with longitudes east of the prime meridian, and the Western Hemisphere on Diagram 21.1.

On a map or globe, meridians can be drawn at any interval.

18. How many degrees of longitude separate each of the meridians on your globe?

_____ degrees of longitude between each meridian

19. Keep in mind that meridians are farthest apart at the equator and converge at the poles. Sketch and label several meridians on Diagram 21.1.
20. Use the diagram that illustrates meridians of lon-

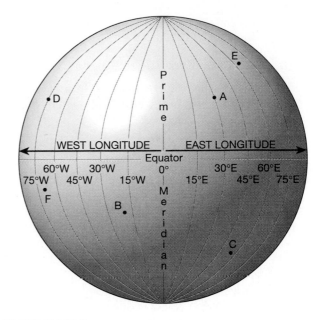

DIAGRAM 21.3

gitude, Diagram 21.3, to answer questions 20a and 20b.

a. Accurately draw and label the following additional meridians of longitude on Diagram 21.3.

35°W longitude
70°E longitude
10°W longitude

b. Refer to Diagram 21.3. Write out the longitude for each designated point as was done for points A and B. Remember to indicate whether the point is east or west of the prime meridian by writing an E or W and include the word "longitude."

Point A: _(30°E longitude)_   Point D: _____

Point B: _(20°W longitude)_   Point E: _____

Point C: _____   Point F: _____

21. Use a globe or atlas to locate the cities listed below and give their longitude to the nearest degree. Indicate either E or W and include the word "longitude."

Wellington, New Zealand: _____

Honolulu, Hawaii: _____

Your home city: _____

Your college campus city: _____

22. Using a globe or atlas, give the name of a city, feature, or country that is at the same latitude as you

home city but equally distant from the prime meridian in the opposite hemisphere.

---

23. The farthest a place can be directly east or west of the prime meridian is (45, 90, 180) degrees of longitude. Circle your answer.

Longitude and Time. Time, while independent of latitude, is very much related to longitude. This fact allows for time to be used in navigation to accurately determine one's location. Knowing the difference in time between two places, one with known longitude, the longitude of the second place can be determined.

Time on the earth can be kept in two ways. **Solar,** or sun, **time** uses the position of the sun in the sky to determine time. **Standard time,** the system used throughout most of the world, divides the globe into 24 standard time zones. Everyone living within the same standard time zone keeps the clock set the same. Of the two, solar time is used to determine longitude.

The following basic facts are important to understanding time.

The earth rotates on its axis from west to east (eastward) or counterclockwise when viewed from above the North Pole (Figure 21.3).

It is noon, sun time, on the meridian that is directly facing the sun (the sun has reached its highest position in the sky, called the *zenith*) and midnight on the meridian on the opposite side of the earth.

The time interval from one noon by the sun to the next noon averages 24 hours and is known as the *mean solar day.*

The earth turns through 360° of longitude in one mean solar day, which is equivalent to 15° of longitude per hour or 1° of longitude every 4 minutes of time.

Places that are east or west of each other, regardless of the distance, have different solar times. For example, people located to the east of the noon

FIGURE 21.3 The noon meridian and solar time.

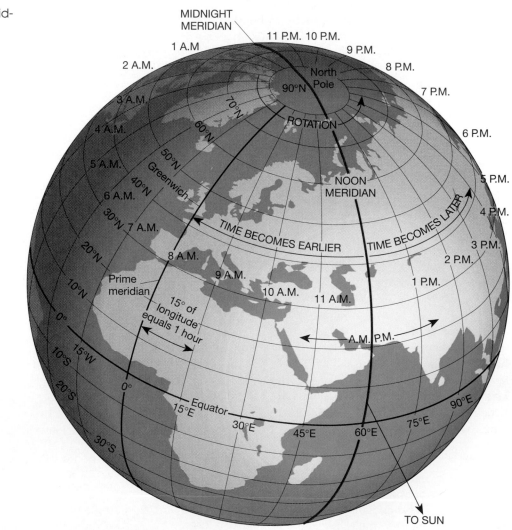

meridian have already experienced noon; their time is afternoon [P.M.— *post* (after) *meridiem* (the noon meridian)]. People living west of the noon meridian have yet to reach noon; their time is before noon [A.M.— *ante* (before) *meridiem* (the noon meridian)]. Time becomes later going eastward and earlier going westward.

Use the basic facts of time to answer questions 24–26.

▬▬▬▬

24. What would be the solar time of a person living 1° of longitude west of the noon meridian? Be sure to indicate A.M. or P.M. with your answer.

Solar time: _____ (A.M., P.M.)

25. What would be the solar time of a person located 4° of longitude east of the noon meridian?

Solar time: _____ (A.M., P.M.)

26. If it is noon, solar time, at 70°W longitude, what is the solar time at each of the following locations?

Solar Time

72°W longitude: _____

65°W longitude: _____

90°W longitude: _____

110°E longitude: _____

Early navigators had to wait for the invention of accurate clocks, called *chronometers*, before they could determine their longitude. Today most navigation is done using satellites, but ships still carry chronometers as a back-up system.

The ship-board chronometer is set to keep the time at a known place on the earth, for example, the prime meridian. If it is noon by the sun where the ship is located, and at that same instant the chronometer indicates that it is 8 A.M. on the prime meridian, the ship must be 60° of longitude (4 hours difference × 15° per hour) east (the ship's time is later) of the prime meridian (see Figure 21.3). The difference in time need not be in whole hours. Thirty minutes difference in time between two places would be equivalent to $7\frac{1}{2}°$ of longitude, twenty minutes would equal 5°, and so forth.

▬▬▬▬

27. It is exactly noon by the sun at a ship's location. What is the ship's longitude if, at that instant, the time on the prime meridian is the following. (*Note:*

Drawing a diagram showing the prime meridian, the ship's location east or west of the prime meridian, and the difference in hours may be helpful.)

6:00 P.M.: _____

1:00 A.M.: _____

2:30 P.M.: _____

Using the Earth's Grid System. Using both parallels of latitude and meridians of longitude, you can accurately locate any point on the surface of the earth.

▬▬▬▬

28. By using Figure 21.4, determine the latitude and longitude of each of the lettered points and write your answers in the following spaces. As a guide, Point A has already been done. Remember to indicate whether the point is N or S latitude and E or W longitude. The only exceptions are the equator, prime meridian, and 180° meridian. They are given no direction because each is a single line and cannot be confused with any other line. Convention dictates that latitude is always listed first.

Point A: (30°N) latitude, (60°E) longitude

Point B: _____ latitude, _____ longitude

Point C: _____ latitude, _____ longitude

Point D. _____ latitude, _____ longitude

Point E: _____ latitude, _____ longitude

29. Locate the following points on Figure 21.4. Place a dot on the figure at the proper location and label each point with the designated letter.

Point F: 15°S latitude, 75°W longitude
Point G: 45°N latitude, 0° longitude
Point H: 30°S latitude, 60°E longitude
Point I: 0° latitude, 30°E longitude

30. Use a globe, map, or atlas to determine the latitude and longitude of the following cities.

Kansas City, MO: _____

Miami, FL: _____

Oslo, Norway: _____

Auckland, New Zealand: _____

Quito, Ecuador: _____

Cairo, Egypt: _____

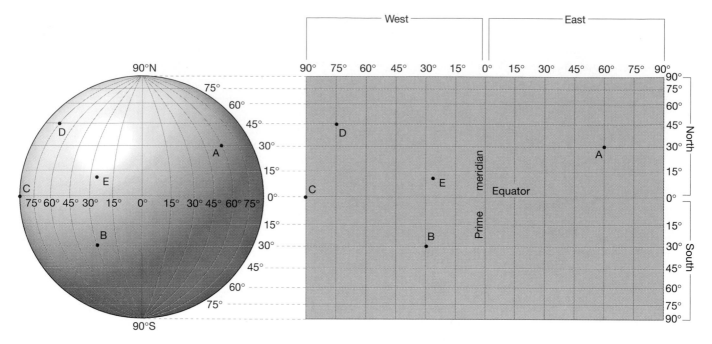

FIGURE 21.4 Locating places using the grid system. (After J. B. Hoyt, *Man and the Earth*, 3rd ed., © 1966. Adapted by permission of Prentice-Hall, Inc., Englewood Cliffs, N.J.)

31. Beginning with a globe or world wall map, and then proceeding to an atlas, determine the city or feature at the following locations.

19°28'N latitude, 99°09'W longitude:

_____

41°52'N latitude, 12°37'E longitude:

_____

1°30'S latitude, 33°00'E longitude:

_____

When you study the earth sciences, it is important to be familiar with the major physical features of the earth's surface. Identifying the features on a map will help acquaint you with their location for future reference.

32. Use a wall map of the world or world map in an atlas to find the following water bodies, rivers, and mountains. Examine their latitudes and longitudes, and then label each on the world map, Figure 21.5. To conserve space, mark only the number or letter of the feature at the appropriate location on the map.

Water Bodies

A. Pacific Ocean
B. Atlantic Ocean
C. Indian Ocean
D. Arctic Ocean
E. Gulf of Mexico
F. Mediterranean Sea
G. Caribbean Sea
H. Persian Gulf
I. Red Sea
J. Sea of Japan
K. Black Sea
L. Caspian Sea

Rivers

North America
a. Mississippi
b. Colorado
c. Missouri
d. Ohio

South America
e. Amazon

Europe and Asia
f. Volga
g. Mekong
h. Ganges
i. Yangtze

Africa and Australia
j. Nile
k. Congo
l. Darling

Mountains

North America
1. Rocky Mountains
2. Cascade Range
3. Sierra Nevada
4. Appalachian Mountains
5. Black Hills
6. Teton Range
7. Adirondack Mountains

South America
8. Andes Mountains

Europe and Asia
9. Pyrenees Mountains
10. Alps
11. Himalaya Mountains
12. Ural Mountains

Africa and Australia
13. Atlas Mountains
14. MacDonnell Ranges

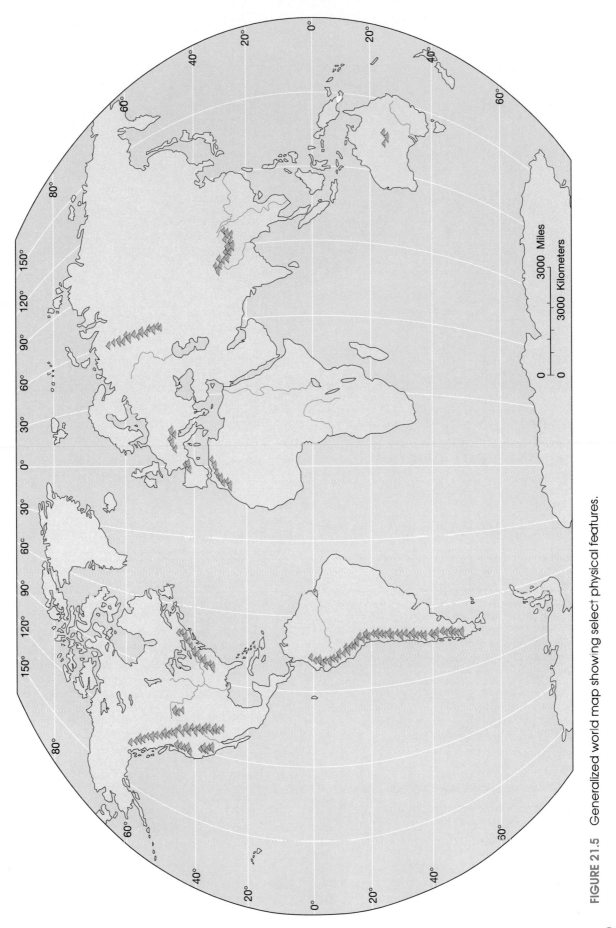

FIGURE 21.5 Generalized world map showing select physical features.

LOCATION AND DISTANCE ON THE EARTH **303**

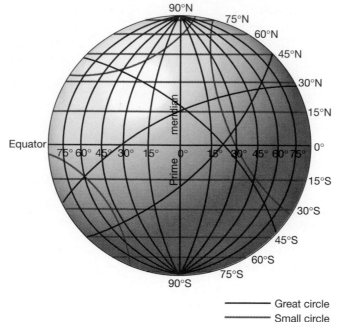

FIGURE 21.6 Illustrated are a few of the infinite number of great circles and small circles that can be drawn on the globe.

——— Great circle
········· Small circle

GREAT CIRCLES, SMALL CIRCLES, AND DISTANCE

Great Circles. A **great circle** is the largest possible circle that can be drawn on the earth (Figure 21.6). Some of the characteristics of a great circle are

A great circle divides the earth into two equal parts, called *hemispheres.*

An infinite number of great circles can be drawn on a globe. Therefore, a great circle can be drawn that passes through any two places on the earth's surface.

The shortest distance between two places on the earth is along the great circle that passes through those two places.

If the earth were a perfect sphere, then one degree of angle along a great circle would cover an identical distance everywhere. Because the earth is slightly flattened at the poles and bulges slightly at the equator, there are small differences in the length of a degree. However, for most purposes, *one degree of angle along a great circle on the earth equals approximately 111 kilometers or 69 miles.*

By using Figure 21.6 and by keeping the characteristics of great circles in mind, examine a globe and answer questions 33–35.

33. Remember that great circles do not necessarily have to follow parallels or meridians and estimate several great circles on the globe. Do this by wrapping a piece of string around the globe that divides the globe into two equal halves. You should be able to see that there are an infinite number of great circles that can be marked on the globe.

34. Which parallel(s) of latitude is/are a great circle(s)?

_____

Meridians of longitude are each half circles. If each meridian is paired with the meridian on the opposite side of the globe, a circle is formed.

35. Which meridians that have been paired with their opposite meridian on the globe are great circles?

_____

Small Circles. Any circle on the globe that does not meet the characteristics of a great circle is considered a small **circle** (see Figure 21.6). Therefore, a small circle *does not* divide the globe into two equal parts and *is not* the shortest distance between two places on the earth. By using Figure 21.6 and by keeping the characteristics of small circles in mind, examine a globe and answer questions 36–38.

----

36. In general, which parallels of latitude are small circles?

_____

37. Which two latitudes are actually points, rather than circles?

_____

38. In general, which meridians that have been paired with their opposite meridian on the globe are small circles?

_____

39. Indicate, by placing an "X" in the appropriate column, which of the following pairs of points illustrated on the earth's grid in Figure 21.4 are on a great circle and which are on a small circle.

	Great Circle	Small Circle
Points A–H	_____	_____
Points D–G	_____	_____
Points C–I	_____	_____
Points B–H	_____	_____

40. Now that you know the characteristics of great and small circles, complete the following statements by circling the correct response.

   a. All meridians are halves of (great, small) circles.
   b. With the exception of the equator, all parallels are (great, small) circles.
   c. The equator is a (great, small) circle.
   d. The poles are (points, lines) of latitude, rather than circles.

Determining Distance along a Great Circle. Determining the distance between two places on the earth when both are on the equator or the same great circle meridian requires two steps:

Step 1. Determine the number of degrees along the great circle between the two places (degrees of longitude on the equator or degrees of latitude on a meridian).

Step 2. Multiply the number of degrees by 111 kilometers or 69 miles (the approximate number of kilometers or miles per degree for any great circle).

Use a globe and these steps to answer questions 41 and 42.

----

41. Approximately how many miles would you journey if you traveled from 10°W longitude to 40°E longitude at the equator by way of the shortest route?

_____ miles

42. Approximately how many kilometers is London, England, directly north of the equator?

_____ kilometers

Determining the shortest distance between two places on the earth that are *not* both on the equator or the same great circle meridian requires the four steps (Figure 21.7):

Step 1. On a globe, determine the great circle that intersects both places.

Step 2. Stretch a piece of string along the great circle between the two places on the globe and mark the distance between them on the string with your fingers (see Figure 21.7A).

Step 3. While still marking the distance with your fingers, place the string on the equator with one end on the prime meridian. Determine the number of degrees along the great circle between the two places by measuring the marked string's length in degrees of longitude along the equator, which is also a great circle (see Figure 21.7B).

Step 4. Multiply the number of degrees along the great circle by 69 miles (111 kilometers) to arrive at the approximate distance. (For example, the great circle distance between X and Y in Figure 21.7 would be approximately 2070 miles, 30° × 69 miles/degree, or 3330 kilometers, 30° × 111 kilometers/degree.)

Use a globe, a piece of string, and the four steps to answer questions 43 and 44.

----

43. Determine the approximate great circle distance in

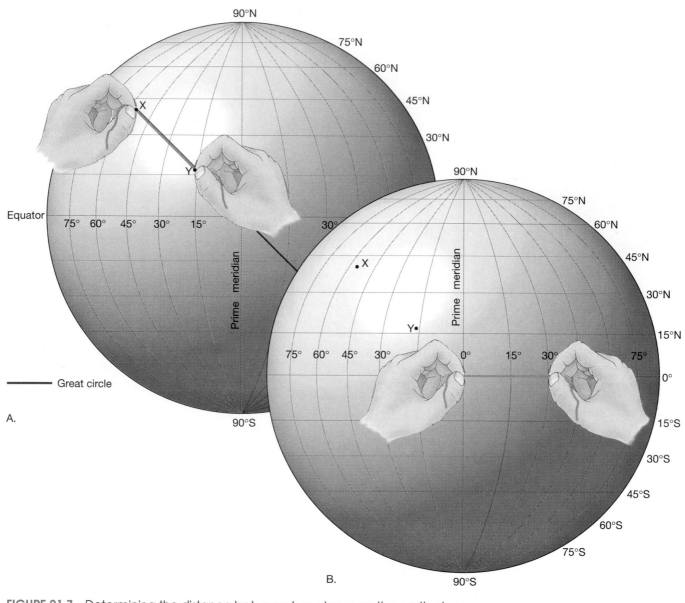

FIGURE 21.7 Determining the distance between two places on the earth along a great circle other than the equator or great circle meridian. In the example illustrated, the distance between X and Y along the great circle is 30°, which is approximately equivalent to 2070 miles (3330 kilometers).

degrees, miles, and kilometers from Memphis, TN, to Tokyo, Japan.

Degrees along the great circle between Memphis

and Tokyo = _____°

Distance along the great circle between Memphis

and Tokyo = _____ miles (_____ km)

44. Describe the flight route, by listing states, countries, etc., that a plane would follow as it flew by

way of the shortest route between Memphis, TN, and Tokyo, Japan.

_____

_____

Determining Distance along a Parallel. Since all parallels except the equator are small circles, the length of one degree of longitude along a parallel, other than the equator, will always be less than 69 miles or 111 kilometers. Table 21.1 shows the length of a degree of longitude at various latitudes on the earth.

TABLE 21.1 Longitude as distance

°Lat.	Length of 1° Long. km	Length of 1° Long. miles	°Lat.	Length of 1° Long km	Length of 1° Long miles	°Lat.	Length of 1° Long. km	Length of 1° Long. miles
0	111.367	69.172	30	96.528	59.955	60	55.825	34.674
1	111.349	69.161	31	95.545	59.345	61	54.131	33.622
2	111.298	69.129	32	94.533	58.716	62	52.422	32.560
3	111.214	69.077	33	93.493	58.070	63	50.696	31.488
4	111.096	69.004	34	92.425	57.407	64	48.954	30.406
5	110.945	68.910	35	91.327	56.725	65	47.196	29.314
6	110.760	68.795	36	90.203	56.027	66	45.426	28.215
7	110.543	68.660	37	89.051	55.311	67	43.639	27.105
8	110.290	68.503	38	87.871	54.578	68	41.841	25.988
9	110.003	68.325	39	86.665	53.829	69	40.028	24.862
10	109.686	68.128	40	85.431	53.063	70	38.204	23.729
11	109.333	67.909	41	84.171	52.280	71	36.368	22.589
12	108.949	67.670	42	82.886	51.482	72	34.520	21.441
13	108.530	67.410	43	81.575	50.668	73	32.662	20.287
14	108.079	67.130	44	80.241	49.839	74	30.793	19.126
15	107.596	66.830	45	78.880	48.994	75	28.914	17.959
16	107.079	66.509	46	77.497	48.135	76	27.029	16.788
17	106.530	66.168	47	76.089	47.260	77	25.134	15.611
18	105.949	65.807	48	74.659	46.372	78	23.229	14.428
19	105.337	65.427	49	73.203	45.468	79	21.320	13.242
20	104.692	65.026	50	71.727	44.551	80	19.402	12.051
21	104.014	64.605	51	70.228	43.620	81	17.480	10.857
22	103.306	64.165	52	68.708	42.676	82	15.551	9.659
23	102.565	63.705	53	67.168	41.719	83	13.617	8.458
24	101.795	63.227	54	65.604	40.748	84	11.681	7.255
25	100.994	62.729	55	64.022	39.765	85	9.739	6.049
26	100.160	62.211	56	62.420	38.770	86	7.796	4.842
27	99.297	61.675	57	60.798	37.763	87	5.849	3.633
28	98.405	61.121	58	59.159	36.745	88	3.899	2.422
29	97.481	60.547	59	57.501	35.715	89	1.950	1.211
30	96.528	59.955	60	55.825	34.674	90	0.000	0.000

45. Examine a globe. What do you observe about the distance around the earth along each parallel as you get farther away from the equator?

_____

46. Use Table 21.1, "Longitude as distance," to determine the length of one degree of longitude at each of the following parallels.

Length of 1° of Longitude

15° latitude: _____ km, _____ miles

30° latitude: _____ km, _____ miles

45° latitude: _____ km, _____ miles

80° latitude: _____ km, _____ miles

47. Use the earth's grid illustrated in Figure 21.4 to determine the distances between the following points.

Distance between points D and G:

_____ degrees × _____ miles/degree =

_____ miles

Distance between points B and H:

_____ degrees × _____ km/degree =

_____ km

Memphis, TN, and Tokyo, Japan, are both located at about 35°N latitude.

48. Use a globe or world map to determine how many degrees of longitude separate Memphis, TN, from Tokyo, Japan.

_____ degrees of longitude separate Memphis, TN, and Tokyo, Japan.

49. From the longitude as distance table, Table 21.1, the length of one degree of longitude at latitude 35°N is

_____ km (_____ miles).

50. How many miles is Tokyo, Japan, directly west of Memphis, TN? Show your calculation below.

_____ miles

51. In question 43 you determined the great circle distance between Memphis, TN, and Tokyo, Japan. How many miles shorter is the great circle route between these cities than the east-west distance along a parallel?

The great circle route is _____ miles shorter.

REVIEW

Having completed the exercise, you should know the following:

1. Latitude and longitude are used to locate a point on the surface of the earth.
2. Latitude is position north or south of the equator, and longitude is position east or west of the prime meridian.
3. The angle of Polaris above the horizon can be used to determine latitude in the Northern Hemisphere.
4. Time can be used to determine longitude.
5. Latitude and longitude can be used to determine distance on the earth.
6. The shortest distance between two places on the earth is along a great circle.

Location and Distance on the Earth

SUMMARY/REPORT PAGE

Date Due: _____

Name: _____

Date: _____

Class: _____

After you have finished Exercise Twenty-one, complete the following questions. You may have to refer to the exercise for assistance or to locate specific answers. Be prepared to submit this summary/report to your instructor at the designated time.

1. In the following space, prepare a diagram illustrating the earth's grid system. Include and label the equator and prime meridian. Refer to the diagram to explain the system used for locating points on the surface of the earth.

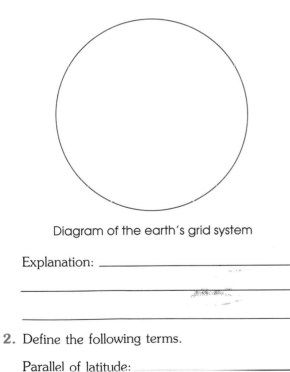

Diagram of the earth's grid system

Explanation: _____

_____

_____

2. Define the following terms.

Parallel of latitude: _____

_____

Meridian of longitude: _____

_____

Great circle: _____

_____

3. Determine whether or not the following statements are true or false. If the statement is false, correct the word(s) so that it reads as a true statement.

T F a. The distance measured north or south of the prime meridian is called latitude.

T F b. All meridians, when paired with their opposite meridian on the earth, form great circles.

T F c. The equator is the only meridian that is a great circle.

4. What is the relation between the latitude of a place in the Northern Hemisphere and the angle of Polaris above the horizon at that place?

_____

5. Approximately how many miles does one degree equal along a great circle?

One degree along a great circle equals _____ miles.

6. What is the latitude and longitude of your home city?

_____ latitude, _____ longitude

7. Use a globe or map to determine, as accurately as possible, the latitude and longitude of Athens, Greece.

_____ latitude _____ longitude

8. From question 51 of the exercise, how many miles shorter is the great circle route between Memphis, TN, and Tokyo, Japan, than the straight east-west distance along a parallel?

_____ miles shorter

9. Approximately how many miles is it from London, England, to the South Pole? (Show your calculation)

_____ miles

10. You are shipwrecked and floating in the Atlantic Ocean somewhere between London, England, and New York, NY. Fortunately you managed to save your globe. You have been in London so your watch is still set for London time. It is noon, by the sun, at your location. Your watch indicates that it is 4 P.M. in London. Are you closer to the United States or to England? Explain how you arrived at your answer.

_____

_____

_____

The Metric System, Measurements, and Scientific Inquiry

Earth science, the study of the earth and its neighbors in space, involves investigations of natural objects that range in size from the very smallest divisions of atoms to the largest of galaxies (Figure 22.1). From atoms to galaxies, objects are each unique in their size, mass, and volume; and yet all are related when it comes to understanding the nature of the earth and its place in the universe.

Almost every scientific investigation requires accurate measurements. One important purpose of this exercise is to examine the metric system as a method of scientific measurement used in the earth sciences. In addition, a few special units of measurement are also examined. The exercise concludes with an activity that focuses on the nature of scientific inquiry.

OBJECTIVES

After you have completed this exercise, you should be able to

1. List the units for length, mass, and volume that are used in the metric system.
2. Use the metric system for measurements.
3. Convert units within the metric system.
4. Understand and use the micrometer and nanometer for measuring very small distances as well as the astronomical unit and light-year for measuring large distances.
5. Determine the approximate density and specific gravity of a solid substance.
6. Conduct a scientific experiment using accepted methods of scientific inquiry.

TEXTBOOK REFERENCE

Introduction and Appendix A

MATERIALS

metric ruler   calculator

Materials Supplied by Your Instructor

metric tape measure or meterstick	paper clip
	nickel coin
metric balance	paper cup
"bathroom" scale (metric)	small rock
large graduated cylinder (marked in milliliters)	thread

TERMS

metric system	micrometer (or	light-year
meter	micron)	density
liter	nanometer	specific gravity
gram	astronomical unit	hypothesis
Celsius degrees		

INTRODUCTION

To describe objects, earth scientists use units of measurement that are relative to the particular feature being studied. For example, centimeters or inches, instead of kilometers or miles, would be used to measure the width of this page; and kilometers or miles, rather than centimeters or inches, to measure the distance from New York to London, England.

Most areas of science have developed units of measurement that meet their particular needs. However, regardless of the unit used, all scientific measurements are defined within a broader system so that they may be understood and compared. In science, the fundamental units have been established by the *International System of Units* (SI).

FIGURE 22.1 The range of
earth science measurements.
(Spiral galaxy photo courtesy
of U.S. Naval Observatory)

Earth science investigations extend from the study of

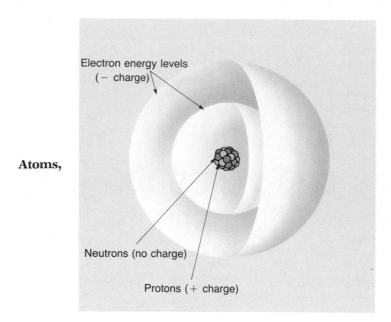

Atoms,

whose diameters can be measured using the *nanometer*. One nanometer =
0.000000001 meter, to the study of

Galaxies,

with diameters measured in hundred thousands of *light-years*. One light-year equals
about 10,000,000,000,000,000 meters.

THE METRIC SYSTEM

The **metric system** is a decimal system (based on fractions or multiples of ten) that uses only one basic unit for each type of measurement: the **meter** (m) as the unit of length (Figure 22.2), the **liter** (l) as the unit of volume (Figure 22.3), and the **gram** (g) as the unit of mass (Figure 22.4). In the English system, the units used to express the same relations are feet, quarts, and ounces.

Working with the Metric System. In the metric system the basic units of weights and measures are in "tens"

relations to each other. It is similar to our monetary system where 10 pennies equal one dime and ten dimes equal one dollar. However, in the English system of weights and measures no such regularity exists; for example, 12 inches equal a foot and 5280 feet equal a statute mile. Thus, the advantage of the metric system is *consistency*.

Table 22.1 illustrates the prefixes that are used in the metric system to indicate how many times more (in multiples of 10) or what fraction (in fractions of ten) of the basic unit you have. Therefore, from the information in Table 22.1, a *kilo*gram (kg) means one thousand

FIGURE 22.2 The SI unit of length is the meter (m), which is slightly longer than a yard. Originally described as one ten–millionth of the distance from the equator to the North Pole, it is currently defined as the distance traveled by light in a vacuum in 0.0000000033 of a second.

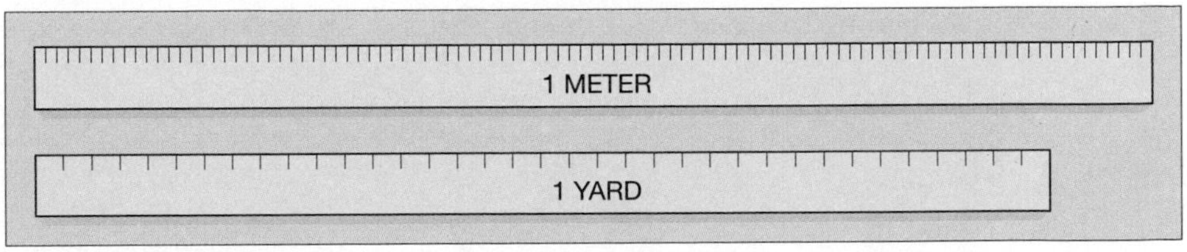

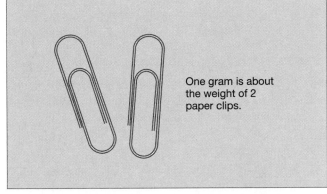

1 METER

1 YARD

1 LITER

1 QUART

MILK MILK MILK

One gram is about the weight of 2 paper clips.

FIGURE 22.4 The basic unit of mass in the metric system is the gram (g), approximately equal to the mass of one cubic centimeter of pure water at 4°C (39.2°F). A gram is about the weight of two paper clips; while an ounce is about the weight of 40 paper clips.

FIGURE 22.3 The basic unit of volume in the metric system is the liter (l), equal to the volume of one kilogram of pure water at 4°C (39.2°F). A liter is 1.06 quarts.

TABLE 22.1 Metric prefixes and symbols

Prefix[1]	Symbol[2]	Meaning
mega-	M	one million times base unit (1,000,000 × base)
kilo-	k	one thousand times base unit (1000 × base)
hecto-	h	one hundred times base unit (100 × base)
deka-	da	ten times base unit (10 × base)
BASE UNIT	m (meter)—base unit of length l (liter)—base unit of volume g (gram)—base unit of mass	
deci-	d	one-tenth times base unit (.1 × base)
centi-	c	one-hundredth times base unit (.01 × base)
milli-	m	one-thousandth times base unit (.001 × base)
micro-	μ	one-millionth times base unit (.000001 × base)
nano-	n	one-billionth times base unit (.000000001× base)

[1] A prefix is added to the base unit to indicate how many times more, or what fraction of, the base unit is present. For example, a kilometer (km) means one thousand meters and a millimeter (mm) means one-thousandth of a meter.

[2] When writing in the SI system, periods are not used after the unit symbols and symbols are not made plural. For example, if the length of a stick is 50 centimeters, it would be written as "50 cm" (not "50 cm." or "50 cms").

grams, while a *milligram* (mg) is one one-thousandth of a gram.

To familiarize yourself with metric units, determine the following measurements using the equipment provided in the laboratory.

▬▬▬▬

Measuring length:

1. Use a metric measuring tape (or meterstick) to measure your height as accurately as possible to the nearest hundredth of a meter (called a centimeter).

   _____._____ meters (m)

2. Use a metric ruler to measure the length of this page as accurately as possible to the nearest tenth of a centimeter (called a millimeter).

   _____._____ centimeters (cm)

3. Accurately measure the length of your shoe to the nearest millimeter.

   _____ millimeters (mm)

Measuring volume:

4. Use a graduated measuring cylinder to measure the volume of the paper cup to the nearest milliliter.

   _____ milliliters (ml)

Measuring mass:

5. Weigh the following and record your results. Follow the directions of your instructor for using a metric balance.

   Sample of rock: _____ grams (g)

   Paper clip: _____ grams (g)

   Nickel coin: _____ grams (g)

(*Note*: Two terms that are often confused are *mass* and *weight*. Mass is a measure of the amount of matter an object contains. Weight is a measure of the force of gravity on an object. For example, the mass of an object would be the same on both the earth and the moon. However, because the gravitational force of the moon is less than that of the earth, the object would weigh less on the moon. On the earth, mass and weight are directly related, and often the same units are used to express each.)

6. Use the metric "bathroom" scale. Weigh yourself as accurately as possible to the nearest tenth of a kilogram. (*Note*: If a metric scale is not available,

convert your weight in pounds to kilograms by multiplying your weight (in pounds) by 0.45.)

_____._____ kilograms (kg)

Metric Conversions. As stated earlier, one important advantage of the metric system is that it is based on "tens." As shown on the metric conversion diagram, Figure 22.5, conversion from one unit to another can be accomplished simply by *moving the decimal point to the left* if going to larger units, or by moving the decimal point to the right if going to smaller units.

For example, if you measure the length of a piece of string and it is 1.43 decimeters long, in order to convert its length to millimeters, start with 1.43 on the "deci-" step of the diagram. Then move the decimal two places (steps) to the right (the "milli-" step). The length, in millimeters, becomes 143.0 millimeters.

▬▬▬▬

7. Use the metric conversion diagram, Figure 22.5, to convert the following:

   a. 2.05 meters (m) = _____ centimeters (cm)

   b. 1.50 meters (m) = _____ millimeters (mm)

   c. 9.81 liters (l) = _____ deciliters (dl)

   d. 5.4 grams (g) = _____ milligrams (mg)

   e. 6.8 meters (m) = _____ kilometer (km)

   f. 4214.6 centimeters = _____ meters (m)

   g. 321.50 grams = _____ kilogram (kg)

   h. 70.73 hectoliters = _____ dekaliters (dal)

8. Use a metric tape measure (or meterstick) to determine the length of your laboratory table as accurately as possible to the nearest hundredth of a meter. Then convert the length to each of the units in question 8b.

   a. Length of table: _____._____ meters

   b. Length of table equals:

      _____ millimeters (mm)

      _____ centimeters (cm)

      _____ km

Metric-English Conversions. Because the change to the metric system will occur gradually over the next few decades, we will be forced to use both systems simultaneously. If we are not able to convert from one

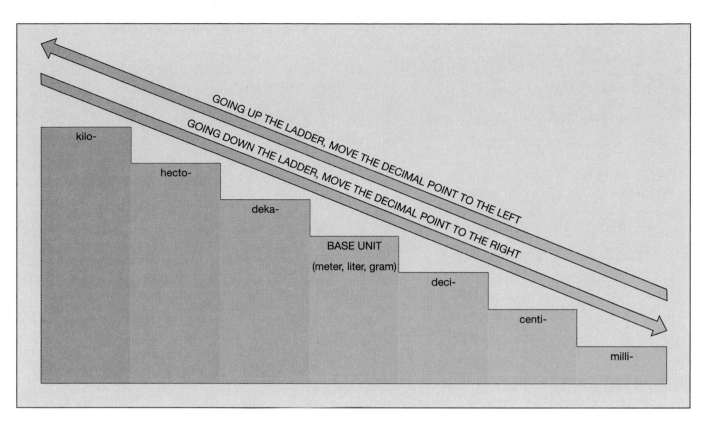

FIGURE 22.5 Metric conversion diagram. Beginning at the appropriate step, if going to larger units (left), move the decimal to the left for each step crossed. When going to smaller units (right), move the decimal to the right for each step crossed. For example, 1.253 meters (base unit step) would be equivalent to 1253.0 millimeters (decimal moved three steps to the right, the milli- step).

system to the other, we will be inconvenienced or mildly frustrated at times.

By using the conversion tables in Appendix A of your textbook or the inside-back cover of this manual, show the metric equivalent for each of the following units.

▬▬▬▬

Length conversion:

9. 1 inch = _____ centimeters

10. 1 meter = _____ feet

11. 1 mile = _____ kilometers

Volume conversion:

12. 1 gallon = _____ liters

13. 1 cubic centimeter = _____ cubic inch

Mass conversion:

14. 1 gram = _____ ounce

15. 1 pound = _____ kilogram

Temperature. Temperature represents one relatively common example of using two different systems of measurement. On the Fahrenheit temperature scale, 32°F is the melting point of ice and 212°F marks the boiling point of water (at standard atmospheric pressure). On the Celsius scale, ice melts at 0°C and water boils at 100°C.

Conversion from one temperature scale to the other can be accomplished using either an equation or graphic comparison scale. To convert Celsius degrees to Fahrenheit degrees, the equation is °F = (1.8)°C + 32°. To convert Fahrenheit degrees to Celsius degrees, the equation is °C = (°F − 32°)/1.8.

▬▬▬▬

16. Convert the following temperatures to their equivalents. Do the first four conversions using the ap-

propriate equation, and the others using the temperature comparison scale in Appendix A of your textbook or the inside-back cover of this manual.

a. On a cold day it was 8°F = _____ °C

b. Ice melts at 0°C = _____ °F

c. Room temperature is 72°F = _____ °C

d. A hot summer day was 35°C = _____ °F

e. Normal body temperature is 98.6°F = _____ °C

f. A warm shower is 27°C = _____ °F

g. Hot soup is 72°C = _____ °F

h. Water boils at 212°F = _____ °C

17. Using the temperature comparison scale, answer the following:

a. The thermometer reads 28°C. Will you need your winter coat? _____

b. The thermometer reads 10°C. Will the outdoor swimming pool be open today? _____

c. If your body temperature is 40°C, do you have a fever? _____

d. The temperature of a cup of cocoa is 90°C. Will it burn your tongue? _____

e. Your bath water is 15°C. Will you have a scalding, warm, or chilly bath? _____

f. "Who's been monkeying with the thermostat? It's 37°C in this room." Are you shivering or perspiring? _____

Metric Review. Use what you have learned about the metric system to determine whether or not the following statements are *reasonable*. Write "yes" or "no" in the blanks. *Do not* convert these units to English equivalents, only *estimate* their value.

▬▬▬

18. A man weighs 90 kilograms. _____

19. A fire hydrant is a meter tall. _____

20. A college student drank 3 kiloliters of coffee last night. _____

21. The room temperature is 21°C. _____

22. A dime is 1 millimeter thick. _____

23. Sugar will be sold by the milligram. _____

24. The temperature in Paris today is 80°C. _____

25. The bathtub has 80 liters of water in it. _____

26. You will need a coat if the outside temperature is 30°C. _____

27. A pork roast weighs 18 grams. _____

SPECIAL UNITS OF MEASUREMENT

Scientists often use special units to measure various phenomena. Most of them are defined using the units of the International System of Units. Throughout this course you will encounter several of these units in your reading and laboratory studies. Only a few are introduced here.

Very Small Distances. Two units commonly used to measure very small distances are the **micrometer** (symbol, μm), also known as the **micron,** and the **nanometer** (symbol, n).

By definition, one micrometer equals .000001 m (one millionth of a meter). There are one million micrometers in one meter and 10,000 micrometers in a centimeter. A nanometer equals .000000001 m (one billionth of a meter).

▬▬▬

28. There are (10, 100, 1,000) nanometers in a micrometer. Circle your answer.

29. What would be the length of a 2.5 centimeter line expressed in micrometers and nanometers?

_____ micrometers in a 2.5 cm line

_____ nanometers in a 2.5 cm line

30. Some forms of *radiation* travel in very small waves with distances from crest to crest of about 500 nanometers (0.5 μm). How many of these waves would it take to equal one centimeter?

_____ waves in one centimeter

Very Large Distances. On the other extreme of size, astronomers must measure very large distances, such as the distances between planets or to the stars and beyond. To simplify their measurements, they have developed special units including the **astronomical unit** (symbol, AU) and the **light-year** (symbol, LY).

The astronomical unit is a unit for measuring distance within the solar system. One astronomical unit is equal to the average distance of the earth from the sun. This average distance is 150 million kilometers, which is approximately equal to 93 million miles.

▬▬▬

31. The planet Saturn is 1,427 million kilometers from the sun. How many AUs is Saturn from the sun?

_____ AUs from the sun

The light-year is one unit for measuring distances to the stars and beyond. One light-year is defined as the distance that light travels in a vacuum in one year. This distance is about 6 trillion miles (6,000,000,000,000 miles).

▬▬▬

32. Approximately how many kilometers will light travel in one year?

_____ kilometers per year

33. The nearest star to Earth, excluding our sun, is named Proxima Centauri. It is about 4.27 light-years away. What is the distance of Proxima Centauri from Earth in both miles and kilometers?

_____ miles

_____ kilometers

Density and Specific Gravity. Two important properties of a material are its **density** and **specific gravity.** Density is the mass of a substance per unit volume, usually expressed in grams per cubic centimeter (g/cm^3) in the metric system. The specific gravity of a solid is the ratio of the mass of a given volume of the substance to that of an equal volume of some other substance taken as a standard (usually water at 4°C). For example, a specific gravity of 6 means that the substance has six times more mass than an equal volume of water.

The approximate density and specific gravity of a rock, or other solid, can be arrived at using the following steps:

Step 1. Determine the mass of a small rock using a metric balance.

Step 2. Fill a graduated cylinder that has its divisions marked in milliliters approximately two-thirds full with water. Note the level of the water in the cylinder in milliliters.

Step 3. Tie a thread to the rock and immerse the rock into the water in the graduated cylinder. Note the new level of the water in the cylinder.

Step 4. Determine the difference between the beginning level and after-immersion level of the water in the cylinder.

Step 5. Calculate the density and specific gravity using the following information and appropriate equations.

A milliliter of water has a volume approximately equal to a cubic centimeter (cm^3). Therefore, the difference between the beginning water level and the after-immersion water level in the cylinder equals the volume of the rock in cubic centimeters. Furthermore, *a cubic centimeter (one milliliter) of water has a mass of approximately one gram.* Therefore, the difference between the beginning water level and the after-immersion water level in the cylinder is the mass of a volume of water equal to the volume of the rock.

Using the four steps for determining density and specific gravity, complete questions 34 and 35.

▬▬▬

34. Determine the density and specific gravity of a small rock sample by completing questions 34a–34f.

a. Mass of rock sample: _____ grams

b. After-immersion level of water: _____ ml

Beginning level of water in cylinder: −_____ ml

_____

Difference: _____ ml

c. Volume of rock sample: _____ cm^3

d. Mass of a volume of water equal to the volume of the rock: _____ g

e. Density of rock:

$$\text{Density} = \frac{\text{mass of rock (g)}}{\text{volume of rock } (cm^3)} = \underline{\quad} \; g/cm^3$$

f. Specific gravity of rock:

$$\text{Specific gravity} =$$
$$\frac{\text{mass of rock (g)}}{\text{mass of an equal volume of water (g)}} = \underline{\quad}$$

35. As a means of comparison, your instructor may require that you determine the density and/or specific gravity of other objects. If so, record your results in the following spaces.

a. Object: _____

   Density: _____ g/cm³

   Specific gravity: _____

b. Object: _____

   Density: _____ g/cm³

   Specific gravity: _____

36. If you have investigated the densities and/or specific gravities of several objects, write a brief paragraph comparing the objects.

_____

_____

_____

METHODS OF SCIENTIFIC INQUIRY

Scientists utilize many methods in an attempt to understand natural phenomena. Some scientific discoveries represent purely theoretical ideas, while others may occasionally occur by chance. However, scientific knowledge is often gained by following a sequence of steps which involve:

Step 1. Establishing a **hypothesis**—a tentative, or untested, explanation.
Step 2. Gathering data and conducting experiments to validate the hypothesis.
Step 3. Accepting, modifying, or rejecting the hypothesis on the basis of extensive data gathering and/or experimentation.

The following simple inquiry should help you understand the process.

Step 1.—Establishing a Hypothesis.   Observe all the people in the laboratory and pay particular attention to each individual's height and shoe length.

▬▬▬

37. Based on your observations, write a hypothesis that relates a person's height and shoe length.

Hypothesis: _____

_____

_____

Step 2.—Gathering Data.   Previously, in questions 1 and 3 of the exercise, each person in the laboratory measured his or her height using a metric tape measure (or meterstick) and shoe length.

▬▬▬

38. Gather your data by asking ten or fifteen people in the lab for their height and shoe length measurements. Enter your data in Table 22.2 by recording height to the nearest hundredth of a meter and shoe length to the nearest millimeter.

Step 3.—Evaluating the Hypothesis Based Upon the Data.   Plot all your data from Table 22.2 on the height vs. shoe length graph, Figure 22.6, by locating a person's height on the vertical axis and his or her shoe length on the horizontal axis. Then place a dot on the graph where the two intersect.

▬▬▬

39. Describe the pattern of the data points (dots) on the height vs. shoe length graph, Figure 22.6. For example, are the points scattered all over the graph, or do they appear to follow a line or curve?

_____

_____

40. Draw a single line on the graph that appears to average, or best fit, the pattern of the data points.
41. Describe the relation of height to shoe length that is illustrated by the line on your graph.

_____

_____

42. Ask several people, whose height and shoe length you have not used to prepare the graph, for their height. Then see how accurately your line predicts what their shoe length should be. Do this by marking each person's height on the vertical axis and then follow a line straight across to the right until you intersect the line on the graph. Read the predicted shoe length from the axis directly below the point of intersection.
43. Summarize how accurately your graph predicts a

TABLE 22.2 Data table for recording height and shoe length measurements of people in the lab.

Person	Height (nearest hundredth of a meter)	Shoe Length (nearest millimeter)
1	_____ . _____ m	_____ mm
2	_____ . _____ m	_____ mm
3	_____ . _____ m	_____ mm
4	_____ . _____ m	_____ mm
5	_____ . _____ m	_____ mm
6	_____ . _____ m	_____ mm
7	_____ . _____ m	_____ mm
8	_____ . _____ m	_____ mm
9	_____ . _____ m	_____ mm
10	_____ . _____ m	_____ mm
11	_____ . _____ m	_____ mm
12	_____ . _____ m	_____ mm
13	_____ . _____ m	_____ mm
14	_____ . _____ m	_____ mm
15	_____ . _____ m	_____ mm

person's shoe length, knowing only his or her height.

_____

_____

_____

44. Using your graph's ability to make predictions as a guide, do you think you should accept, reject, or modify your original hypothesis? Explain the reason(s) for your choice.

_____

_____

_____

Your study has been restricted to people in your laboratory.

▬▬▬▬

45. Why would your ability to make predictions have

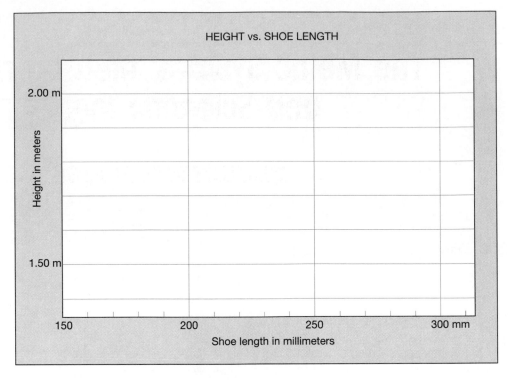

FIGURE 22.6 Height vs. shoe length graph.

been more accurate if you had used the heights and shoe lengths of ten thousand people to construct your graph?

_____

_____

_____

Drawing hasty conclusions with limited data can often cause problems. In science you can never have "too much" data. Experiments are repeated many times by many different people before the results are accepted by the scientific community.

REVIEW

Having completed the exercise, you should know the following:

1. The metric system is a method of measurement used in scientific studies.
2. The metric system is a decimal system that uses the meter, liter, and gram to measure length, volume, and mass.
3. How to use the metric system and convert units within the system.
4. Scientists use special units of measurement when necessary. These units can often be defined using standard units such as the meter.
5. How to conduct a scientific investigation using a hypothesis.

The Metric System, Measurements, and Scientific Inquiry

SUMMARY/REPORT PAGE

Name: _____

Date Due: _____

Date: _____

Class: _____

After you have finished Exercise Twenty-two, complete the following questions. You may have to refer to the exercise for assistance or to locate specific answers. Be prepared to submit this summary/report to your instructor at the designated time.

1. List the basic metric unit and symbol used for these measurements:

    Length: _____

    Mass: _____

    Volume: _____

2. Convert the following units:

    a. 2 liters = _____ deciliters

    b. 600 millimeters – _____ meter

    c. 72°F = _____ °C

    d. 0.32 kilograms = _____ grams

    e. 12 grams = _____ milligrams

3. Indicate by answering "yes" or "no" whether or not the following statements are reasonable:

    a. A person is 600 centimeters tall. _____

    b. A bag of groceries weighs 5 kilograms. _____

    c. It took 52 liters of gasoline to fill the car's

    empty gasoline tank. _____

4. List your height and shoe length using the metric system.

    a. Height: _____ . _____ meters

    b. Shoe length: _____ millimeters

5. How many micrometers are there in 3.0 centimeters?

    _____ micrometers in 3.0 centimeters

6. How many waves, each 500 nanometers wide, would fit along a two centimeter line?

    _____ waves along a two centimeter line

7. What would be the distance in kilometers of a star that is 6.5 light-years from the earth?

    _____ kilometers from the earth

8. Uranus, one of the most distant planets, is 2,870 million kilometers from the sun. What is its distance from the sun in astronomical units?

    _____ astronomical units from the sun

9. Explain the difference between the two terms, *density* and *specific gravity*.

    _____

    _____

    _____

10. At the conclusion of your height–shoe length experiment, in question 44 you (accepted, rejected, modified) your original hypothesis. Circle your answer and explain the reason for this decision.

    _____

    _____

    _____

ISBN 0-02-419011-X

9 780024 190116

90000>